VULTURES

Their Evolution, Ecology and Conservation

VULTURES

Their Evolution, Ecology and Conservation

Michael O'Neal Campbell

Simon Fraser University
British Columbia, Canada

CRC Press
Taylor & Francis Group
Boca Raton London New York

CRC Press is an imprint of the
Taylor & Francis Group, an **informa** business

A SCIENCE PUBLISHERS BOOK

CRC Press
Taylor & Francis Group
6000 Broken Sound Parkway NW, Suite 300
Boca Raton, FL 33487-2742

First issued in paperback 2020

© 2015 by Taylor & Francis Group, LLC
CRC Press is an imprint of Taylor & Francis Group, an Informa business

No claim to original U.S. Government works

ISBN-13: 978-1-4822-2361-3 (hbk)
ISBN-13: 978-0-367-73810-5 (pbk)

Library of Congress Cataloging-in-Publication Data

Campbell, Michael O'Neal, 1965-
    Vultures : their evolution, ecology, and conservation / Michael O'Neal Campbell.
        pages cm
    Includes bibliographical references and index.
        ISBN 978-1-4822-2361-3 (hardcover : alk. paper) 1. Vultures. I. Title.

QL696.F32C35 2015
598.9'2--dc23   2014035370

Visit the Taylor & Francis Web site at
http://www.taylorandfrancis.com

and the CRC Press Web site at
http://www.crcpress.com

Preface

In current times, a science project, far from being a rudimentary exercise in laboratory methods techniques, is an acute business of timely investigations, popular and/or adequately promoted unpopular positions, political lobbying and replicable experiments that in concert may gain traction. In other words, the project must catch the big wave, ride it and keep it going. Fundamental questions for any scientific endeavour are: 'Why is this important?' 'Are you sure?' and finally 'So what?'

Vultures are a group of birds that have been at least until recently been relatively ignored by scientists. The question is why should we study vultures? And what would be the relevance of our results? Are vultures relics from the past, irrelevant in the modern context of glass, concrete, highways, farms and small, rare mammals, destined for extinction with their former food, the huge mammals of the Pleistocene? Do they still serve any useful function, considering modern sanitary and landscape management may make their role in cleaning the environment redundant? Considering the historical fascination, quasi-religious and spiritual status, criminal association, repugnance and awe vultures elicit, do we even know enough about them, to assess what we will be missing when they are gone? Finally, what is a vulture—just any scavenging bird with a bald head?

Five key, incomplete components have emerged in current studies of vultures: (1) changing evidence, based on new, chemical and biochemical techniques, has waded into the already murky knowledge concerning the differences between the so called *New World* and *Old World* vultures; (2) vultures are variably susceptible to particular chemicals and many species have gone from being among the commonest birds of prey in the world to critically endangered species; (3) local vulture extinctions have serious negative consequences for both people and animals, in ways not foreseen; (4) captive breeding, unexpectedly can rescue nearly extinct species; and consequently (5) it is quite evident that we know so little about the background of vultures and their ecological importance, that the study of vulture is necessary.

The book searches current knowledge of all the vulture species, using literature sources covering the biology, ecology, evolutionary history, cultural appraisal and current conservation status of both *New World* and *Old World* vultures. It examines recent developments in knowledge on the classification of these two groups, and assesses the importance of these changes. The book differs from older texts by creating ecological and conservation possibility networks around vulture species, examining their relations with other species, including avian competitors (eagles, storks and hawks), mammalian competitors (hyenas, felids and canids), disease-causing agents (that have recently seriously decimated vulture populations in Eurasia and Africa) and environmental change (deforestation, desertification, urbanization, agricultural intensification, hunting and public attitudes).

While these issues have been mentioned in earlier literature, and are scattered among many journal articles, there is insufficient information available on the overall scenario for all the vulture species, comparisons of different factors of vulture extinction and hence the socio-environmental networks within which these variables are embedded. In-depth attention is paid to the trends of these networks, in relation to the evolution, resilience, vulnerability and ecology for the future of vulture species, and to the necessary ameliorative actions. Throughout the book, as many references to relevant literature sources are cited as possible, to allow the reader to use this volume as a source for further reading and a guide to further studies.

The book is divided into three parts. Part 1 (Vulture Classification, Genetics and Ecology) introduces the reader to the ecology and forms of the vultures. Topics include anatomy, scavenging and predation. Three chapters examine in a systematic mode, the eight species of Griffon vultures of the Genus Gyps, the eight species of other *Old World* vultures and the seven species of *New World* vultures. Griffons are given their own chapter, as their genus comprises half of the *Old World* vultures and they have traits absent in the other species. This part may be used for field identification and study, and also serve as a background to the scientific analysis of vulture evolution, ecology and geography in the later chapters. Where the scientific name of the species is first mentioned, the naming authority (i.e., the person(s) who gave the animal its scientific name) is placed after the scientific name. For example, the Turkey Vulture's scientific name is *Cathartes aura* (Linnaeus 1758). Here *Cathartes* is the generic name or the name of the Genus to which the bird belongs. *Aura* is the species name. Linnaeus is the name of the person who gave the bird its classification, in 1958. The same would apply to the Griffon vulture, Gyps fulvus (Hablizl 1783). The name Linnaeus, which is a common name for classification, is usually shortened to L.

Part 2 (Vulture Evolution and Ecology) starts with an introduction that briefly describes the prehistoric evolution of animal species and land cover change and how this affects vulture ancestors and current species. After this there are three chapters. Chapter 4 deals with the biological evolution of vultures. This chapter illustrates the evolution of the New and Old World vultures, their close relations among other species of birds (the storks and raptors), the species they fed on (other mammals and birds) and the land cover of the Earth. It also describes the different systems used for the classification of vultures; bone structure, other body features and DNA. This provides the background for Chapter 5, which looks at the relationship between modern vultures and other species, i.e., mammal and bird scavengers and predators, and Chapter 6, which looks at the climatic relations and landscape ecology of modern vultures.

Part 3 (Vulture Ecology and Conservation) describes the human attitudes and actions, including mysticism, preservation, conservation, extermination, urbanization, agricultural development and industrialization. This is followed by four chapters. Chapter 7 looks at Vultures, Cultural Landscapes and Environmental Change, concerning the impacts of urbanization, agriculture and human hunting practices on vulture populations. These events have had devastating effects on vulture ecology and biogeography, both positively (e.g., food supply increases in urban areas) and negatively (e.g., killing of the animals that provide food, and direct killing of vultures). Chapter 8 looks at a crucial aspect of this engagement; the pollutants and diseases that have ravaged the vulture populations in Asia and Africa, these effects being so important as to merit a separate chapter. Vultures formerly considered among the commonest raptors or large birds in some regions are now severely threatened, sometimes with local extinction. Solutions to such problems depend on social, economic and political factors, these largely depend on the social attitude towards vultures in both the local context, and among the international financial actors who may fund conservation and protection schemes. For this reason, Chapter 9 examines public attitudes and conservation policies related to both New and Old World vultures, through history, as historical views, both positive and negative are variably embedded in current attitudes to vultures in many parts of the world, and as such must be examined as a component of both current and future scenarios for vultures. Finally, Chapter 10 looks at the future of vultures. This future is assessed, using all the information presented in the preceding chapters, as vulture survival and wellbeing will depend on their biological and ecological resilience, their relations with other animals and environmental dynamics and ultimately, the will of humans.

Contents

PART 1

Vulture Classification, Genetics and Ecology

This part comprises three chapters that give a detailed summary of the seven species of New World vultures and the sixteen species of Old World vultures. The New World vultures are now found only in the New World, and Old World vultures are found only in the Old World. However, this was not always so; if we were classifying these birds to include prehistoric ancestors it would be better to describe the New World vultures as Cathartid vultures and the Old World vultures as Accipitrid vultures. Part 2 of this book will examine the location of ancestral fossils. As will be discussed in detail in later chapters, the main differences between the two groups are that Old World vultures are more similar to the Accipitrid raptors (hawks and eagles), because they have stronger, more raptor-like bills, feet and talons than the New World vultures. For these reasons, the balance of debate is shifting towards two positions, i.e., either the Old World vultures are related to hawks and eagles and New World vultures are more distantly related to these species, or Old World vultures remain related to the hawks and eagles, and the New World vultures are placed with or near the stork family. Further complications arise, because among the Old World vultures, different species appear to have different links with other raptors. Some authorities consequently speculate on the possibility that these vulture species emerged from similar lifestyles among different raptor species and eventually attained similar appearances and habits. This is termed convergent evolution. Considering the differences between New and Old World vultures, 'vulture is an ecologically but not a systematically meaningful term' (Wink and Sauer-Gürth 2004; see also Wink 1995 and Storch et al. 2001). However, before we start with the issues and systematics of these species, it is necessary to elaborate on some terms and phenomena that are basic for the understanding of a vulture's lifestyle, such as definitions of scavengers and scavenging, and the chemical bases of scavenging such as putrification.

Scavenging is a behavior practiced by animals classified as scavengers, predators and omnivores. Therefore, the terms "scavenger" and "scavenging" are not easily defined. Most predatory vertebrates also scavenge on dead flesh they did not kill themselves. This behavior may influence the evolution of scavenging behavior in vertebrates (DeVault et al. 2003). However, among vertebrates, only the 23 species of New and Old World vultures have been described as obligate or permanent full-time scavengers (Ruxton and Houston 2004; Ogada et al. 2011). Their bodies are designed for scavenging and they require this lifestyle for their survival. Physical adaptations for this activity include: bare skin or down covered heads and sometimes necks (with the exception of only the Bearded vulture and Palm-nut vulture), facilitating the thrusting of heads and necks into meat, while avoiding matted blood covered feathers; feet adapted to walking on the ground, rather than grasping prey; and large wings designed for soaring rather than pursuing the prey. Other avian scavengers, such as gulls, corvids, eagles and storks are more suited than vultures to alternate carnivorous or herbivorous lifestyles. Hence, when these species forage for carrion they are called facultative scavengers or temporary, part-time scavengers. An introduction to the ecology of avian scavengers is therefore a complex exercise, including species that compete with, evolve into, or may be mistaken for scavengers.

One definition is that a scavenger is 'an animal (such as a vulture or coyote) that eats carcasses abandoned by predators, digs through trash cans for food, etc., true scavengers seldom kill their own prey (but many animals are not exclusively scavengers)' (Biology Online 2013). Another definition is that a scavenger is either 'one that scavenges, as a person who searches through refuse for food' or 'an animal, such as a bird or insect, that feeds on dead or decaying matter.' (The American Heritage Dictionary of the English Language 2009), Getz (2011) gives an incisive description of scavengers:

> *Scavenging is both a carnivorous and herbivorous feeding behavior in which the scavenger feeds on dead animal and plant material present in its habitat. The eating of carrion from the same species is referred to as cannibalism. Scavengers play an important role in the ecosystem by consuming the dead animal and plant material. Decomposers and detritivores complete this process, by consuming the remains left by scavengers*

Hamilton (2003) also describes the roots of the term "scavenger": *Scavenger* is an alteration of "scavager", from Middle English "skawager" meaning "customs collector", from skawage meaning 'customs', from Old North French "escauwage" meaning "inspection", from "escauwer" meaning "to inspect", of Germanic origin; akin to Old English "sceawian" meaning "to look at", and modern English "show".

Several points emerge from these definitions. First scavengers may be animals that are not exclusively scavengers, but may also derive sustenance using other methods. Second, true scavengers do indeed derive their food entirely or almost entirely from scavenging. Third, the food of the scavenger should be dead or decaying animal or plant matter. Fourth, scavengers may be birds, mammals, insects or even humans. Fifth, decomposers and detritivores, which consume the remains left by scavengers are not scavengers.

As mentioned above, scavengers may be further described as facultative scavengers (i.e., they are also predators, omnivores or herbivores that supplement their diet with decaying animal or plant or animal matter) or obligate scavengers that eat only carrion (Schmitz et al. 2008; Cortés-Avizanda et al. 2009a). The latter may be lesser in number, as carrion is not very common in natural contexts (Wilson and Wolkovich 2011). Cold climates may limit the availability of decaying matter, as decomposing bacteria are restricted. Hot temperatures in warmer climates may increase the action of decomposers, limiting carrion due to rapid decomposition (Selva et al. 2003; DeVault et al. 2004). Decomposers compete with scavengers by consuming carrion rapidly (Shivik 2006). Competition between scavengers, carnivores and decomposers may increase with limited edible matter (Schmitz et al. 2008).

Scavenger removal of animal carcasses is considered beneficial for the ecosystem (Sekercioglu 2006). In this setting, the avian scavenger, whether an eagle, vulture, stork or gull, possesses a unique role, due to its ability to cover large areas, and its visibility to associated ground dwellers, be they people, canines or felines, that may be able to interpret the avian scavenger's behavior and location for self-benefit. The avian and mammalian scavengers have a long history in the ecology of the Earth. They have close ties with the emergence of the age of large mammals and later extinctions, and the consequent proliferation of large carcasses. After this period, in prehistoric and historic times, scavengers and carnivores have had a strong relation with people, within an ecological setting in which people were important players and roles of carnivores overlapped with those of people.

Scavengers may also be described for their roles in the ecosystem. One role is that of energy transfer. Energy is transferred when scavengers eat carrion, sometimes accounting for as much as 50 per cent of the total energy transfer in the ecosystem (DeVault et al. 2003; Shivik 2006). Decaying meat has been described as a 'high quality nutritional source', as compared to decaying plant material it has a high protein content (i.e., high nitrogen:carbon ratio (Wilson and Wolkovich 2011). Scavengers also may reduce diseases that may result from the decomposition of dead animals, by consuming the meat before bacterial colonization during later putrefication (Von Dooren 2011).

The most important issue in scavenging is the food that is actually eaten. Such food is composed of dead plant or animal matter, and hence it is at a particular stage of decomposition. The science of decomposition is usually called taphonomy. Decomposition is the process of disintegration and tissue change that starts with the death of the organism and ends when it is either a skeleton or entirely destroyed. There are two main types of decomposition, abiotic (which occurs through chemical or physical processes) and biotic (also termed biodegradation, which involves the metabolic breakdown by organisms of matter into simpler components). Decomposition is crucial to the existence of the ecosystem, because it is the method by which living materials are returned to the inanimate earth, eventually to be recycled back into the living world through chemical uptake and food.

After death, when the heart of the animal stops beating and pumping blood, the body of the organism is in the fresh stage, when blood moves according to gravity. The muscles then harden (rigor mortis), the body cools (algor mortis), and during the chemical decomposition termed autolysis, cells lose their structure and enzymes are released. Oxygen within the body is depleted by the aerobic microbes in the body, creating a suitable environment for anaerobic organisms. These anaerobic organisms, from the respiratory system and gastrointestinal tract change carbohydrates, lipids, and proteins into organic acids (propionic acid, lactic acid) and gases (methane, hydrogen sulfide, ammonia). This leads to a situation termed bloating when the carcass is bloated with gases. Most tissues of multicellular organisms are putrescible, or subject to putrefaction upon death. The organic tissues that comprise living organisms, being stores of chemical energy, are broken down and simplified when death occurs, with the loss of the biochemical processes of life.

Putrefaction starts with the decomposition of the proteins in the tissue, with the result that the cohesion of the tissues is gradually destroyed, and some tissues are liquefied. The process of putrefaction may be accelerated in the case of animals by the action of microorganisms that were present within the digestive system during life, which after death, digest the proteins and excrete simpler products, including gases such as hydrogen sulphide, carbon dioxide, and methane, and amines including putrescine and cadaverine, the cause of the odor of decaying flesh. Intestinal anaerobic bacteria change haemoglobin to sulfhemoglobin, and the gases transport the sulfhemoglobin through the circulatory and lymphatic systems. The gases spread through the body cavities and then the blood vessels and tissues, bloating the body. When there is sufficient breakdown of the body tissues, the gases may escape, and the tissues further disintegrate, leading to the skeletonization of the body. The rate of putrefaction may vary greatly, the main determining factor being temperature, which either retards bacterial action (cold temperatures, as in refrigeration) or increases it (warm temperatures, as in tropical climates).

After the bloating and putrefaction stages, there is more active decay, the stage of greatest loss of mass to the body, which occurs when insects lay eggs in the flesh, and maggots (larvae) feed on the tissue. This results in the decomposition of the tissue under the skin, contributing to skin slippage and hair losses, and ruptures in the skin that allow more gases and fluids to escape, while oxygen enters through the ruptures, encouraging more development of aerobic microbes and insect larvae. After this active stage, there is the advanced decay stage, when insect larvae move from the body to pupate, and decomposition declines due to the loss of tissue. There may be loss of vegetation in the area around the body, and an increase in soil carbon, phosphorus, potassium, calcium, magnesium and nitrogen. After this stage, the former body is in a dry remains stage, merely bones, cartilage and possibly skin. It may be partially skeletonised if there is still some skin, or completely skeletonised if there is only bones remaining.

Carcasses may be recognized by the presence of insects, mammals or birds on the body, the stationary posture, the partial or total dismemberment of the body, or the loss of several body parts, which will be the norm when the death was due to carnivores. The amount of flesh remaining on a carcass depends on the stage of abandonment by the carnivores. As will be seen later, only vultures of the Genus Cathartes (the Turkey vulture, Greater Yellow-headed vulture and Yellow-headed vulture) are able to detect carrion by smell (although in most cases they probably use sight); all other vultures must rely on sight, using the indicators mentioned above. As will also be discussed later, different vulture species have different abilities at carcasses. The larger species are able to tear through the thick skin of large mammalian carcasses, while the smaller, weaker billed species either pull morsels from the carcass after it has been opened by carnivores or by the larger tearing species, or wait and pick small morsels from the bones. It is at this stage, the acquisition of food from carcasses, and how this varies among Species, Genera, Families and possibly Orders, and the enabling processes of foraging, reproduction and competition that the story of vultures begins.

1

Systematic List of Old World Vultures: The Griffons

1 INTRODUCTION

This chapter examines the forms, foraging and feeding habits, breeding patterns and statuses of the eight species of Griffon vultures of the Genus Gyps (Savigny 1809). The Gyps vultures are a genus of Old World vultures in the family Accipitridae, which also includes eagles, kites, buzzards and hawks. Compared to other vultures, the heads of Gyps species are more feathered, with a characteristic downy cover. The eight Gyps species are: the White-backed vulture (*Gyps africanus*); White-rumped vulture (*Gyps bengalensis*); Cape Griffon (*Gyps coprotheres*); Griffon vulture (*Gyps fulvus*); Himalayan vulture (*Gyps himalayensis*); Indian vulture (*Gyps indicus*)—formerly Long-billed vulture; Rüppell's vulture (*Gyps rueppellii*); and Slender-billed vulture (*Gyps tenuirostris*)—formerly included in *G. indicus*. A prehistoric species, *Gyps melitensis* is known only from fossil remains found in Middle to Late Pleistocene sites all over the central and eastern Mediterranean.

Gyps species are unique among Old World vultures in that they feed exclusively as scavengers, whereas other vultures are also known to kill their prey on occasions or, rarely, to feed on fruits (i.e., the Palm-nut vulture *Gypohierax angolensis* (Houston 1983; Mundy et al. 1992; del Hoyo et al. 1994; Johnson et al. 2006). Specialization in feeding behavior among Gyps vultures is thought to have evolved due to their close association with ungulate populations, particularly migratory herds in Africa and Asia. In fact, the observed temporal and geographic diversification of Gyps vultures coincides with the diversification of Old World ungulates, especially those of the family Bovidae (Vrba 1985; Arctander et al. 1999; Hassanin and Douzery 1999; Matthee and Davis 2001), and the expansion of grass-dominated ecosystems in Africa and Asia (Jacobs et al. 1999). These close associations likely played a significant role in the adaptation and rapid diversification of Gyps vultures. Indeed, Houston (1983) proposed that their large body size and ability to soar over large distances in search for food are related to the associated migrant distributions and seasonal fluctuations in the mortality of ungulates, and that they have consequently become incapable of actually killing their own prey (Ruxton and Houston 2004).

The Indian vulture, Rüppell's Griffon vulture and the Common Griffon vulture have been argued to be polytypic or descended from several sources (Mayr and Cottrell 1979; Sibley and Monroe 1990; del Hoyo et al. 1994; Ferguson-Lees and Christie 2001). Wink and Sauer-Gürth (2000) found that *G. rueppellii* and *G. himalayensis* may be closely related in a monophyletic group with *G. coprotheres* and *G. fulvus,* based on nucleotide squences of the cytochrome b gene. Arshad et al. (2009a) supported this position and argued that there was a recent, rapid diversification among the Gyps vultures, possibly related with the diversification of wild ungulates in the Old World. *G. bengalensis* and *G. africanus* were formerly classified together as a separate species, *Pseudogyps*, as unlike other Gyps species they shared a smaller body size and a reduced number of rectrices (wing feathers, 12 vs. 14 for the other species) (Mundy et al. 1992; Sharpe 1873, 1874; Peters 1931). In addition, proposals were made to consider the 'long-billed' vultures as two separate species (*G. indicus* and *G. tenuirostris*) based on morphological differences (Grey 1844; Hulme 1869, 1873; Rasmussen et al. 2001, 2005), and the taxonomic status of the two subspecies of the Eurasian vulture (*G. f. fulvus* and *G. f. fulvescens*), as well as their characteristics and geographic distribution are unclear. Currently, taxonomic relationships among Gyps species, including subspecies relationships, have yet to be fully studied with molecular sequence characters, and the validity of fulvescens has not been considered in recent times (Hulme 1869, 1873; Jerdon 1871).

1.1 White-rumped Vulture (*Gyps bengalensis* Gmelin 1788)

Physical appearance

The White-rumped vulture (*Gyps bengalensis*), a typical griffon, is the smallest of the griffons, but is still a very large bird. As a medium-sized vulture it weighs between 3.5 and 7.5 kg (7.7–16.5 lbs), is between 75 and 93 cm (30–37 in) in length, and has a wingspan of 1.80–2.6 m (6.3–8.5 ft) (Alstrom 1997; Ferguson-Lees et al. 2001; Rasmussen and Anderton 2005).

The dominant plumage is blackish, with slate-grey to silvery secondaries (Fig. 1.1a; see also Fig. 1 for an illustration of the locations of the various feathers). The underside is marked with whitish streaks and the back, rump

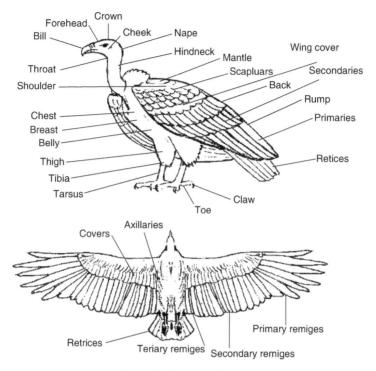

Fig. 1. Feathers on a Vulture.

and underwing coverts are whitish. The neck is unfeathered, partly down covered, slightly pink or maroon tinted, with a white neck ruff slightly open at the front, the bill is large with slit nostrils. The wings have dark edges and white linings, and black undertail coverts are visible on an adult bird in flight (Rasmussen and Anderton 2005). The coloration of this species therefore differs strongly from the White-backed vulture of Africa.

The juveniles acquire adult plumage at the age of four to five years (Fig. 1.1b). The juvenile is described as dark brown, with thin whitish streaks. Some feathers, especially the mantle and scapulars, lesser and median coverts are dark brown, while the greater coverts are blackish brown. The lower back and rump area are brown rather than white as in the adult (Rasmussen and Anderton 2005). The bare skin of the head is colored greyish-white, and the neck may also be tinged pinkish or bluish, with a dark grey throat. Both the head and neck are covered with scattered pale brown, grey and/or offwhite feathers. The cere and the bill are blackish, and the top of the upper mandible is pale bluish in the subadult. The legs,

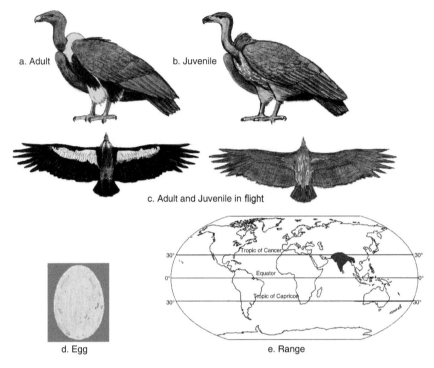

a. Adult

b. Juvenile

c. Adult and Juvenile in flight

d. Egg

e. Range

Fig. 1.1. White-rumped Vulture.

feet and iris of the juvenile are the same as those of the adult. The subadult is intermediate between the juvenile and the adult (Alstrom 1997).

In flight from below, white wing-linings on the wings, and black undertail coverts, are visible on an adult bird (Rasmussen and Anderton 2005). There are white underwing-coverts, with dark leading edges, contrasting with the blackish body (Fig. 1.1c).

Classification

The White-rumped vulture was once classified as closely related to the African White-backed vulture. *G. bengalensis* and *G. africanus* have been classified by some as a separate genus, Pseudogyps. As mentioned above, this was based on physical attributes; the smaller body size compared to other species, and the smaller number of rectrices or tail feathers (12 vs. 14) compared to other Gyps vultures (Mundy et al. 1992; Sharpe 1873, 1874; Peters 1931). Although not all scientists agree with this classification, Lerner et al. (2006: 169) found a very close genetic match between the two white-backed vultures. For example, Seibold and Helbig (1995) argue that molecular data do not support the split from Gyps to a separate genus. As can be seen in the diagrams, its coloration differs strongly from the white-backed vulture of Africa (Fig. 1.1d). Wink (1995: 877) noted 'we suggest that the name Pseudogyps should be omitted and the respective taxa included in the Genus Gyps. *G. africanus* clearly differs from *G. bengalensis*, indicating that both vultures represent distinct species' (see also Dowsett and Dowsett-Lemaire 1980).

Foraging

The White-rumped vulture, in common with other vultures, begins foraging in the morning, when the thermals are strong enough to allow extensive soaring flight. Due to their comparatively smaller size, they are dominated by the larger Asian vultures, such as the Red-headed vulture. The main foraging and nesting areas are open plains and sometimes hilly areas, with grass, shrubs and light forest. A highly social vulture, it nests and roosts in large numbers, usually on trees near human habitation (Cunningham 1903; Morris 1934; Ali et al. 1978).

Breeding

For the nesting and egg-laying period, a study by Baral and Gautam (2007) found that the egg-laying period occurred from late September to late October, the incubation period from December to January and the nesting period from February to May. This was supported by a study by Sharma (1969: 205) in Jodhpur, India which noted that 'fuller nesting behavior gets under way only during December, when minimum temperatures drop to about 11°C and relative humidity to about 50%.' These nesting activities peak in January when the minimum temperatures are reached (about 8°C), the relative humidity is around 50% and the days are the shortest (10.24 hours). In February into March, the temperatures rise to 14°C and above, and the relative humidity is around 60%. The researchers conclude that 'low temperatures, short days and high relative humidity favor breeding in *G. bengalensis*, while rising temperatures and falling humidities are adverse.' Eggs are elliptical, white, sometimes with a few reddish brown marks, with a chalky surface (Fig. 1.1e). General dimensions are 87.9 x 69.7 and 84.4 x 66.4 mm (Wells 1999).

Different nesting locations have been recorded (Baral and Gautam 2007; Sharma 1970). Baral and Gautam state that trees are favored in the Rampur Valley, Nepal. This study located 42 nesting trees, comprising 33 *kapok* trees, 2 *khair* and 1 each of the other seven tree species (*barro, kavro, ditabark, tuni, padke, saj* and *karma*). The *kapok* (*Ceiba pentandra* Linnaeus) Gaertn. is a very large tree that grows 60–70 m (200–230 ft) tall with a butressed trunk of up to 3 m (10 ft) in diameter (Gibbs and Semir 2003). The *barro* (*Terminalia chebula* Retz.) is also a large deciduous tree growing up to 30 m (98 ft) tall, and a trunk up to 1 m (3 ft 3 in) thick (Saleem et al. 2002). Similar species are the *kavro* (*Ficus religiosa* L.) (Singh et al. 2011), ditabark *Alstonia scholaris* L.R. Br. (World Conservation Monitoring Centre 1998), the *saj* (*Terminalia alata* Heyne ex Roth, or *Terminalia tomentosa* (Roxb.) Wight & Arn.) and the *karma* (*Adina cordifolia*) (Willd. ex Roxb.) Benth. and Hook.f. ex Brandis. The *khair* (*Senegalia catechu* (L.f.) P.J.H. Hurter & Mabb) is slightly smaller, growing to about 15 m in height. The Padke, also known as the Persian Silk tree or the Pink Silk tree (Albizia julibrissin Durazz., 1772 non sensu Baker), 1876 is a smaller deciduous tree growing to only 5–12 m (Gilman and Watson 1993). Sharma's study (1970: 205) states that vultures use both cliffs and trees, cliffs being favored because they require fewer twigs and are a safer refuge from predators. For trees, favored species include the very large banyan tree (*Ficus begalensis* L.) and in more arid areas the large *Prosopis spicigera* L., this latter species described as the only 'tree that meets the vultures' nesting requirements.'

Population status

The distribution of the White-rumped vulture is shown in Fig. 1.1e. Recently, the White-rumped vulture was considered one of the commonest raptors worldwide, but its numbers have greatly declined (Gilbert et al. 2003, 2006; Green et al. 2004; Baral et al. 2005; Gautam et al. 2005; Arshad et al. 2009b; BirdLife International 2001, 2012; Cuthbert et al. 2011a,b; Chaudhary et al. 2012). This species was common in Myanmar in the nineteenth century, especially near the Gurkha cattle-breeding villages. Extensive populations were also recorded by Macdonald (1906), Hopwood (1912) and Stanford and Ticehurst (1935, 1938–1939).

1.2 White-backed Vulture, *Gyps africanus* (Salvadori 1865)

Physical appearance

The White-backed vulture is a medium-sized vulture, slightly larger than the White-rumped vulture. The body mass is 4.2 to 7.2 kg (9.3–16 lbs), it is 78 to 98 cm (31 to 39 in) long and has a 1.96 to 2.25 m (6 to 7 ft) wingspan (Ferguson-Lees et al. 2001). The dark tail and flight feathers, contrast with the lighter brown to cream-colored body feathers. The rump is white. The head is paler than the neck. The only other similar birds within its range are *G. fulvus*, and *G. rueppellii*, but these species are larger and have less contrast between the flight feathers and underwing-coverts (Fig. 1.2a,b,c).

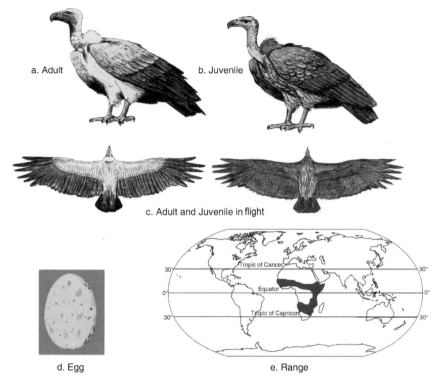

a. Adult

b. Juvenile

c. Adult and Juvenile in flight

d. Egg

e. Range

Fig. 1.2. White-backed Vulture.

Classification

The White-backed vulture, as described above, is closely related to the Indian White-rumped vulture, to the extent that they were (and still are in some publications) classified together, but 'grouping of bengalensis and africanus together in the Genus Pseudogyps, as historically proposed, is not upheld based on mitochondrial data' (Johnson et al. 2006: 65).

Foraging

The White-backed vulture frequents open wooded savanna, particularly areas of acacia. Phipps (2011) notes that the foraging preferences of this species are poorly understood, but its occurrence is linked to free ranging ungulates in open land. The Serengeti ecosystem in Tanzania has been cited as a good foraging ecosystem for this and other species of African vultures, based on the habitat of large wild ungulates and the carnivores that kill them to provide food for vultures (Houston 1974).

Breeding

For the nesting and egg-laying period, a study by Virani et al. (2010) in the Serengeti National Park of Tanzania found the peak period to be mid-April. Vultures may deliberately select this period for fledgling as it coincides with the peak of ungulate carcass availability (Houston 1976). Another study by Herholdt and Anderson (2006) in Kgalagadi Transfrontier Park, Botswana, found that most White-backed vultures laid eggs in June. Nesting periods were observed to start in May to June in Zimbabwe, and June to July in Swaziland and KwaZulu-Natal (Monadjem 2001). Only one egg is laid (Fig. 1.2d) and incubated for about 56 days. The chick is fledged after about four months. However, Virani et al. (2012) note that despite the single annual breeding season in southern Africa (as also noted by Mundy et al. 1992), in East Africa there is a bimodal breeding season, with nesting in April/May and December/January (as also reported by Ferguson-Lees and Christie 2001), and peak egg laying between March and May (Houston 1976).

Colonial nesting usually involves ten or fewer pairs, with one or occasionally two nests per tree (Mundy et al. 1992; Monadjem and Garcelon 2005). Nesting is usually in large, crowned trees in loose colonies, near rivers (Mundy et al. 1992; Bamford et al. 2009a, 2009b). Malan (2009) observed, nesting White-backed vultures predictably selected the tallest trees available. Studies by Varani et al. (2010) in the Masai Mara in Kenya, and by Monadjem and Garcelon (2005) and Bamford et al. (2009a) in Swaziland found that the White-backed vulture nests mostly in tall trees in riparian vegetation. Other studies found that nesting trees were a minimum of 11 m tall (Houston 1976; Monadjem 2003a; Herholdt and Anderson 2006). A study by Comba and Simuko (2013) found that the average height of nests in Zambia was 16.6 m. Other heights for nests were 19 m in Zimbabwe (Mundy 1982), 14 m in the Kruger National Park (Tarboton and Allan 1984), 13 m in Swaziland (Monadjem 2003) and 7 m in Kimberly (Mundy 1982). Common tree species for nests are *Faidherbia albida* [(Delile) A. Chev.], which grows from 6 to 30 m tall; *Vachellia xanthophloea* [(Benth.) P.J.H. Hurter], which grows up to 15–25 m and *Senegalia nigrescens* [(Oliv.) P.J.H. Hurter] which grows up to 18 m in height. Occasionally, these vultures nest on pylons in South Africa (Anderson and Hohne 2007; Malan 2009).

Population status

The distribution of the White-backed vulture is shown in Fig. 1.2e. This species has been described as the most widespread and common vulture in Africa, but in recent times it has declined (Thiollay 2006; McKean and Botha 2007; Ogada and Keesing 2010; Otieno et al. 2010). The population reduction in Western Africa is over 90% in some areas (Thiollay 2006). There have been population reductions in the Sudan (Nikolaus 2006), Kenya (Virani et al. 2011) and southern Africa (Hockey et al. 2005) Declines in Tanzania and Ethiopia are disputed (Nikolaus 2006). The conservation status has been upgraded from Least Concern to Near Threatened (BirdLife International 2007).

Habitat conversion to agro-pastoral systems, loss of wild ungulates leading to a reduced availability of carrion, hunting for trade, persecution and poisoning are factors for the declining population (Virani et al. 2011). The diclofenac compound, used to reduce pain and inflammation in livestock, acts as a poison as African and Asian vultures are vulnerable to its effects (Swan et al. 2006; Naidoo and Swan 2009; Ogada and Keesing 2010; Otieno et al. 2010; Phipps 2011). In Kenya, the toxic Carbamate-based pesticide Furadan™ has killed many vultures (Maina 2007; Mijele 2009; Otieno et al. 2010). In southern Africa, some birds are killed and eaten for perceived medicinal and psychological benefits (McKean and Botha 2007). Electrocution from powerlines is common in some areas (Bamford et al. 2009). In addition, the ungulate wildlife populations on which this species relies have declined precipitously throughout East Africa, even in protected areas (Western et al. 2009).

1.3 Slender-billed Vulture (*Gyps tenuirostris*)

Physical appearance

The Slender-billed vulture is another medium-sized vulture. Compared to the other Gyps species, this species appears smaller-headed, larger-eyed, longer-billed, longer-legged, ragged, dingy, and graceless with a less feathered head and neck, and 'large prominent ear canals that are noticeable even at a distance, not like the smaller ones in the Indian vulture and other Gyps vultures' (Rasmussen et al. 2001: 25).

Seen from a perch, adults have a black, nearly featherless head and neck, a dark bill with a pale culmen and a black cere, and a dirty white ruff. Brown is the dominant feather color with lighter colored streaked underparts, and white thigh patches. Juveniles differ in having white down on the upper neck and nape, and streaked upperparts (BirdLife International 2014). Subadults are intermediate between adults and juveniles (Rasmussen et al. 2001) (Fig. 1.3a,b,c).

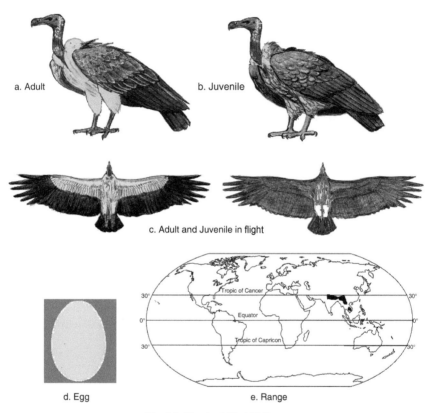

a. Adult

b. Juvenile

c. Adult and Juvenile in flight

d. Egg

e. Range

Fig. 1.3. Slender-billed Vulture.

Classification

The Slender-billed vulture was formerly classified with the Indian vulture *Gyps indicus* as the Long Billed vulture *Gyps indicus tenuirostris*, but is now recognized as a different species (Rasmussen and Parry 2000). Arshad et al. (2009a) studied the phylogeny and phylogeography of the Gyps species, using nuclear (RAG-1) and mitochondrial (cytochrome b) genes, and concluded that *G. indicus* and *G. tenuirostris* are separate species. The Slender-billed vulture differs from the Indian vulture in having a slenderer bill and darker brown plumage (Hall et al. 2011). The two species are also found in different regions. The Slender-billed vulture, a tree nester, is found in Southeast Asia north to the Sub-Himalayan regions. The Indian vulture, mostly a cliff nester, is found south of the Ganges in India.

Foraging

The Slender-billed vulture shares similar habitat with the White-rumped vulture, i.e., open grassland, savanna or mixed dry forest with open patches, near or far from human habitation (Baker 1932–1935; Lekagul and Round 1991; Robson 2000; Satheesan 2000a,b; BirdLife International 2001). This is reflected in its diet, which comprises carcasses of domestic animals, and wild deer and pigs killed by tigers (Sarker and Sarker 1985). In Nepal, their main diet comprises domestic livestock rather than wild ungulates (Baral 2010).

Breeding

The breeding season is December–January, recorded in studies in India (Baker 1932–1935), Myanmar (Smythies 1986) and in Kamrup district, Assam (Saikia and Bhattacharjee 1990c). Only one egg is laid in regularly used nests (Brown and Amadon 1968), sometimes in groups of up to 16 birds (Baker 1932–1935) (Fig. 1.3d).

   The Slender-billed vulture usually nests in large trees, such as the larger woody trees of the Genus Ficus, e.g., *Ficus religiosa*. This is a large deciduous or semi-evergreen tree up to 30 m (98 ft) tall, with a trunk diameter of up to 3 m (9.8 ft). Nests are located near or far from human settlement, usually 7 to 14 m (23–46 ft) high (Baker 1932–1935; Ali and Ripley 1968–1998; Brown and Amadon 1968; Grubh 1978; del Hoyo et al. 1994; Alström 1997; Grimmett et al. 1998; Rasmussen and Parry 2000). Nesting trees are the mango (*Mangifera indica* L.) and *kadam* (*Anthocephalus indicus* A. Rich.) in Kamrup district, Assam (Saikia and Bhattacharjee 1990). Other large trees were used for nesting in Myanmar (Smythies 1986) and in Khardah, Calcutta (Munn 1899, cited in BirdLife International 2001).

Population status

The distribution of the Slender-billed vulture is shown in Fig. 1.3e. It was one of the most numerous vultures in Southeast Asia during the first half of the twentieth century (Ferguson-Lees and Christie 2001; Pain et al. 2003; Hla et al. 2010; Prakash et al. 2012). The population of this species, in combination with the Indian vulture, declined to 3.2% of its former level in 2007 in India (Prakash et al. 2007) with similar declines in Nepal (Chaudhary et al. 2012). Diclofenac, the anti-inflammatory drug for the treatment of livestock, contributed to renal failure, visceral gout and death in vultures (Oaks et al. 2004a; Shultz et al. 2004; Swan et al. 2005; Gilbert et al. 2006). The veterinary drug ketoprofen, was also toxic in concentrations (Naidoo et al. 2009). Processing of dead livestock also reduced vulture access to carcasses (Poharkar et al. 2009).

1.4 The Indian Vulture (*Gyps indicus*) (Scopoli 1786)

Physical appearance

The Indian vulture is closely related to the Slender-billed vulture; in fact as noted above, both species were once considered one species, the Long-billed vulture. It is also smaller and more lightly-built than the Eurasian Griffon *Gyps fulvus*, with a weight of 5.5 to 6.3 kg (12–13.9 lbs), a body and tail length of 80–103 cm (31–41 in) and a wingspan of 1.96 to 2.38 m (6.4 to 7.8 ft) (Ferguson-Lees and Christie 2001).

This species has a pale yellowish bill and cere, a large fluffy white ruff, buff back and upperwings (with larger feathers containing dark centers that give the back 'a broadly scalloped appearance'); the neck is blackish with pale down on the upper hind neck, and pale yellow feet; in flight its thighs are similar in color to the underparts (Rassmussen et al. 2001: 24). Juveniles have a darker bill and cere, a pinkish head and neck covered with light colored down and heavily streaked plumage (Fig. 1.4a,b,c). Subadults are intermediate between the adults and juveniles (Rassmussen et al. 2001).

Classification

The physical differences between this species and the Slender-billed vulture, and the fact that their ranges do not overlap, contributed to the reclassification of the Long-billed vulture into the two species (Rasmussen and Parry 2001). Compared with the Slender-billed vulture, it has a shorter, deeper bill, a thicker partly down-covered neck (the other species has no down), a fluffier ruff, a clean buff rather than 'dingy' buff coloration of the back and upper wing coverts, lower feathers and uppertail coverts lacking dark centers, pale tipped upper- and uppertail-coverts and less downy outer leg feathers (Rassmussen et al. 2001: 25).

Foraging

This species is found in cities, towns and villages near cultivated areas, and in open and wooded areas. It also forages in open and wooded areas, and often in association with the White-rumped vulture when scavenging (BirdLife International 2012).

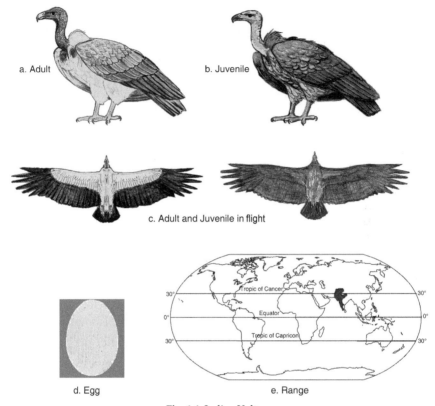

a. Adult

b. Juvenile

c. Adult and Juvenile in flight

d. Egg

e. Range

Fig. 1.4. Indian Vulture.

Breeding

The Indian vulture is mostly a colonial cliff nester (Ali and Ripley 1968–1998; Brown and Amadon 1968; del Hoyo et al. 1994; Alström 1997; Grimmett et al. 1998). A survey of the vultures present in Sindh, Pakistan by Iqbal et al. (2011) found all the vultures' nests on cliffs. In India, nests are usually built by small colonies of 2 to 12 pairs on cliff-face ledges and/or rock outcrops (Baker 1932–1935). Exceptions on record included a large colony of 50 nests at Taragurh hill, Rajasthan reported by Baker (1932–1935) and one of 30 pairs mentioned recorded by Hume and Oates (1889–1890). In Pakistan, colonies of 3–16 nests existed on outcrops up to the altitude of 325 m (Roberts 1991–1992). Subramanya and Naveein (2006) describe a nest 35 m up a steep-sided cliff, on the south side of the Ramadevarabetta hills, south-west of Bangalore (Karnataka, India).

However, in some areas, such as Rajasthan, tree nesting has been recorded (Naoroji 2006). Chhangani (2009: 65) notes that the Indian vultures in the Great Indian Thar desert of Rajasthan nested in both cliffs and trees, 'an aspect requiring more intensive work.' Mukherjee (1995), reported trees for nesting sites, but Collar, Andreev, Chan, Crosby, Subramanya and Tobias (the editors of Threatened Birds of Asia; The Bird Life International Red Data, Book, Bird Life International 2001: 618) argue that as there are no other reports of tree-nesting, this assertion 'may therefore be doubtful; it certainly requires verification.'

The breeding season in India is November–March, mostly concentrated between December and January for both India (Baker 1932–1935) and Pakistan (Roberts 1991–1992). Only one oval white egg is laid (Barnes 1885; Hume and Oates 1889–1890; Brown and Amadon 1968; Roberts 1991–1992). The incubation period for the egg is estimated to be about 50 days (Brown and Amadon 1968) (Fig. 1.4d).

Population status

The distribution of the Indian vulture is shown in Fig. 1.4e. As noted by Chaudhry et al. (2012) and Thiollay (2000) Long-billed vultures (including the current Indian vulture) and White-rumped vultures were the most abundant vultures in South Asia during the first half of the twentieth century. Now, they have severely declined in numbers.

1.5 Rüppell's Vulture (*Gyps rueppelli*) (Alfred Brehm 1852)

Physical appearance

The Rüppell's Griffon vulture, or the Rüppell's vulture, named after Eduard Rüppell, a nineteenth century German explorer, collector, and zoologist, is found mainly in the Sahel Savanna region of western and central Africa, between the Sudanian Savannas to the south and the Sahara Desert to the north (see Part 2, Chapter 6, 6.2. for a discussion of these terms). It is a large vulture, having a body length of 85 to 103 cm (33 to 41 in), a wingspan of 2.26 to 2.6 metres (7.4 to 8.5 ft), and a weight of 6.4 to 9 kg (14 to 20 lb) (Ferguson-Lees and Christie 2001). It is similar to the White-backed vulture, *Gyps bengalensis*, but has a yellowish bill and is considerably larger. The head and neck are covered with white down feathers, and the base of the neck has a white collar. The plumage is mottled brown to dark brown or black with a mixed white-brown underbelly (Fig. 1.5a,b,c).

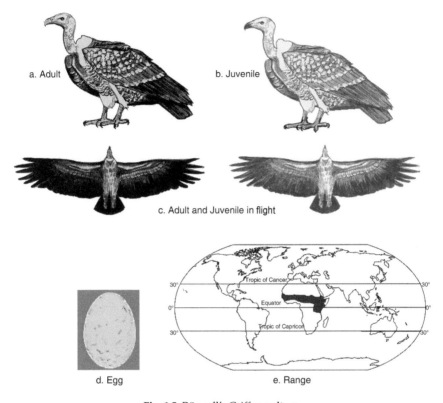

Fig. 1.5. Rüppell's Griffon vulture.

Classification

The Rüppell's Griffon vulture is closely related to the other species of the Genus Gyps. A study by Johnson et al. (2006: 65) using phylogenetic results from mitochondrial cytB, ND2 and control region sequence analysis 'supported a sister relationship between the Eurasian Vulture (*G. f. fulvus*), and Rüppell's Vulture (*G. rueppellii*), with this clade being sister to another consisting of the two taxa of "Long-billed" vulture (*G. indicus indicus* and *G. i. tenuirostris*), and the Cape Vulture (*G. coprotheres*)'.

Foraging

The foraging of this species is dependent on the large ungulate herds on the savannas. Migratory ungulates within the vultures' range, especially in East Africa include Blue Wildebeest (*Connochaetes taurinus*, Burchell

1823) Burchell's Zebra (*Equus burchelli* Gray 1824) and Thomson's Gazelle (*Eudorcas thomsonii*, Günther 1884) (Boone et al. 2006). A study by Kendall et al. (2012) found that the average number of Rüppell's vulture was higher during the ungulate migration season, enabling a fast response to wildlife density changes (Mundy et al. 1992; Kendall et al. 2012).

The Rüppell's Griffon is the commonest vulture of the Sahel savanna of Chad and Niger in West-Central Africa, Wacher et al. (2013). Here it feeds mostly on the carcasses of livestock (mostly camels, cattle, goats, horses and donkeys) rather than the outnumbered wild ungulates such as Dorcas Gazelle (*Gazella dorcas*, L. 1758), and to a lesser extent Dama Gazelle (*Nanger dama*, Pallas 1766) Barbary Sheep (*Ammotragus lervia*, Pall., 1777) and Addax (*Addax nasomaculatus*, de Blainville 1816). Rüppell's griffons in the Serengeti have been recorded as flying up to 150 km from the nest site to food source among the herds of migratory ungulates (Houston 1974b). Pennycuick (1972) suggested that the average foraging radius may be as far as 110 km.

The Rüppell's Griffon vulture is the world's highest-flying bird. An individual was involved in collision with an airplane over Abidjan, Côte d'Ivoire, at 11,000 metres or 36,100 ft. This species has a specialized variant of the hemoglobin alpha^D sub-unit which has a high affinity for oxygen which allows the species to absorb oxygen efficiently despite the low partial pressure in the upper troposphere (Laybourne 1974; Hiebl et al. 1988; Weber et al. 1988; Storz and Moriyama 2008).

Breeding

The breeding season of the Rüppell's Griffon varies greatly. For example, Virani et al. (2012) describe a colony in Kwenia, southern Kenya, composed of 150 to 200 adults (from 2002 and 2009) with a maximum of 64 nests occupied at any time. Egg laying dates varied each year; in some cases there were two egg-laying periods in one year (Fig. 1.5d). The number of nests was correlated with the previous year's rainfall. Ungulate populations in this area may have influenced breeding. The study concludes that nesting in Rüppell's vultures 'may be triggered' by rainfall and geared to producing fledged young at the end of the dry season (July–October) when carrion is most 'abundant' (ibid. 267). This point is also noted by Houston (1976). In another study, Houston (1990) found that the breeding time for colonies in the Serengeti region of Tanzania changed by 5 months between 1969–1970 and 1985, possibly correlated to the changes in ungulate populations. Food availability may have influenced two alternate breeding seasons, the choice of each period depending on the food available (Bouillault 1970; Mendelssohn and Marder 1989; Schlee 1988).

Nests may be in cliffs or rock outcrops or in trees. Tree nesting, considered atypical, has been recorded in West Africa (Rondeau et al. 2006). Wacher et al. (2013) give a detailed survey of Rüppell Griffon nesting and foraging presence in Chad and Niger in West-Central Africa (see also Scholte 1998). Of the 572 Rüppell's vultures recorded in the survey 47 nested on rocky inselbergs and 24 in tree crowns. Also, there were 24 cases of Rüppell's vultures using treetop stick nests, mostly on the crowns of the flat-topped thorny trees. Common tree species for nesting were the desert date (*Balanites aegyptiaca* (L.) Delile 1812) a medium-sized tree and to a lesser extent jiga (*Maerua crassifolia* Forssk), a slightly smaller tree (Wacher et al. 2013).

Population status

The distribution of the Rüppell's Griffon vulture is shown in Fig. 1.5e. Rüppell's Griffon vulture is considered near threatened by conservationists. In West Africa there have been severe declines in Mali, South Sudan, Burkina Faso, Mali, Niger, Cameroon, Uganda, Kenya, Somalia and Malawi and may be extinct in Nigeria (Rondeau and Thiollay 2004; Thiollay 2001, 2006; Nikolaus 2006; Virani 2006; BirdLife International 2014). One contributory factor may be the conversion of natural landcover to agriculture and urban landscapes (Buij et al. 2012). Other factors for the decline of the population of this species are poisoning from the toxic pesticide carbofuran, mostly in East Africa (Ogada and Keesing 2010; Otieno et al. 2010; Kendall and Virani 2012), the reduction of the ungulate wildlife populations (Western et al. 2009), diclofenac (BirdLife International 2007) and a substantial trade in vulture flesh and body parts, mostly in West Africa (Rondeau and Thiollay 2004; Nikolaus 2006). Possibly, there are also impacts from the actions of human climbing expeditions near the rocky outcrops in the Hombori and Dyounde massifs of Mali during the breeding season (Rondeau and Thiollay 2004).

1.6 Cape Vulture (*Gyps coprotheres*) (Forster 1798)

Physical appearance

The Cape Griffon vulture or Cape vulture is a large to very large vulture, larger than the Rüppell's or White-backed vultures. It occurs only in southern Africa, i.e., South Africa, Lesotho, Botswana, and is labelled as 'Critically Endangered' in Namibia (Simmons and Brown 2007). The length from bill to tail end is about 96–115 cm (38–45 in), the wingspan about 2.26–2.6 m (7.4–8.5 ft) and the weight 7–11 kg (15–24 lb) (Ferguson-Lees and Christie 2001). The coloration for an adult is creamy-buff, which contrasts with dark flight and tail feathers. The neck ruff is pale buff to dirty white.

Light silvery feathers and a black alula are visible in the underwing during flight. A black bill and its slightly larger size distinguish this species from the Rüppell's Griffon vulture. The juveniles have a darker, streaked plumage and a reddish neck (Fig. 1.6a,b,c).

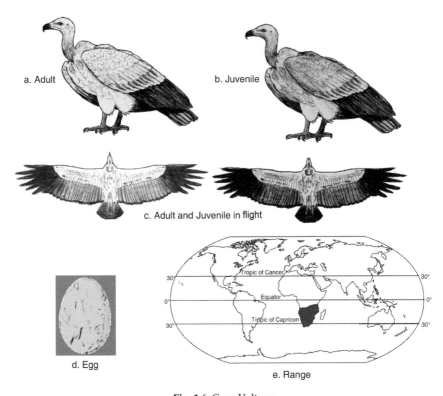

Fig. 1.6. Cape Vulture.

Classification

Johnson et al. (2006: 72) in a molecular study of the Gyps genus, noted a strong relation between *G. coprotheres*, *G. i. indicus* and *G. i. tenuirostris* in a clade, with these related to another clade of *G. f. fulvus* and *G. rueppellii*. Older publications held that the Cape vulture formed a superspecies with the Eurasian Griffon and Rüppell's vulture. In this classification, the Cape vulture would be a subspecies of *Gyps fulvus* (Stresemann and Amadon 1979; Amadon and Bull 1988). A close relation between *G. coprotheres* and *G. fulvus* was also supported by Wink (1995). This was based on the divergence in nucleotide sequences in the cytochrome b gene. Wink suggested that the two species diverged from a common ancestor about half a million years ago.

Foraging

Cape vultures in Namibia were described as denizens of open habitat such as grassland and open woodland savanna, with the majority (79%) of prey animals being wild ungulates (Schultz 2007). In that study, Greater Kudu (*Tragelaphus strepsiceros* Pallas 1766) were the most common prey item (54.2% of all carcasses) followed by cattle, eland and horses. The Cape vulture is a heavy bird with a high wing loading (112 N/m). It practices fast cross-country soaring (Pennycuick 1972). In this it contrasts with the commoner, smaller White-backed vulture in Namibia (a lighter bird with a wing loading of 76 N/m) (Pennycuick 1972; Bridgeford 2004). Possibly the White-backed vulture may have evolved in the denser wooded savanna, and hence is more adapted to bush encroachment than the Cape vulture, which may favor a more open grassland (Brown 1985; Schultz 2007). The White-backed vulture frequents alpine grassland, followed by moist woodland, sour grasslands, arid woodland, and mixed and sweet grasslands (Mundy et al. 2007).

The Cape vulture is the only vulture in southern Africa south of 28° and thus has little foraging competition (Mundy et al. 2007). North of this area it is in competition with four other species: the common White-backed vulture, the Hooded vulture, the Lappet-faced vulture and the White-headed vulture.

Breeding

The Cape vulture is a colonial cliff-nester (Bamford et al. 2007; Schultz 2007). Some individuals nest in trees (Bamford et al. 2007). Breeding usually starts in April to June, with the laying of one egg (Fig. 1.6d), with fledging between the end of October and mid-January. Juveniles normally remain in the vicinity of the colony until the following breeding season, followed by dispersal to start their own breeding some six years later (Pickford 1989; Mundy et al. 1992; Piper 1994; Hockey et al. 2005; Boshoff and Anderson 2006).

Population status

The distribution of the Cape vulture is shown in Fig. 1.6e. This species has been declining across its range (Monadjem et al. 2004; Shultz 2007). For example, in Namibia its range declined precipitously after the Second World War. All its known breeding colonies and roosts were abandoned, except for a handful of individuals by the 1990s (Mundy and Ledger 1977; Collar et al. 1994; Simmons and Bridgeford 1997; Simmons 2002). The incidence of the rinderpest disease was a factor for this decline (1886–1903) which killed many herbivorous animals, including domestic cattle. Other factors were the Anglo-Boer War, the destruction of game herds, the replacement of wild ungulate herds with domestic stock, the conversion of grazing land to cultivation and poisoning (Boshoff and Vernon 1980). Recent threats include electrocution and collision with power lines, persecution, killings for traditional medicine, drowning in farm reservoirs and human disturbances (Anderson 2000; Monadjem et al. 2004).

Bush encroachment after 1950 is a factor for the decline in the range of the Cape vulture (Schultz 2007). This involves the conversion of grassland and woodland savannas to dense acacia-dominated vegetation with minimal grass cover (Barnard 1998; Muntifering et al. 2006). This may result from changes in the incidence of bush fires and increased grazing pressure on grass (Ward 2005). There is no clear evidence that bush encroachment negatively affects vultures (Smit 2004). However, Schultz (2007) argues that it may reduce the visibility of carcasses for foraging vultures in dense bush (Houston 1974; Mundy et al. 1992). In some other studies, however, vultures have been recorded as finding non-visible food by following other avian scavengers such as the Bateleur eagle, Milvus kites, corvids and jackals (Mundy et al. 1992; Camina 2004). Bush encroachment may also indirectly reduce the livestock stocking rates and affect food sources (Bester 1996; Dean 2004; Smit 2004) which in turn may affect scavenger bird populations (Schultz 2007). Satellite image-based studies have shown that vultures prefer commercial farmland to communal areas or the protected Etosha National Park (Mendelsohn et al. 2005).

Vulture numbers in Botswana peak between protected and grazing land, as the birds may breed and roost inside conservation areas and feed on livestock in non-protected areas (Herremans and Herremans-Tonnoeyr 2000). This strategy is also observed among Griffon vultures *Gyps fulvus* in Israel (Bahat 1995; Schultz 2007).

1.7 Griffon vulture (*Gyps fulvus*, Hablizl 1783)

Physical appearance

The Griffon vulture, also called the Eurasian Griffon vulture or Common Griffon vulture is a large to very large vulture, and very closely related to the Cape Griffon vulture when molecular phylogeny of the mitochondrial cytochrome b (cyt) gene is used (Wink 1995). The Griffon vulture is one of only a few vulture species resident in Africa, Asia and Europe; the others are the Bearded vulture *Gypaetus barbatus*, the Cinereous vulture *Aegypius monachus* and the Egyptian vulture *Neophron percnopterus* (Houston 1983; Mundy et al. 1992; Clark 2001).

The body length and tail is 93–122 cm (37–48 in), while the wingspan ranges from 2.3 to 2.8 m (7.5–9.2 ft). Males usually weigh 6.2 to 10.5 kg (14 to 23 lbs) and females about 6.5 to 11.3 kg (14 to 25 lbs). This species has a yellow bill and white neck ruff. The buff body and wing coverts contrast with the dark flight feathers (Ali 1996; Ferguson-Lees and Christie 2001) (Fig. 1.7a,b,c).

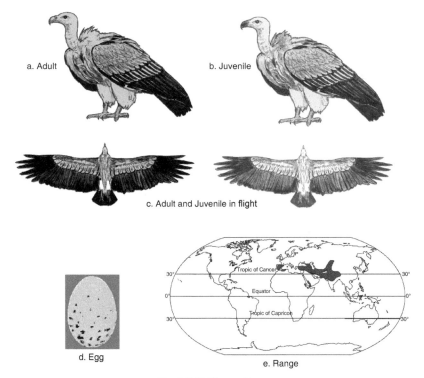

Fig. 1.7. Griffon vulture.

Classification

There are currently two recognized subspecies: *Gyps fulvus fulvus* and *Gyps fulvus fulvescens*. The former is found in Spain, Southern France and Morocco, eastwards to the eastern Mediterranean. The latter is found in northern India, north into Kyrgyzstan, Russia and the western Himalayas. Johnson et al. (2006: 66) note that their study represents the 'first attempt to ascertain Gyps systematics based on samples of all recognized species using molecular techniques'. The results show that the 'taxonomic status of the two subspecies of the Eurasian vulture (*G. f. fulvus* and *G. f. fulvescens*), as well as their characteristics and geographic distribution are unclear', possibly as *G. f. fulvescens* has not been included in a study of molecular phylogenetics, it may even be a different species. The evidence for this assertion is that the results for the two subspecies *G. f. fulvus* and *G. f. fulvescens* were 'phylogenetically distinct'; they were 'not placed as sister taxa', and importantly the samples of the *G. f. fulvescens* clustered with the Himalayan Giffon vulture (*G. himalayensis*) rather than with *G. f. fulvus*. Therefore the two subspecies, *fulvus* and *fulvescens*, were found to be phylogenetically distinct, not sister taxa, and further study might elevate both to different species.

Foraging

García-Ripollés et al. (2011: 127) state that 'little is known about the spatial ecology and ranging behavior of vultures in Europe', therefore they use GPS satellite telemetry to assess home-ranges of Eurasian Griffon vultures in Spain. The results showed that the birds ranged mainly in areas with traditional stock-raising practices, pasturing and to a lesser extent vulture restaurants. Griffon vultures may use temporary communal roosts where there is high food availability (Donazar 1993; Xirouchakis 2007; Olea and Mateo-Tomás 2013).

The Griffon vulture is almost exclusively a carrion-eater, with few records of killing injured and weak sheep or cattle (Camiña et al. 1995; Camiña 2004a). The diet mainly comprises livestock species (sheep, goat, cattle and horses) (Fernández 1975; Camiña 1996). Wild ungulate carcasses, such as those of *Rupicapra rupicapra*, *Cervus elaphus*, and *Capreolus capreolus* are eaten in mountain areas such as the Pyrenees and the Alps.

Many records show a dependence on domestic livestock (Donazar 1993; Tucker and Heath 1994). For example, Camiña in Slotta-Bachmayr et al. (2006) writes that in Spain, they are strongly dependent upon domestic pigs, cattle and sheep. A similar situation obtains in Portugal, where there is a tradition of putting carcasses out for vultures. Fernández y Fernández-Arroyo (2012) point out that the recent compulsory removal of animal carcasses in Spain has reduced vulture numbers (Camiña 2004b; Tella 2006; Camiña 2007; Camiña and López 2009; Fernández 2007a, 2007b, 2009; Melero 2007; Pérez de Ana 2007; FAB 2008; González and Moreno-Opo 2008; Suárez Arangüena 2008).

The carcass removal programme against the Bovine Spongiform Encephalopathy (BSE) in the Ebro of Aragon, Spain, decreased the food availability for vultures from about 3 carcasses per farm a day in 2004 to

0–0.5 carcasses in 2005 (Camina and Montelio 2006). Also, after the removal program, the proportion of adult birds increased relative to immature birds. The authors conclude that the removal program significantly affected the Griffon populations and hence requires more study. Zuberogoitia et al. (2010: 53) point out that the 'systematic removal of livestock from the mountains and the closing of vulture feeding stations in neighboring provinces have caused the local population to decline, with a simultaneous decrease in breeding parameters. Moreover, mortality due to starvation has become increasingly common in recently fledged birds of some vulture colonies' (see also Zuberogoitia et al. 2009).

The reduction in vulture numbers has also been seen in the French Pyrenees (LPO 2008; Razin et al. 2008). In Armenia, the situation is similar (Ghasabian and Aghababian 2005; Slotta-Bachmayr et al. 2005), but there is greater access to wild animal carcasses, including wild ibex and bears (Adamian and Klem 1999).

Griffon vultures in Spain and southern France may kill livestock, to the anger of local farmers (Margalida et al. 2011). Changes in European sanitary and conservation policies in 2002, to control the spread of bovine spongiform encephalopathy (commonly termed mad cow disease), and regulations for animal husbandry, which required that all carcasses be removed from farms are considered factors for these killings. Griffon vulture populations in Western Europe have also increased in the last decade, while those in the Eastern Mediterranean have declined (Parra and Telleria 2004; Slotta-Bachmayr et al. 2005). Margalida et al. (2011) write that there were 1,165 reported cases of vultures killing livestock in 2006 to 2010 in northern Spain. The compensation cost was nearly $350,000. In retaliation, stakeholders poisoned 243 vultures during this period.

Breeding

The Griffon vulture is a cliff nester, usually nesting on cliffs below 1,500 m in elevation and rarely in trees (sometimes in those of Cinereous vultures) (Traverso 2001; Xirouchakis and Mylonas 2005). It is also a colonial breeder, with colonies numbering up to 100 pairs (Del Moral and Martí 2001). Gavashelishvili and McGrady (2006), using the results of a study in the Caucasus, found that the probability of a cliff ledge being occupied by Griffon vulture nests was positively correlated with the percentage of open areas and the annual biomass of dead livestock, and was negatively correlated with annual rainfall. The breeding season extends from December/January until July, and the brood consists only of one chick (Donázar 1993). The egg is white, with light brown spots (Adamian and Klem 1999) (Fig. 1.7d).

Population status

The distribution of the Griffon vulture is shown in Fig. 1.7e. This species was once classified as threatened, but has recovered its numbers and is now classified as of least concern (BirdLife International 2007). Large populations are found in the Iberian Peninsula (Del Moral and Martí 2001). Crete 'supports the largest insular Griffon vulture population in the world' (Xirouchakis and Mylonas 2005:229). The very large range of the Griffon vulture is possibly a factor for its survival, as it encounters different circumstances in each of the regions that it inhabits (Parra and Telleria 2004). For example, in Italy in 2005 the total population stood at 320–390, with supports such as supplementary feeding, and problems such as poisoned baits, new power-lines and wind-power generation projects (Genero 2006). In Croatia, they are rare and possibly declining, with less than 200 birds in 2005. Problems include illegal poisoning, the end of traditional sheep farming in breeding areas, and disturbances from tourism or recreation near nest sites (Pavokovic and Susic 2006). In France, intensive work on conservation and re-introduction has resulted in increased populations (Tarrasse et al. 1994; Terrasse 2004; Terrasse 2006). In Cyprus, the Griffon vulture is one of the most threatened birds (Iezekiel and Nicolaou 2006). In Sicily, it was common throughout the nineteenth and early twentieth centuries, but it became extinct in 1969, due to poisoning (Priolo 1967). It has since been reintroduced, and a small population of about nine are breeding (Vittorio 2006).

In the countries of the former Soviet Union, Katzner et al. (2004a: 235) note that although there is 'no sign of severe decline in vulture populations', there are few or no records of Griffon breeding and population status. Russia and Uzbekistan are recorded as having the largest number of this species, and there are no records of breeding in Belarus, or in the Baltic States (Estonia, Latvia, Lithuania). The Ukraine has a small population in the Carpathians, and it appears the breeding population in Moldova. Populations in the Caucasus (Armenia, Azerbaijan, Georgia and Russia) are listed as declining. The most important factor for these declines is the 'massive declines in the number of livestock herded throughout this region' (Katzner et al. 2004a: 239).

1.8 Himalayan Griffon vulture (*Gyps himalayans*, Hume 1869)

Physical appearance

The Himalayan Griffon vulture or Himalayan vulture is a very large vulture, rivalling the Cinereous vulture as the largest of the Old World vultures (Mundy et al. 1992; Ferguson-Lees and Christie 2001). The adult length is 109 to 115 cm (43–45 in). The weight may vary from 8 to 12 kg (18–26 lbs) and the wingspan varies from 2.6 to 3.1 m (8.5 to 10 ft) (Schlee 1989; Rasmussen and Anderton 2005; Namgail and Yoram Yom-Tov 2009). The Himalayan Griffon compared with the Cinereous vulture has a slightly longer body length due to the longer neck, but the largest of the latter species are larger and heavier than the largest Himalayan Griffons (Thiollay 1994; Grimmett et al. 1999; Ferguson-Lees and Christie 2001; Chandler 2013).

In coloration, adults have a pale brown, white streaked ruff, a yellowish-white down-covered head, pale brown to whitish underside, under-wing coverts and leg feathers, and white to grayish feet (Fig. 1.8a,b,c). The upper body is pale buff, with no streaks, this color contrasting with the dark brown outer greater coverts and wing quills. The inner secondary feathers are paler on the tips. The facial skin is pale blue, a lighter color than the blue facial skin of *Gyps fulvus*. The bill is pale bluish-grey with darker tip, unlike the yellowish bill of *Gyps fulvus*. In flight it differs from *Gyps fulvus* in having dark wings and tail feathers that contrast with the pale body

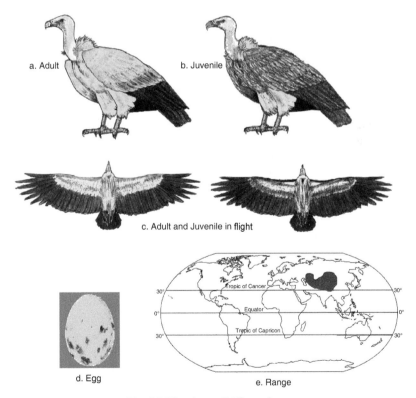

a. Adult b. Juvenile

c. Adult and Juvenile in flight

d. Egg e. Range

Fig. 1.8. Himalayan Griffon vulture.

and coverts. It also differs from *Gyps indicus* in being larger with a larger bill. In juveniles, the down on the head is whiter, with whitish streaks on the scapulars and wing coverts, with dark brown underparts (Brown and Amadon 1986; Alström 1997; Ferguson-Lees and Christie 2001; Rasmussen and Anderton 2005).

Classification

The Himalayan Griffon vulture is closely related to the European Griffon vulture (*G. fulvus*) and was considered a subspecies. In a study billed as the first inclusion of *G. himalayensis* and *G. f. fulvescens* in a molecular phylogenetic study, Johnson et al. (2006) found a close relation between *G. himalayensis* and *G. f. fulvescens*, in fact closer than the relation between *G. f. fulvus* and *G. f. fulvescens*, 'suggesting a topic for further analysis' (Johnson et al. 2006: 74).

Foraging

The Himalayan Griffon vulture forages in small groups over open meadows, alpine shrublands and open, partly forested landcover. The commonest food source are carcasses of domestic yak (64%), followed by human corpses (2%) and wild ungulates (1%) (Lu et al. 2009). Himalayan Griffon vultures are usually dominant over all other avian scavengers, except for the slightly larger Cinereous vulture (Thiollay 1994; Grimmett et al. 1999; Ferguson-Lees and Christie 2001). This species forages widely; in fact there is evidence of its range spreading southwards into southeast Asia, possibly due to 'the decline of large mammals, leading to food shortage in the breeding range, and resulting in long-distance dispersal of the species', and also 'climate change, deforestation and hunting, coupled with natural patterns of post-fledging dispersal and navigational inexperience may be contributing to this change' (Li and Kasorndorkbua 2008: 57, 60).

Breeding

The Himalayan Griffon is one of the world's least-known vultures and opinions differ regarding their nesting (Flint et al. 1989). While some writers state that they are not colonial breeders, others state that they are semi-colonial breeders (Ferguson-Lees and Christie 2001). The Himalayan vulture is mostly a cliff nester, sometimes colonial, utilizing inaccessible ledges. Some nests in northeastern India were located between 1,215 and 1,820 m (3,986 and 5,970 ft) in altitude and in Tibet at 4,245 m (13,927 ft), with colonies of about 10 to 14 birds in pairs (Brown and Amadon 1989).

Nesting may be in association with other species, such as Bearded Vultures, with variable levels of conflict (Brown and Amadon 1989; Katzner et al. 2004b). The breeding season has been recorded as beginning in January, but variable dates have been recorded for egg-laying (for example in India from December 25 to March 7). The single white egg is marked with red irregular spots (Fig. 1.8d). In captivity, the general incubation period was about 54–58 days, and is similarly variable in the wild.

Population status

The distribution of the Himalayan Griffon vulture is shown in Fig. 1.8e. The main range of this species is in the higher altitudes of the Himalayas, the Pamirs, Kazakhstan and the Tibetan Plateau. In older sources, the breeding range was from Bhutan in the south to Afghanistan in the north (Peters 1931). However, as noted above, more recent literature sources have recorded young, vagrant and foraging birds in Southeast Asia, as far south as Thailand, Myanmar, Singapore and Cambodia (Li and Kasorndorkbua 2008).

The Himalayan Griffon vulture has not experienced the major population decline of the other Gyps vultures in Asia (Lu et al. 2009). The remote location of the Himalayan Plateau and protection by Tibetan Buddhism are possible factors for this. The total population in the Tibetan Plateau is estimated at 229,339 (+/–40,447). The maximum carrying capacity of the plateau, based on food possibilities, may be 507,996 Griffons. The main habitat would be meadow habitats, which would support 76% of this population (Lu et al. 2009).

Although similar to the other Gyps species, the Himalayan Griffon vulture is recorded as less susceptible to the effects of diclofenac. There is no evidence of strong decline in their numbers over their range, despite population reductions in Nepal (Virani et al. 2008; Acharya et al. 2009). This may partly be the result of their remote range and individuals dispersing into the lowlands of southeast Asia might be more vulnerable (Li and Kasorndorkbua 2008).

1.9 Conclusions

This chapter has looked at the species of the Genus Gyps, which is currently seen as unique among the Old World vultures, because they are exclusively scavengers, unlike some other species that kill for food, or even eat fruits (Houston 1983; Mundy et al. 1992; del Hoyo et al. 1994). The specialized feeding habits are thought to have evolved with the development of migratory herds of large ungulates and associated savanna landcover

(Vrba 1985; Arctander et al. 1995; Hassanin and Douzery 1999, 2003; Jacobs et al. 1999; Matthee and Davis 2001). These developments may have contributed to their inability to take live prey (Houston 1983; Ruxton and Houston 2004). All the Gyps species have become increasingly dependent on domestic animals and human modified landscapes (Mundy et al. 1992). This increased the negative impact of introduced chemical compounds on their health (Pain et al. 2003; Arun and Azeez 2004; Green et al. 2004; Oaks et al. 2004; Shultz et al. 2004; Swan et al. 2004). Also, laboratory work has shown that the various Gyps species are closely related (Johnson et al. 2006), thus supporting their obvious physical similarities.

From an ecological perspective, the next question concerns the extent to which the Gyps species may resemble or differ from the non-Gyps vultures. Therefore the next chapter examines the other Old World vultures in order of size, and in terms of their biological relationships with each other and with the Griffons. Many of the issues discussed in this chapter will be repeated in a comparative mode. The objective is to give the reader a detailed perspective on all the Old World vultures, for later comparison with the New World vultures, and other scavengers and carnivores in the later chapters.

2

Systematic List of Old World Vultures: Huge and Small Non-Griffons

2 INTRODUCTION

This chapter looks at the eight species of Old World vultures which are not classified within the Genus Gyps. They are not necessarily closely related, but are separated for simple classification. They are, in approximate order of size from smallest to largest: the Egyptian vulture (*Neophron percnopterus*); the Hooded vulture (*Necrosyrtes monachus*); the Palm-nut vulture (*Gypohierax angolensis*); the White-headed vulture (*Trigonoceps occipitalis*); the Red-headed vulture (*Sarcogyps calvus*); the Lappet-faced vulture (*Torgos tracheliotus*); the Lammergeier or Bearded vulture (*Gypaetus barbatus*); and the Cinereous vulture (*Aegypius monachus*). This group varies greatly in size, appearance and habits. The Egyptian vulture and the Bearded vulture have been linked in classifications, sometimes with the Palm-nut vulture (Wink and Sauer-Gürth 2004: 489). These three species have feathered necks and in the case of the latter two, feathered heads (Wink and Sauer-Gürth 2004). The rest of the vultures in this chapter are usually grouped with the Griffons; they however differ markedly from each other, sharing mostly large tearing bills and generally solitary behavior, but widely varying coloration in exposed skin and plumage.

2.1 The Egyptian Vulture (*Neophron percnopterus*, Linnaeus 1758)

Physical appearance

The Egyptian vulture is also called the White Scavenger vulture or Pharaoh's Chicken and is the smallest of the Old World vultures. Despite its name, it is not found only in Egypt. Its range extends from the Cape Verde and Canary Islands, and Morocco in the West, across Algeria, Tunisia, Libya, to Northern Egypt and southwards to West and East Africa, including the Sahelian parts of Niger, northern Cameroon, Chad, northern Sudan and Ethiopia (Ferguson-Lees et al. 2001). In the north, it breeds in southern Europe, from Spain in the west, through the Mediterranean to Turkey. In Asia, it ranges from the Caucasus and central Asia to Northern Iran, Pakistan, northern India and Nepal (Cramp and Simmons 1980; Del Hoyo et al. 1994; Levy 1996; Angelove et al. 2013; BirdLife International 2014).

The three widely-recognized subspecies vary in size, with slight graduation due to migration, dispersal and intermixing. The medium-sized, nominate subspecies *Neophron percnopterus percnopterus* is found from Southern Europe to central Asia and northwestern India, and also in northern Africa south through Tanzania, into Angola and Namibia. The smaller *Neophron percnopterus ginginianus* occurs in Nepal and India (except for northwestern India). The largest, *Neophron percnopterus majorensis* is found in the Canary Islands, where it is may have been resident for more than 2000 years (Donázar et al. 2002a; Agudo et al. 2010). In the nominate species, the average length is 47–65 cm (19–26 in) from bill to tail. The weight is about 1.9 kg (4.2 lbs). The long, slender bill is dark grey. *N. p. majorensis* is about 18 percent larger in body mass, with a weight of 2.4 kg (5.3 lb) (Donázar et al. 2002a,b). *N. p. ginginianus* is 10 to 15 percent smaller than *N.p. pernopterus*, as males are 47–52 cm (19–20 in) long and females are 52–55.5 cm (20–21.9 in) long (Rasmussen and Anderton 2005). The bill is pale yellow. The wingspan is about 155 to 170 cm (Ferguson-Lees and Christie 2001; Donázar et al. 2002a).

The face and bill base are yellow in both sexes, but males may have deeper orange skin during the breeding season. The feathers of the neck are long. The tail in flight is wedge shaped and the primaries and secondaries are black, contrasting with the white underwing coverts and white centres to the primaries above. The wings are pointed and the legs are pink in adults, but grey in juveniles. Females are slightly larger than males. Juveniles are brownish black to brown, patched with black and white, until they attain adult plumage at five years (Ali and Ripley 1978; Clark and Schmitt 1998; Ferguson-Lees and Christie 2001) (Fig. 2.1a,b,c).

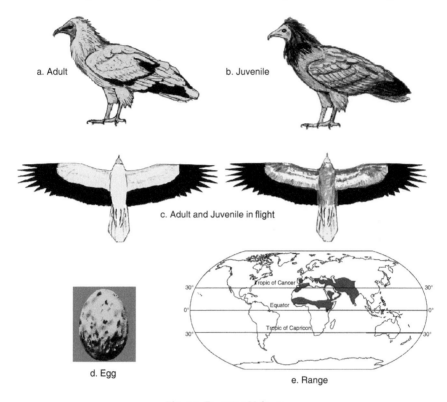

a. Adult

b. Juvenile

c. Adult and Juvenile in flight

d. Egg

e. Range

Fig. 2.1. Egyptian Vulture.

Classification

The Egyptian vulture is the only member of the Genus Neophron (Savigny 1809). This genus is recorded as the oldest branch within the evolutionary tree of vultures (Wink et al. 1996). Regarding the different subspecies within the species of *Neophron percnopterus*, Kretzmann et al. (2003) used microsatellite loci to determine the genetic difference between the sedentary populations on the Canary and Balearic Islands and the migratory populations on the Iberian Peninsula. This and other studies examining both molecular and morphological differences discovered the new subspecies *N. p. majorensis*, proving limited gene flow between Canarian and other populations (Donázar et al. 2002a). Donázar et al. (2002a) note that the morphological differences between *N. p. majorensis* and *N. p. percnopterus* in western European and African populations (including Cape Verde Islands) are as significant as those between the subspecies *ginginianus* and *percnopterus* in Central Asia.

In addition to the three subspecies mentioned above (i.e., *Neophron percnopterus percnopterus, Neophron percnopterus majorensis* and *Neophron percnopterus ginginianus*) two investigators named Nikolai Zarudny and Härms also described another possible subspecies, named *N. p. rubripersonatus*, from Baluchistan region in 1902. This was identified by a dark bill with a yellow tip and darker reddish-orange skin on the head (Hartert 1920). This coloration was not recognized as grounds for a different subspecies, but rather an intermixing of subspecies due to its intermediate coloration (Rasmussen and Anderton 2005).

The Egyptian vulture has also been studied within the evolutionary lineages of Old World vultures (Wink and Hedi Sauer-Gurth 2004). There are two main evolutionary lineages in Old World vultures (Wink 1995; Wink and Seibold 1996; Wink et al. 1998; Seibold and Helbig 1995). One assemblage, which 'shares many biological characters' includes the Egyptian vulture (*Neophron percnopterus*) and the Bearded vulture (*Gypaetus barbatus*) (Wink and Sauer-Gürth 2004: 489). The Bearded vulture is considered the nearest evolutionary relative of the Egyptian vulture, and the two are often placed together in the separate subfamily of the Gypaetinae (Seibold and Helbig 1995). There are three Genera in this subfamily: Genus Neophron for the Egyptian vulture, *Neophron percnopterus*; Genus Gypaetus for the Lammergeier or Bearded vulture; and Genus Gypohierax for the Palm-nut vulture. The second lineage includes the genera Gyps and Necrosyrtes which are a monophyletic clade with Sarcogyps and Aegypius/Torgos/Trigonoceps, and Sarcogyps. The first group appears physically different as, while the second group have downy or bare heads, the Egyptian, Bearded and Palmnut vultures have feathered necks and for the latter two, the heads are also feathered (Wink and Sauer-Gürth 2004).

Foraging

The Egyptian vulture is ecologically adaptable, eating carrion, faeces, waste, insects and eggs. It lives mainly in open arid and rugged landscapes (Donázar et al. 2002b). It is usually marginalized by the larger vultures at a carcass, due to its small size. Consequently, it feeds after the larger birds have opened the carcass and consumed most of the flesh, leaving small morsels for the pickers (Ferguson-Lees and Christie 2001).

Egyptian vultures, similar to some other Old and New World vultures, roost communally (Ceballos and Donazar 1990). Roosts have been reported throughout the entire range of the species (Brown and Amadon 1968; Cramp and Simmons 1980). Roosting trees include the European white poplar (*Populus alba* L.). A study by Ceballos and Donazar (1990) in northern Spain using fecal pellets in roosting areas found the principal food species to be European Rabbit (*Oryctolagus cuniculus*, Linnaeus 1758), domestic cat (*Felis catus*, Linnaeus 1758), domestic dog (*Canis familiaris* Linnaeus 1758), European badger (*Meles meles*, Linnaeus 1758), red fox (*Vulpes vulpes*, Linnaeus 1758), domestic horse (*Equus caballus*, Linnaeus 1758), domestic and wild pig (*Sus scrofa*, Linnaeus 1758), domestic cattle (*Bos primigenius*, Bojanus), domestic sheep (*Ovis aries*, Linnaeus 1758) and domestic goat (*Capra aegagrus hircus*, Linnaeus 1758). The preponderance of each species was linked to the seasonal changes in the death rate of these mammals due to disease. There was also much plant matter in the food, mostly the stems and seeds of both cultivated and wild plants, e.g., watermelon (*Citrullus lanatus* (Thunb.) Matsum. and Nakai), cherry (*Prunus avium* (L.) L. 1755) and melon (*Cucumis melo* L.). Most of the plant material was associated with wool and hair, as the plant matter was eaten to aid pellet formation (Ceballos and Donazar 1990).

Breeding

Typically the Egyptian vulture nests on ledges or in caves on cliffs (Sarà and Di Vittorio 2003). Nest sites were on huge boulders, and ledges, crevices and caves on cliffs in Turkey (Vaassen 2000), Israel (Shirihai 1996), the Canary Islands (Clarke 2006), the Cape Verde Islands (Clarke 2006) and Eritrea (Smith 1957). Donázar et al. (2002b) record breeding in the Canaries Islands taking place in holes in cliffs of variable size. Angelove et al. (2013: 141) in study in Oman, found that all the recorded nests were 'in holes or crevices on very steep slopes or cliffs that had an abundance of potential nesting cavities', these being located 'high up (mean elevation 119 m, n = 32) on ridges and hills that were remote from human habitation.' Also favored are cliffs located near the bottoms of valleys (Bergier and Cheylan 1980). This may be due to 'the necessity of minimizing the energy investment in finding and carrying food to the nest', as this species carries food to the nest in its beak, 'so for the small amount it is able to carry long trips may not be worthwhile' (Cerballos and Donazar 1989: 358). Cerballos and Donazar (1989: 358) further record that 'the placement of nests at the bottom of valleys shows that the Egyptian vulture is a species notably tolerant towards human activities, a factor which has an extremely variable effect on other species.'

Records exist of wider choices in nest locations, especially in the past when the population of this species was much higher. Rarely, ground nesting is recorded. 'To our knowledge, there are three published records on pairs occupying ground nests under extreme conditions' (Nikolov et al. 2013:

418). One was in the Canary Islands, where two cave nesters moved to flat shrubland, reared a juvenile and returned to cave nesting the next year (Gangoso and Palacios 2005). Another case was on the Island of Farasan, in the Red Sea near Saudi Arabia, where a pair nested in an old Osprey *Pandion haliaetus* nest (Jennings 2010). The third case was in India, where there was a nest at the base of a tree (Paynter 1924).

Other nesting sites are in towns and even the Egyptian pyramids (see also Heuglin 1869; Nikolaus 1984; Baumgart et al. 1995; Nikolov et al. 2013). Tree nesting was recorded in India, Pakistan, Somalia, Sudan and Oman (Butler 1905; Archer and Godman 1937; Gallagher 1989; Naoroji 2006). Other records include those of Wadley (1951) who described a breeding colony of forty birds in trees from 1943 to 1946 in Western Turkey; and also those of (Prostov 1955) near the village of Veselie, and of Arabadzhiev (1962) near the town of Yambol, both in Bulgaria (Nikolov et al. 2013). In northern Spain, many birds nested in pines (Aleppo Pine, *Pinus halepensis* Miller; Scots Pine, *P. sylvestrics* L. and European Black Pine, *P. nigra* J.F. Arnold). Ceballos and Donazar (1990: 23) note that 'The choice of pines for roosting may only be a consequence of abundance in the study area and of the fact that pines in Spain reach taller heights than broad-leafed trees.'

The egg-laying period varies in different countries. In Spain it was from March to April (Ceballos and Donazar 1990). In Georgia, eggs were laid in the first half of April (Gavashelishvili 2005), in Armenia May to June (Adamian and Klem 1999), in Israel from March to April (Shirihai 1996), in the Canary Islands February to mid-April (Clarke 2006), the Cape Verde Islands from November to April (Clarke 2006), Morocco from late March to early May, starting with desert areas (Thévonot et al. 2003); in Algeria in late March to late April (Heim de Balsac and Mayaud 1962); in Tunisia from late March to late April (Heim de Balsac and Mayaud 1962; Isenmann et al. 2005); and in Saudi Arabia during the first few months of the year (Jennings 1994, 1996). Angelove et al. (2013: 143) reported that vultures on Masirah Island, Oman laid eggs October–March ($n = 25$), with most laying in January and February. They however speculate on the possibility that there would be a laying period May–September, 'but the timing of our surveys precluded determining this.'

The records for Africa are more scanty, however in Eritrea and Ethiopia, eggs were laid in the months of January to May, and October to November (Ash and Atkins 2009). In Somalia, laying took place from January–April (Archer and Godman 1937; Ash and Miskell 1998). In Uganda, laying occurred from October to December (Carswell et al. 2005). In South Africa, the critical months are August to November or December, beginning with the spring rains (Brooke 1982; Simmons and Brown 2006a).

'The Egyptian vulture is, with the Lammergeier *Gypaetus barbatus*, the only vulture whose clutch usually has two eggs' (Donazar and Ceballos 1989: 217). In about 70% of breeding attempts, one chick survives (see Cramp and Simmons 1980; Del Hoyo et al. 1994). Eggs are oval to round oval, rough dull white, dark orange-brown spotted and streaked eggs (Fig. 2.1d) (Isenmann et al. (op cit.) Heim de Balsac and Mayaud 1962; Shirihai 1996; Adamian and Klem 1999). The incubation periods and nesting duration vary. For example, in Armenia the incubation period was 42 days and the nesting period 69 to 90 days (Gavashelishvili 2005). In Israel it was 39 to 45 days, and the nesting period 70 to 90 days (Shirihai 1996).

Population status

The distribution of the Egyptian vulture is shown in Fig. 2.1e. Some populations, however have remained stable or slightly increased; for example in Spain (Del Moral 2009; Kobierzycki 2011; Yemen (Porter and Suleiman 2012) and Oman (Angelove et al. 2013). Egyptian vulture populations have declined across its wide range, especially outside Spain, in the Mediterranean islands (e.g., Cyprus, Greece, Crete and Malta, and in the southern parts of Africa and Central Asia (Mundy et al. 1992; Tucker and Heath 1994; Levy 1996; Xirouchakis and Tsiakiris 2009; Angelove et al. 2013). In the Macaronesic archipelagos off the Western European and northwestern African coasts, it declined from common status during the latter part of the twentieth century (Bannerman 1963; Bannerman and Bannerman 1968). It is also extinct or very rare from several islands in the Canary Islands (Martın 1987; Delgado 1999), with reduced populations only on the islands of Fuerteventura and Lanzarote 1980 (Palacios 2000).

The Canary population is particularly important, due to its reclassification as a new subspecies (*N. p. majorensis*) (White and Kiff 2000; Donazar et al. 2002b). In the Canaries, as in some parts of Western Europe, the reasons for the population decline have been described as 'illegal persecution, poisoning, electrocution, habitat destruction and reduction of food supplies' (Donazar et al. 2002b: 90). These authors note that in the Canaries at least Egyptian vultures seem frequently to select power lines for roosting, making them 'extremely vulnerable to accidents by collision or electrocution' (p. 95).

Egyptian vultures in temperate Europe migrate to Africa during winter, mostly avoiding large water bodies (Spaar 1997; Yosef and Alon 1997). In Italy, they migrate through Sicily and the islands of Marettimo and Pantelleriato to Tunisia (Agostini et al. 2004). In the Iberian Peninsula, they migrate across the Strait of Gibraltar to North Africa, and others move south through Lebanon, Syria, Palestine, Jordan, Israel (Ferguson-Lees and Christie 2001; Meyburg et al. 2004; García-Ripollés et al. 2010).

2.2 The Hooded Vulture *Necrosyrtes monachus* (Temminck 1823)

Physical appearance

The Hooded vulture (*Necrosyrtes monachus*) is another small vulture, only marginally larger than the Egyptian vulture. Anderson and Horwitz (2008) describe the Hooded vulture and the Egyptian vulture as the same length (66 cm) and weight (1.9 kg). The largest individuals are found in South Africa, with smaller birds found in the more northerly parts of its range (Snow 1978; Mundy et al. 1992). The average length is 62–72 cm (25–28 in) long, have a wingspan of 155–165 cm (61–65 in) and a body weight of 1.5–2.6 kg (3.3–5.7 lbs) (Ferguson-Lees and Christie 2001). The coloration is brown, with brownish to whitish down on the crown and the back of the neck, and pinkish to greyish skin on the face and throat. There is also a brown feather ruff on the lower neck. The underparts are whitish, streaked with brown (Fig. 2.2a,b,c).

Classification

The Hooded vulture is the only species of the Genus *Necrosyrtes* (Gloger, 1841). Some investigators recognize a second subspecies, *N. m. pileatus*, occurring throughout the range outside West Africa, with the nominate subspecies classified as *Necrosyrtes monachus monachus*, and limited to West Africa. Kemp (1994: 126) noted that 'Two subspecies sometimes recognized, with nominate race in W. Africa and rest of range covered by race pileatus, but variation only in size... from small in W to large in S'. Other authors write that it should be monotypic (Brown and Amadon 1968). In the past, it was placed with the Egyptian vulture in the Genus Neophron due to perceived morphological similarities (e.g., Britton 1980; Louette 1981; Nikolaus 1987; Gore 1990). This classification was however rejected by cytochrome b gene studies (Wink 1995), and studies using the sequences of two mitochondrial genes and one nuclear intron (Lerner and Mindell 2005). The results of these studies were that the Hooded vulture is more likely to be related to the clade of true Old World vultures, Aegypiinae, including Aegypius, Gyps, Sarcogyps, Trigonoceps and Torgos, and less likely related to the Neophron-Gypaetus clade, Gypaetinae. Studies have not clarified the link of Necrosyrtes to Aegypiinae, but it is possible it diverged 3 to 5 million years BP, from the Genus Gyps, commonly considered its nearest relatives (Wink 1995).

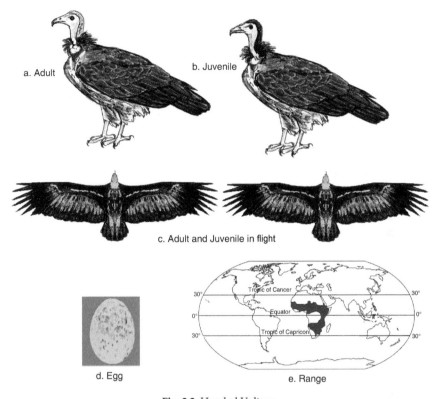

a. Adult

b. Juvenile

c. Adult and Juvenile in flight

d. Egg

e. Range

Fig. 2.2. Hooded Vulture.

Foraging

Hooded vultures, unlike some other larger vultures are often associated with human settlements. They are also common in desert, coastal land, grassland, wooded savanna and forest edge (Ferguson-Lees and Christie 2001). Wacher et al. (2013: 195) write that 'the distribution of human activity is clearly significant to Hooded Vultures in West Africa, but the interaction is less clear-cut for larger species.' Buij et al. (2013: 240) write of 'the greater abundance of Hooded vultures in the areas with little natural habitat and high human populations.' This species has been recorded in high altitudes up to 4,000 m, but is commonest below 1,800 m (BirdLife International 2012). In the Masai Mara National Park, the Hooded vultures, similar to the Egyptian vulture, compared to the larger vultures 'have a more varied diet, using several food sources other than carrion, but when feeding at a carcass tend to peck on scraps' (Kendall et al. 2012: 525). Hooded vultures were also more likely to associate with Egyptian vultures at kills, despite their similar feeding habits, than with the other larger vultures.

Breeding

The Hooded vulture is usually a tree nester, commonly nesting in stick nests in colonies on palm trees. Daneel (1984) gives examples of nesting from Kruger National Park South Africa. These were located in trees, namely Jackalberry or African Ebony *Diospyros mespiliformis* Hochst. ex A. DC. and sycamore fig or fig-mulberry *Ficus sycomorus* L. trees. Both of these trees may reach over 20 m in height.

The breeding period of the Hooded vulture varies regionally. In Northeast Africa most birds breed from October to June. Breeding in West Africa and Kenya occurs mainly from November to July, but may take place all year. In southern Africa, most breeding takes place from May to December. One egg is laid, and the incubation period lasts 46–54 days, with a fledging period of 80–130 days (Ferguson-Lees and Christie 2001) (Fig. 2.2d).

Population status

The population distribution of the Hooded vulture is shown in Fig. 2.2e. The Hooded vulture is mostly resident across most if its range. It is fairly common in the Sahelian countries (Senegal, southern Mauritania, Niger, Chad, and Sudan, South Sudan, Ethiopia and western Somalia), southward to northern Namibia and Botswana, Zimbabwe, southern Mozambique and South Africa (Ferguson-Lees and Christie 2001).

Recently, the population has declined across its range (Ogada and Buij 2011). Indirect poisoning, killings for capture for human food or 'bushmeat' or traditional medicine, and deliberate persecution are the main factors for this decline (Ogada and Buij 2011). The indirect poisoning is usually due to the use of carbofuran pesticides in dead livestock to poison mammalian predators (Otieno et al. 2010; Kendall 2012).

Hooded vulture meat is reportedly sold as chicken in some places. Intentional poisoning of vultures may be carried out by poachers to hide the locations of their kills. Secondary poisoning occurs when vultures ingest meat from carcasses laced with carbofuran pesticides (Otieno et al. 2010; Kendall 2012). Improved abattoir hygiene and garbage disposal, and land conversion to more developed uses also affect vultures (Ogada and Buij 2011). The Hooded vulture may also acquire and die from avian influenza (H5N1), possibly from feeding on dead domestic poultry; 'Hooded vultures could potentially be vectors or sentinels of influenza subtype H5N1, as are cats and swans elsewhere' (Ducatez et al. 2007: 611).

2.3 Palm-nut Vulture, *Gypohierax angolensis* (Gmelin 1788)

Physical appearance

The Palm-nut vulture is one of the smaller Old World vultures. The head is feathered and the plumage is mostly white, with black primaries on the wings and a red face and eye patch. The head, body and tail is 60 cm (24 in) long and the wingspan is about 150 cm (59 in). The average weight is 1.2–1.5 kg (2.6–4 lbs) (Fig. 2.3a,b,c). The adult Palm-nut vulture is similar to the African Fish Eagle and the Egyptian vulture, but lacks the chestnut body feathers of the eagle and white tail of the Egyptian vulture. Juveniles are predominately brown with partially black wings and a yellow eye patch (Fig. 2.3a,b,c).

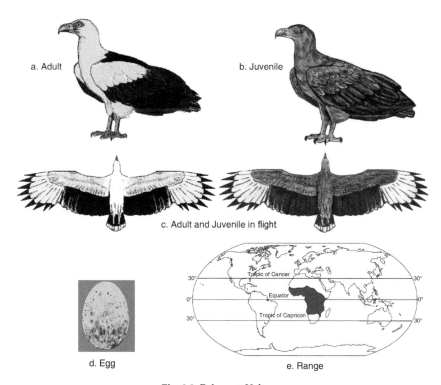

Fig. 2.3. Palm-nut Vulture.

Classification

The Palm-nut vulture, also called the Vulturine Fish Eagle, is the only member of the Genus *Gypohierax* Rüppell, 1836. As cited earlier, molecular studies have identified two clades of old world vultures; one comprising the hooded vulture Necrosyrtes and Gyps, with a sister group of Aegypius/Torgos/Trigonoceps; and a second clade of the Bearded vulture (*Gypaetus barbatus*) and the Egyptian Vulture (*Neophron percnopterus*). The Palm-nut vulture of Africa is usually classified near the second clade (Wink and Sauer-Gürth 2004).

Foraging

Adults are largely sedentary, but juveniles are more mobile, and may wander beyond their breeding range into the northern Sahel Savanna and to the extreme south of Southern Africa, foraging on both plants and animals (Ferguson-Lees and Christie 2001). The prefered habitat is dense forest and treed savanna, usually near water, with and without oil and raffia palms, and also with and without human proximity.

The Palm-nut vulture eats fruit of the oil palm (*Elaeis guineensis* Jacq.), raffia palm (Genus Raphia P. Beauv.), and fruits and grains of other plants, which may comprise as much as 65% of its diet (del Hoyo et al. 1994; Ferguson-Lees and Christie 2001). However, other food sources exist; Thomson and Moreau (1957) describe it eating living fish in the Gambia. In East Africa, it occurs in the absence of oil-palms, subsisting partly on Raffia fruit. These authors argue that based on this evidence, and also on the 'survival of Gypohierax in zoos, it is concluded that the vitamin-rich oil-palm fruit is not essential to the bird; but one that had been in captivity for 11 years, when presented with the fruit, preferred that to its usual meat' (Thomson and Moreau 1957: 608). It also eats living small mammals, birds, reptiles, fish, invertebrates and amphibians, and scavenges on carcasses (Ferguson-Lees and Christie 2001; del Hoyo et al. 1994).

Breeding

The Palm-nut vulture nests in tall trees, using large stick nests. The breeding season is usually from October to May in West and Central Africa, June to January in East Africa, August to January in southern Africa and May to December in Angola (Ferguson-Lees and Christie 2001). It lays one white and chocolate-brown egg, which is incubated for four to six weeks, and the chick fledges in 85 to 90 days (Fig. 2.3d).

Population status

The distribution of the Palm-nut vulture is shown in Fig. 2.3e. It is a tropical African species, found from west and central Africa, where it is common, south to north east South Africa where it is rarer. In the northern and central latitudes of its range it is common to abundant, but it is rarer in the south and east. Due to its wide range and stable population, the Palm-nut vulture is not classified as vulnerable like other vultures (BirdLife International 2013). It is not known to be affected by pesticides or poisons (Del Hoyo et al. 1994). It is not persecuted by humans for food or medicine, but there are issues related to habitat loss in West Africa. Its range is related to the distribution of the oil palm, which has expanded in West Africa, Angola and Zululand for commercial reasons even if harvesting disturbs potential nesting sites (Ferguson-Lees and Christie 2001).

2.4 White-headed Vulture, *Trigonoceps occipitalis* (Burchell 1824)

Physical appearance

The White-headed vulture is a medium-sized vulture, 72–85 cm (28–34 in) in length and a wingspan of 207–230 cm (82–91 in). Females are slightly larger than males; they usually weigh around 4.7 kg (10.4 lbs), while males weigh 4 kg (8.8 lbs) or less. It has a rear-peaked, downy white topped head with pinkish skin, with a large dark tipped red bill and blue cere, a black collar ruff, ruffed white legs, and a black feathered breast. The female has white inner secondaries, but in the male they are grey. In the juvenile, the plumage is dark, with a white head and white edges to wing linings similar to those of the adult (Ferguson-Lees and Chrsitie 2001; BirdLife International 2014) (Fig. 2.4a,b,c).

Classification

Wink (1995) classifies this species with the clade Aegypius complex (Aegypius, Torgos, Trigonoceps and Sarcogyps). Snow (1978) noted that the White-headed vulture has similar morphological similarities to the Asian Red-headed vulture (*Sarcogyps calvus*), despite the neck wattles of the latter species and considered the possibility of merging the two genera Trigonoceps and Sarcogyps. This species is often described as eagle-like. For example, Hancock (2013) (Birdlife of Botswana) writes 'On a continuum from predatory eagles to scavenging vultures, the White-headed vulture sits near the middle, rubbing shoulders with its larger relative the Lappet-faced vulture on the one side, and the Bateleur on the other.'

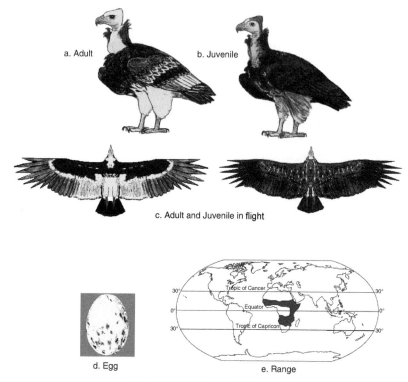

a. Adult

b. Juvenile

c. Adult and Juvenile in flight

d. Egg

e. Range

Fig. 2.4. White-headed Vulture.

Foraging

The White-headed vulture is the most solitary forager among the African vultures, a loner, often the first at a carcass, or feeding alone on small carcasses (Steyn 1982). Ash and Atkins (2009) taking a case study of foraging birds in Ethiopia, recorded 56% flying single, 24% in pairs, 18% in small groups of 2–5 birds, and only once in a large group (10 individuals). Dowsett et al. (2008) in Zambia and Dowsett-Lemaire and Dowsett (2006) in Malawi also report only individuals or small groups of three or four appear at carcasses, unlike the social Griffon vultures. The distribution of this species is also related to its main nesting tree, the Baobab *Adansonia digitata* (Palgrave 1977). However, preferred habitats include lightly wooded savanna, bush savanna, montane forest-grassland mosaic, mixed deciduous and broad-leaved woodland in low elevations, and even semi-desert (Mundy et al. 1992; Penry 1994; Dowsett-Lemaire and Dowsett 2006; Dowsett et al. 2008;

Ash and Atkins 2009; Borrow and Demey 2010). It is occasionally recorded in the arid thornveld areas of western Etosha National Park in Namibia and the Kalahari (Mundy 1997). It is usually rare near dense forests, human habitation and cultivation (Penry 1994; Dowsett et al. 2008).

Reinforcing its eagle-like status, the White Headed vulture predates lizards, snakes, insects, even mongooses, piglets and possibly *Queleas quelea quelea* (Linnaeus 1758) and occasionally takes the prey of other avian predators (Attwell 1963; Biggs 2001a,b; Dowsett-Lemaire 2006; Dowsett et al. 2008). Despite its slightly smaller size, it is usually dominant over single White-backed vultures, but can be driven away by flocks of the species (Mundy 1997). In a study in Botswana, Herremans and Herremans-Tonnoeyr (2000) found the foraging patterns of the White Headed vulture more similar to the Wahlberg's Eagle *Aquila wahlberg*i and steppe eagle *A. nipalensis*, which foraged in both conservation areas and unprotected land, rather than those of the other vultures, which clustered at the interface between conservation areas and unprotected land.

Breeding

The White-headed vulture is a non-colonial nester; individual nests may be up to 8 to 15 km apart, in the crown of tall, usually solitary baobab or acacia trees (e.g., *Acacia nigrescens*). Its distribution is closely linked to that of the baobab (Palgrave 1977; Irwin 1981; Dowsett-Lemaire 2006; Hartley and Hulme 2005).

Different breeding dates have been recorded in different countries. It usually lays one egg (Dowsett et al. 2008) (Fig. 2.4d). For example, in Ethiopia, egg-laying was reported from October to December (Ash and Atkins 2009), while in Somalia the breeding season was from late October to March, with most of the egg laying occurring in December and January (Ash and Miskell 1998). In Uganda, there are variable egg-laying dates recorded; for example April (1949 and 1969), March 1964 and November 1949 (East Africa Natural History Society Nest Records) and also November and January (Kinloch 1956).

In Southern Africa, egg-laying occurs mostly in the winter (i.e., May to July) (Mundy et al. 1992, 2008). In Zambia, egg-laying was documented from May to October, with the largest number of records for June and July (Dowsett et al. 2008). Dowsett-Lemaire and Dowsett (2006) gave examples of nesting in May and August in Malawi. In Namibia, there are two records for June and July (Jarvis et al. 2001). In Bostwana, there are records from August (Penry 1994) and May to July (Skinner 1997). In Zimbabwe, egg-laying dates from June, July and August (Irwin 1981; Hartley and Hulme 2005).

Population status

The distribution of the White-headed Vulture is shown in Fig. 2.4e. The evidence points to a small population in a large range within sub-Saharan Africa generally outside the dense forest and human habitation (from Senegal, Gambia and Guinea-Bissau east to Eritrea, Ethiopia and Somalia, and south to easternmost South Africa and Swaziland) (Harrison et al. 1997). The population of White-headed vultures has declined in West Africa from the 1940s (Hall 1999; Thiollay 2006a,b,c, 2007a,b, 2012), and is currently declining in East Africa (Virani et al. 2011). In southern Africa, it now occurs mostly in protected areas. Hancock (2008a) reported that it was the rarest vulture recorded in a survey of Botswana. Other depressing news concerns Niger (Brouwer 2012) and Mozambique (Parker 2005a).

Factors for the reduction in the populations of White-headed vultures include reduced populations of medium and large wild ungulates, habitat conversion, indirect poisoning (poisoned carcass baits to kill jackal predators of small livestock, and for lions and hyenas that kill larger livestock), secondary poisoning from carbofuran, exploitation for the international trade in raptors, killings for traditional medicine (especially in South Africa) and high human presence disturbing breeders (Mundy et al. 1992; Hall 1999; Genero 2005; Baker 2006; Davies 2006; Simmons and Brown 2006b; Hancock 2008; Otieno et al. 2010; Kendall 2012). There is also a potential threat from the use of the non-steroidal anti-inflammatory drug diclofenac for veterinary purposes, which fatally affects vultures (BirdLife International 2007; Woodford et al. 2008). For these reasons, this species is commoner in protected areas in southern and East Africa (Dowsett-Lemaire 2006).

2.5 Red-headed Vulture, *Sarcogyps calvus* (Scopoli 1786)

Physical appearance

The medium sized Red-headed vulture, is also named the Asian King vulture, Indian Black vulture or Pondicherry vulture (Ali 1993). It has a length of about 76 to 86 cm (30 to 34 in), a wingspan of about 1.99–2.6 m (6.5–8.5 ft) and a weight 3.5–6.3 kg (7.7–14 lb) (Ferguson-Lees and Christie 2001). In the adult, the head has deep red to orange skin, with scattered light whitish down. The iris is white in the male and dark brown in the female. The heavy bill is black with a reddish pink cere, the feet are a similar reddish to orange color to the head and the plumage is mainly dark with a grey band at the base of the flight feathers (Naoroji 2006) (Fig. 2.5a,b,c).

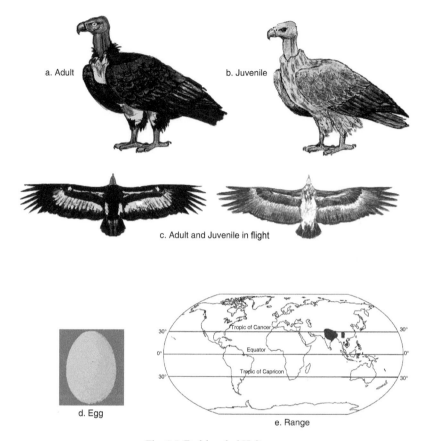

Fig. 2.5. Red-headed Vulture.

Classification

The Red-headed vulture was formerly named the King Vulture *Torgos calvus*, and is named by some as *Aegypius calvus*, e.g., Ferguson-Lees and Christie (2005) and Rasmussen and Anderton (2005). This the only species of the Genus Sarcogyps (Lesson 1842). Currently, no subspecies have been recorded. Based on nucleotide sequences of the mitochondrial cytochrome b gene, Seibold and Helbig (1995) found that within the clade of the Old World vultures including the genera Aegypius, Gyps, Sarcogyps, Torgos, and Trigonoceps, this species was consistently excluded from the clade including the other three genera. This hinted that it might be the most primitive of the group. Wink (1995) also found that this species belongs to this clade of Old World vultures.

Other authors, namely Amadon (1977), Stresemann and Amadon (1979) and Amadon and Bull (1988) argued that the four monotypic genera of large Old World vultures, due to morphological closeness should be placed within one genus, Aegypius (as also noted by Seibold and Helbig 1995; Wink 1995). However, not all authors agree, and sometimes the Red Headed vulture is placed in the Genus Torgos Kaup, 1828, with the Lappet-faced vulture *Torgos tracheliotos* Forster 1791.

Foraging

The Red-headed vulture is usually more solitary than other vultures (Ali and Ripley 1983; Kazmierczak 2000). It forages in open, mixed vegetation landcover, often close to human habitation, and also more forested landcover such as hilly and dry deciduous forest, usually as a single forager rather than as a flock like the Gyps vultures (Roberts 1991; Grimmett et al. 1998; Robson 2000). Chhangani (2007a) in a study of Rajasthan, India, noted 'a strong correlation' between the presence of vultures and mammalian carnivores such as tigers (Panthera tigris Linnaeus 1758), leopards (Panthera pardus Linnaeus 1758) and Indian wolves (Canis lupus pallipes Sykes 1831) in their study area, as vultures fed on the predators' kills.

Despite the common perception that this species prefers mixed vegetation and dry forest, Nadeem et al. (2007) recorded a pair in the Khairpur area in the Tharparker Desert, a location covered by desert, shrub vegetation such as *Calligonum polygonoides* (L.), *Aerva javanica* ((Burm.f.) Shult), *Cymbopogon jwarancusa* (Jones), *Haloxylon salicornicum* ((Moq.) Bunge ex Boiss), *Dipterygium glaucum* (Decne), *Leptadenia pyrotechnica* ((Forssk.) Decne), *Calotropis procera* ((Aiton) W.T. Aiton) and small to medium-sized trees such as *Prosopis cineraria* ((L.) Druce) and *Salvadora oleiodes* (Decne, Ziziphus mauritiana Lam.) and *Acacia nilotica* ((L.) P.J.H. Hurter and Mabb). They describe this as the first sighting of this species in Pakistan since 1980, and also 'This characteristic plant community has not previously been recorded as a preference of this vulture, which has never before been reported, even as a transient, from such deep desert areas' (Nadeem et al. 2007: 146). It is hypothesized that vulture presence in this desert was due to food shortages and increasing urbanisation in more favored habitats.

Breeding

Red-headed vultures generally nest on large- and medium-sized trees. In the study by Chhangani (2007a) in Rajasthan, India, nests were located in the following tree species: Khejri *Prosopis cineraria* (L.) Druce (3–5 m tall), Babul *Acacia nilotica* (L.) P.J.H. Hurter & Mabb (5–20 m tall), *Rohida Tecomella undulata* (D. Don) (up to 12 metres tall), *Ficus Ficus benghalensis* L. (20–25 m tall) and *Godal Lannea coromandelica* (Houtt.) and Merr (up to 14 m tall). Nesting density may be low, as unlike many other vulture species it is very territorial and is not gregarious (Naoroji 2006). It usually lays a single egg (Fig. 2.5d).

Population status

The distribution of the Red-headed vulture is shown in Fig. 2.5e. Red-headed vultures are described as rare; even historically and globally it was 'nowhere very abundant' (Blanford 1895; see also Bezuijen et al. 2010). Chhangani (2007a: 220) noted that in Rajasthan 'sightings of Red-headed vulture were fewer than those of Long-billed, White-backed or Egyptian vultures. Most sightings of Red-headed vultures were of solitary birds or breeding pairs, or of birds at feeding sites.' Other authors report similar results from Gujarat (Khachar and Mundkur 1989), and recently with population declines it is increasingly rare (Chhangani 2004, 2005, 2007b; Chhangani and Mohnot 2004).

The population reduction has been very rapid, due to feeding on animal carcasses treated with the veterinary drug diclofenac, and possibly other contributory factors, leading to its classification as Critically Endangered (BirdLife International 2014). Cuthbert et al. (2006) calculated a decline of over 90% within 10 years in India. Factors included habitat loss, disturbance, predation, hunting, road accidents when feeding, food and water scarcity, land use change and poisoning (Chhangani 2002, 2003, 2005). Other authors record a decline in other Asian nations such as Pakistan (Nadeem et al. 2007); Bhutan and Burma (Bezuijen et al. 2010; Hla et al. 2011), China (Chan 2006); Thailand (Round 2006), Laos, Vietnam, peninsular Malaysia, Singapore (Ferguson-Lees et al. 2001). Possibly in Cambodia the population is stable since 2004 (Eames 2007a,b).

Doloh (2007: 79) describes a five-year joint programme by Kasetsart University (KU) and the Zoological Park Organisation (ZPO) to rescue the 'near extinct' Red-headed vulture in the forests of Uthai Thani, Thailand. This concerns a breeding and reintroduction programme to release the birds back into the wild. It is noted the 'the last big flock disappeared in early 1992, after eating a poisoned deer carcass. The carcass had been contaminated with toxic chemicals by tiger hunters using the deer as bait' (Doloh 2007: 80).

2.6 Lappet-faced Vulture, *Torgos tracheliotus* (Forster 1791)

Physical appearance

The Lappet-faced vulture is a very large vulture, rivalling the Cinereous vulture, Cape vulture and Eurasian griffon in size, and almost the same size as the New World condors. It is the most aggressive and powerful African vulture, and dominates other vultures at a carcass. It measures between 78 and 115 cm (31–45 in) in the length of body and tail, with a 2.5–2.9 m (8.2–9.5 ft) wingspan, and a weight of 4.4. to 13.6 kg (9.7–30 lb) (Ferguson-Lees and Christie 2001; Hockey et al. 2005; BirdLife International 2014). It has one of the largest bills (up to 10 cm (3.9 in) long and 5 cm (2.0 in) deep) of all accipitrids, only equaled by that of the Cinereous vulture (Hardy 1947; Ferguson-Lees and Christie 2001). The plumage is largely blackish on the back and on the backs of the wings. In African birds, the black back feathers have brown linings, while the Arabian birds are brown, not black above. The underside is white to pale brown, and the thigh feathers are white. The head varies in color; usually pink on the rear of the head and greyish in front in birds from Arabia, pink in northern African birds and reddish in southern African birds (Fig. 2.6a,b,c). It thus varies greatly from

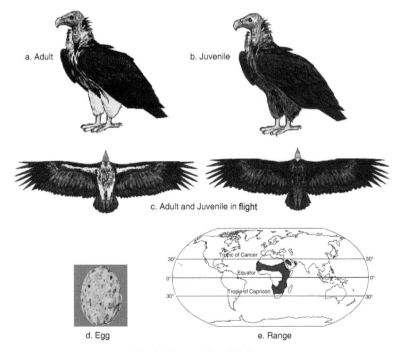

Fig. 2.6. Lappet-faced Vulture.

the Gyps vultures, which are much paler with less white lining in the wings. It also varies from the more similar Hooded vulture by its much larger size, and from the Cinereous vulture (both species overlap in range in the Arabian peninisula) by the latter's all dark plumage (Ferguson-Lees and Christie 2001).

Classification

As described for the other species, the Lappet-faced vulture is classified using molecular sequences of the mitochondrial cytochrome b gene, as belonging to the clade of Old World vultures, which contains the genera Aegypius, Gyps, Sarcogyps, Torgos, and Trigonoceps (Wink 1995). The findings of some studies supported the merging of four monotypic genera of large Old World vultures into one single genus named Aegypius (Stresemann and Amadon 1979; Amadon and Bull 1988; Wink 1995). It has been suggested that this species is a recent offshoot of the Cinereous vulture *Aegypius monachus* (Wink and Sauer-Gurth 2000, 2004).

Further analyses using the sequences of the cytochrome b gene, supported the existence of subspecies, as there was evidence of a small genetic distance between the nominate subspecies *Torgos tracheliotus tracheliotus* and another subspecies *Torgos tracheliotus negevensis* (Wink 1995). The latter had already been described as a separate subspecies (Bruun et al. 1981). Although earlier studies hypothesized that *Torgos tracheliotus negevensis* was a cross between the Lappet-faced vulture and the Cinereous vulture (*Aegypius monachus*), the study by Wink (1995), using the maternal cytochrome b gene showed it could not be a link between a female *A. monachus* and a male Torgos due to the similarity between *T. t. tracheliotus* and *T. t. negevensis*, and the greater differences between *T. t. negevensis* and *A. monachus*. The other possibility, a mating between a male *A. monachus* and a female *T. t. tracheliotus* was considered possible but remote. A divergence between *T. t. tracheliotus* and *T. t. negevensis* was hypothesized at about 350,000 years BP, this estimation being based on the so-called 2% per million years rule (Shields and Wilson 1987; Wink 1995).

As a result of these studies, there are three recognized subspecies of the Lappet-faced vulture (Bruun et al. 1982; Shobrak 1996). *T. t. tracheliotus* from south and east Africa, differs in having white-feathered thighs, a scarlet bald head and larger skin folds or lappets on the head. Mundy et al. (1992) note that East African adult birds have a black bill, while those of southern Africa have yellow bills. *T. t. negevensis* is found mainly in Israel and the Arabian Peninsula (Mendelssohn and Marder 1989). This subspecies has brown rather than black plumage, greyer skin on the face and a pink colored rear of the head. It has a thicker down on the head, and either an absence of lappets or smaller lappets. The thighs are covered with dark brown rather than white feathers, and the bill is blackish (Brunn et al. 1981).

T. t. nubicus, a hybrid of the recognized subspecies may constitute a third subspecies in Egypt and Sudan (Bruun 1981; Mundy et al. 1992). It is identified by very small lappets, a pink head and brown feathers on the thighs. Not all authors recognize this third subspecies (Cramp and Simmons 1980; Brown et al. 1982; Weigeldt and Schulz 1992). In the past, before DNA studies, the classification of subspecies considered the 'size of the lappets, the coloration of the head, color of the thighs, and the degree of development of a white bar on the underwing' (Weigeldt and Schulz 1992: 24; see also Bruun 1981; Bruun et al. 1981; Leshem 1984). Some studies emphasized the lack of geographical separation between the Negev and Arabian populations (Brown et al. 1982; Jennings and Fryer 1984) and variabiity of individual colors from both populations (e.g., variation of thigh colors between brown and white, as reported by different authors, e.g., Bruun et al. 1981; Mendelssohn and Leshenm 1983; Leshem 1984; Shirihai 1987; Mendelssohn and Marder 1989).

Weigeldt and Schulz (1992: 24) argue that 'it appears that initial description and classification of *T. t. negevensis* was based on only a very few birds, and the variability of the characters employed has not been considered in sufficient detail.' Therefore, 'it has to be concluded that the two populations are not morphologically distinct' (*ibid.*). These authors also question the distinction between *T. t. negevensis* and *T. t. nubicus* and *T. t. tracheliotus*, although they acknowledge the differences (using color of head, size of lappets and feather coloration) between the Arabian and African populations are greater than those between the Arabian and Negev populations.

Foraging

This species favors dry open savannas, deserts and open mountain slopes (Ferguson-lees and Christie 2001; Shimelis et al. 2005), up to elevations of 3,500 m (Shimelis 2007). It also enters denser habitats and even human habitated areas, feeding on human discards, dead livestock and road kills. In Ethiopia, it has been recorded near forest-edge vegetation and in the Bonga forest and forest in Bale Mountains National Park in 2007, and in the high altitude alpine habitats of this national park (Shimelis 2007). The Lappet-faced vulture is a wide range forager and although mainly a scavenger, it is also a predator for small reptiles, fish, birds and mammals, and even hunts young flamingos (*Phoenicopterus* spp. Linnaeus 1758) (Mundy 1982; Mundy et al. 1992; McCulloch 2006a, 2006b). It also feeds on dead livestock, as in Saudi Arabia (Newton and Newton 2008).

While Andersson (1872) termed this species 'sociable', Roberts (1963, in Sauer 1973) referred to it as 'less sociable' than other vultures. However, several observers agree that the Lappet-faced vulture is solitary, as foraging pairs usually outnumber groups and it usually attends carcasses in smaller numbers than the griffons (Mundy et al. 1992).

Breeding

The Lappet-faced vulture is principally a tree nester (Brown et al. 1982). Nests are often located in trees of *Maerua* spp. (Forssk.), about 4.5 m high (Weigeldt and Schulz 1992). The mostly solitary nests are commonly in acacia, balanites and terminalia (Boshoff et al. 1997; Shimelis et al. 2005). Bolshoff et al. (1997) gives evidence that in some cases the distribution of the Lappet-faced vulture is limited by the distribution of the acacia. The low density of the nests has been compared with other species. For example, while White-backed vultures were recorded with average nesting densities of 49.8 nests/100 km^2, with areas up to 266 nests/100 km^2 in Swaziland, Lappet-faced vultures had a 'high' nesting density of only 1.5 nests/100 km^2 (Monadjem and Garcelon 2005; see also Mundy et al. 1992), which was higher than those for the Kruger National Park (Tarboton and Allan 1984). Comparatively high figures of 2–7 nests/100 km^2 were reported from Kenya and Zimbabwe (Pennycuick 1976; Hustler and Howells 1988; Mundy et al. 1992; Monadjem and Garcelon 2005).

The nesting period is from November–July/September in the north of the range, throughout the year in East Africa, and May to January in southern Africa (see also Shimelis et al. 2005; Ferguson-Lees and Christie 2001). It mostly lays a single egg (Fig. 2.6d). Newton and Newton (2008), studying birds in western Saudi Arabia, wrote that most eggs were laid in December, the time of the lowest mean daily air temperatures and young birds fledged after 180 days. The clutch is 1–2, incubation 54–56 days, fledging 125–135 days, and juveniles attain breeding age at about six years (Shimelis et al. 2005).

Population status

The distribution of the Lappet-faced vulture is shown in Fig. 2.6e. Although distributed across sub-Saharan Africa and Arabia, Yemen, Oman and the United Arab Emirates, the population of the Lappet-faced vulture has been described as declining, especially in the Sahel and parts of western, northern and southern Africa (Ferguson-Lees and Christie 2001). In Syria, birds have been killed for medicine and have also died after ingesting meat from poisoned carcasses (Serra et al. 2005). It may be extinct in Israel since the 1990s (Yosef and Hatzole 1997), but *T. t. negevensis* is recorded as still present in Israel and the Arabian Peninsula (Mendelssohn and Marder 1989). In the Arabian Peninsula it is recorded as rare, due to human disturbance, shooting and poisoning (Aspinwall 1996). In Morocco it is extinct, due to human predation and pesticides (Sayad 2007).

In the northern Sahel, the Lappet-faced vulture was recorded as the most widespread vulture in the early 1970s, linked to the distribution of acacia trees (about 20° N) (Thiollay 2006a; Newby et al. 1987). In 2001–2004, they were very rare (Clouet and Goar 2003; Thiollay 2006a,b,c). In central West Africa, Rondeau and Thiollay (2004) and Thiollay (2006a,b,c), reported a 97% decline in numbers. Possible reasons are fewer carcasses, killing for medicine and food, contaminant poisoning, powerline electrocutions and captures for zoos (Thiollay 2006a,b,c). In parts of West and East Africa, there are similar declines (Dowsett-Lemaire 2006; Dowsett et al. 2008).

There have been few surveys of this species in southern Africa (Mundy et al. 1992). However, several authors have described population declines, with factors for this including those factors mentioned above for central West Africa (Simmons 1995; Simmons and Bridgeford 1997; van Rooyen 2000; Bridgeford and Bridgeford 2003; Borello 2004; Bridgeford 2004; Diejmann 2004; Monadjem and Garcelon 2005; Simmons and Brown 2006a; Hancock 2008b). In Mozambique, problems include large mammal declines after the colonial period and the wars following independence (Parker 2004, 2005b). It was recorded to be faring better in Zimbabwean cattle areas (Mundy 1997; Bridgeford 2004). In South Africa, it was historically common and reported to be rare towards the end of the twentieth century, but its current status is uncertain, as much of the literature is dated (Steyn 1982; Boshoff et al. 1983; Tarboton and Allan 1984; Brown 1986; Mundy et al. 1992; Anderson and Maritz 1997; Piper and Johnson 1997; Simmons and Bridgeford 1997; Verdoorn 1997a; Oatley et al. 1998; Anderson 2000, 2003; van Rooyen 2000; Bridgeford 2001, 2002a,b, 2003, 2004).

2.7 Lammergeier or Bearded Vulture, *Gypaetus barbatus* (Linnaeus 1758)

Physical appearance

The Bearded vulture, also known as the Lammergeier or Lammergeyer, is the only species of Genus Gypaetus. It is a very large vulture, rivalling the Cinereous vulture and the Himalayan Griffon vulture in size. In length, it is 94–125 cm (37–49 in), and weighs 4.5–7.8 kg (9.9–17.2 lb) with a wingspan of 2.31–2.83 m (7.6–9.3 ft). Two subspecies have been identified, *G. b. barbatus* the nominate, slightly larger subspecies from Eurasia and North Africa (weighing 6.21 kg or 13.7 lb); and *G. b. meridionalis* weighing 5.7 kg (13 lbs) from south of Tropic of Cancer in southwestern Arabia and East and southern Africa (Mundy et al. 1992; Kruger et al. 2013). Some of the largest specimens of *G. b. barbatus* are found in the Himalayan mountains (Hiraldo et al. 1984; Ferguson-Lees and Christie 2001). In coloration, Bearded vultures have light buff colored feathers on the head and body, with dark brown to blackish back and upper wing feathers. The underside of the wings are also dark blackish brown, and the primaries are greyish. The tail feathers are blackish to dark brown. The head, breast and leg feathers may be orange, possibly due to dust-bathing, mud-rubbing or drinking in mineral-rich waters. The juvenile bird is dark black-brown over most of the body, with a buff-brown breast (Fig. 2.7a,b,c). It is five years before it acquires adult plumage (Ferguson-Lees and Christie 2001). Brown (1977) describes juveniles of the African subspecies *G.b. meridionalis* as having a blackish brown head, a paler dull brown breast and belly, and dark brown, pale streaked upper sides. Later, in the more mature subadult, more white feathers grow on the crown and cheeks, and a dark chestnut neck-ring separates the head from the paler, more reddish breast and belly.

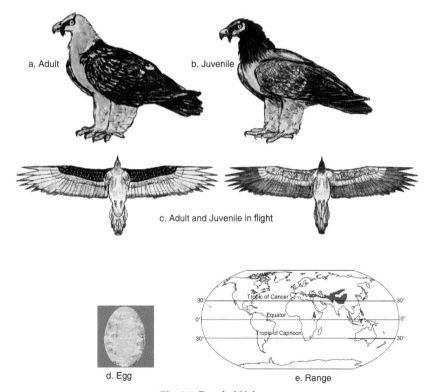

a. Adult b. Juvenile c. Adult and Juvenile in flight d. Egg e. Range

Fig. 2.7. Bearded Vulture.

Classification

As mentioned above, there are two recognized subspecies of the Bearded vulture. It has been argued that this is not a true vulture (see Amadon and Bull 1988); however whether or not it is a true vulture (depending on the definition of vulture), several studies using the nucleotide sequences of the mitochondrial cytochrome b gene have classified the Bearded vulture with the Egyptian vulture as a sister species in a Neophron-Gypaetus clade that is separate and basal to other Old World vultures (Wink 1995; Siebold and Helbig 1995; Wink and Seibold 1996; Wink et al. 1998). Older studies using karyological (De Boer and Sinoo 1994), morphological (Jollie 1976), and embryological data (Thaler et al. 1986) had previously hinted at this conclusion. More recently, Lerner and Mindell (2005) used molecular sequences from one nuclear and two mitochondrial genes, and found that the Bearded vulture and the Egyptian vulture, although highly divergent, are more closely related to each other than to other Accipitridae species. Hence, these two species share a clade called Gypaetinae, including the Palm-nut vulture and possibly the Madagascar Serpent Eagle (Eutriorchis astur, Sharpe 1875).

Regarding the Bearded vulture subspecies, *G. b. barbatus* and *G. b. meridionalis*, research based on phylogenetic analysis of DNA from museum specimens from the Western Palearctic found two divergent mitochondrial lineages, one (lineage A) occurring mainly in western European populations and the other (lineage B) in African, eastern European, and central Asian populations (Godoy et al. 2004). The results also showed that while the two lineages were situated in two different regions, there was marked expansion and mixing especially by the lineage B through central Europe and North Africa, with most of the mixing in western regions of the Alps and Greece. A third subspecies. *G.b. haemachalanus* was not fully recognized (Brown and Amadon 1968; del Hoyo et al. 1994).

Foraging

The Bearded vulture is a scavenger, but is the only scavenger that prefers bone marrow to meat; bone marrow comprises 85–90% of its diet (Boudoint 1976; Hiraldo et al. 1979; Brown 1988; Margalida and Bertran 2001; Ferguson-Lees and Christie 2001). It has a powerful digestive system, including acids with a pH of about 1, for dissolving bones (Brown and Plug 1990; Houston and Copsey 1994). Bones too large to be swallowed [e.g., over 4 kg (8.8 lbs) are broken for their internal marrow by carrying them in flight to a height of over 50 m (160 ft) and dropping them onto rocks below]. This requires several years' practice by juveniles. They may also break small bones by beating them against rocks using the bill. Margalida et al. (2009) record sheep and goats as comprising 61% of the prey remains identified in the Pyrenees. Domestic livestock were 73% of their diet in the Pyrenees and 80% in southern Africa (see also Brown and Plug 1990). Margalida et al. (2009: 240) also note that 'although the Bearded Vulture is considered a bone-eating species, our results suggest that during the breeding season, small dead animals (i.e., prey with a high meat content) are very important prey when parents are feeding young nestlings.'

Particular food preferences were also noted (Margalida et al. 2009), which were not proportionate to their availability. Larger species' bones were generally avoided, possibly because of the greater energy expenditure required for transporting and digesting these larger bones (see also Margalida 2008a, 2008b). The ideal species in terms of size and weight were sheep *Ovis aries* Linnaeus 1758, goats (*Capra aegagrus hircus* Linnaeus), 1758 or Southern Chamois *Rupicapra rupicapra* Linnaeus 1758. The commonest anatomical part favored by Bearded vultures are sheep limbs (Margalida et al. 2007a), the most nutritious part of the sheep limb being the fat content of the extremities (Margalida 2008b).

In the study by Margalida et al. (2009), there was no relationship between the percentage of Ovis/Capra in the diet and the presence of feeding stations supplied with sheep limbs. This demonstrated that Bearded vultures took sheep irrespective of the presence of feeding stations. One fact justifying this assessment was that feeding stations were principally used by the non-breeding population (Heredia 1991; Sesé et al. 2005; Margalida et al. 2009).

These authors also found that during the first month of breeding, adults selected prey items with higher meat biomass (possibly indicating a link between meat consumption and breeding success). This was also linked to a lowered use of the bone breaking sites (termed ossuaries) during the first month of the chick's life (see also Margalida and Bertran 2001). Speculatively, areas with limited meat availability and inexperienced adult foragers, might be linked to the breeding failures that are commoner during the hatching and chick-rearing periods (Margalida et al. 2003). Of course, other factors such as weather, human pressure and parent quality may play a role in breeding success (Margalida et al. 2009).

Bearded vultures also eat living tortoises (Testudinidae, Batsch 1788); these are killed by the same method as bone-breaking; they are carried to a height and dropped on to rocks below. Other species killed in this way include rock hyraxes *Procavia capensis* (Pallas 1766), hares (Genus Lepus, Linnaeus 1758), marmots (Genus Marmota, Blumenbach 1779) and monitor lizards (Genus Varanus, Merrem 1820). Mammals killed on the ground, usually young animals (some driven off cliffs, perhaps accidently) include ibex and wild goats (Genus Capra, Linnaeus 1758), Chamois (*Rupicapra*

rupicapra, Linnaeus 1758) and (*R. pyrenaica*, Bonaparte 1845) and Steenbok (*Raphicerus campestris*, Thunberg 1811) (Ferguson-Lees and Christie 2001).

The range of the Bearded vulture encompases cliffs, crags, precipices, canyons, gorges and inselbergs, in mountainous areas. The vegetation of these areas includes forest clumps, high altitude steppe, alpine pastures and meadows and montane grassland and heath. These high altitude habitats are usually above 1,000 m (3,300 ft). Most birds are recorded above 2,000 m (6,600 ft), near or above the tree line, i.e., up to 2,000 m (6,600 ft) in Europe, 4,500 m (14,800 ft) in Africa and 5,000 m (16,000 ft) in central Asia (Ferguson-Lees and Christie 2001; Gavashelishvili and McGrady 2006a,b, 2007; Kruger et al. 2006). For food, they also require the presence of predatory birds and mammals, such as Golden Eagles and wolves, as these animals kill animals and therefore provide the bones upon which the Bearded vulture feeds. Other feeding areas have also been recorded. For example, in Ethiopia, Bearded vultures are reported at refuse dumps near human settlements, feeding on dead animals and human organic discards (Brown 1977).

Breeding

The Bearded vulture is generally a solitary nester, usually building large stick nests on sheltered ledges of rock cliffs. There are regional variations in nesting. For example, in Iraq, a nest was located on a cliff 2,400 m high (Ararat et al. 2011). Abuladze (1998) writes that nests in Caucasia (parts of Georgia and Russia) were recorded at altitudes ranging from 815 to 2200 m, with the greatest number (76%) of nests found between 1500 and 1900 m above sea level. In another study in Russia, nests were also located on cliffs, but mostly in narrow ravines near rivers, with an average height of 113 m and a range of 10–250 m (Karyakin et al. 2009). In Russia, nests were large in some cases, up to 2–3 m high, and also small in some cases, about 40 cm high. The main determinant was the availability of branches and sometimes bones and animal hide (Karyakin et al. 2009). In the Spanish Pyrenees, nests were recorded as near the midpoint of cliff faces (Donázar et al. 1993). On the Mediteranean island of Corsica, nests were observed at variable elevations, as low as 10 metres above the ground and as high as 200 m, but all were inaccessible at least to people (Grussu 2008). Nearby, in Sardinia, an old nest was located only 4.5 m high in a calcareous cliff surrounded by oaks (Genus Quercus L.) and junipers (Juniperus L.) (Grussu 2008). Also in the Mediteranean, in Crete, nests were built on cliffs facing either the east or south (Vagliano 1984).

The clutch size is usually 1 or 2 light brown or pink, brown spotted eggs (Barrau et al. 1997; Karyakin et al. 2009) (Fig. 2.7d). While several authors describe the clutch as of one egg, Margalida et al. (2003) report that 80% of the clutches comprised two eggs. Generally, the incubation period is

55–60 days and the nestling period is about 105–112 days (Tarboton 1990; Shirihai 1996).

There are regional differences for the egg laying period. In Asia, for example, in the Altai and Tuva regions, nesting begins in January, and chicks fledge in July (Karyakin et al. 2009). In Armenia, most egg laying starts in January (Gavashlishvili 2005). However, Adamian and Klem (1999) report that a young bird about 35 days old was seen in a nest in May 1994. Abuladze (1998: 179) noted that most eggs are laid 'in the first half to middle of January', while hatching takes place early to mid-March and fledgling in the second half of June to the beginning of July. In Israel, the breeding season started in mid-December and ended just before June (Shirihai 1996). In Greece and Macedonia, eggs were laid from December to early February (Reiser 1905; Grubac 1991; Handrinos and Akriotis 1997).

In Europe, in the eastern Pyrenees, Margalida et al. (2003) record the average laying date as January 6 (with a range from 11 December to 12 February) and no significant annual differences. Hatching averaged between 21 February and 3 March and fledgling in May to July.

Concerning Africa, Kruger et al. (2013: 1) note that 'within sub-Saharan Africa, knowledge of the species is poor.' There are two records from Africa: in Uganda, eggs were laid in October and November (Urban and Brown 1971) and December 1949 (EANHS Nest Records).

Population status

The population distribution of the Bearded vulture is shown in Fig. 2.7e. In past decades, the Bearded vulture was described as commonest in Ethiopia. Brown (1977: 50) writes that this species was 'Common, locally abundant' and that 'There is thus no question that the Lammergeier is relatively common in Ethiopia, perhaps more so than anywhere else in the world except Tibet' (Schaefer 1938 is cited in reference to Tibet). In addition, it was 'most numerous around towns and villages, but also numerous in high mountains' and present in most of East and South Africa (ibid.). On reason was noted: 'in Ethiopia, the fact that immatures can readily find scraps of meat, skin, and bones around any village rubbish dump or slaughter house improve survival and helps to explain the apparently higher proportion of immatures in the Ethiopian than in the Lesotho or Spanish populations' (ibid. 52). In Europe, it nearly reached extinction by the early twentieth century, due to hunting and habitat modification. It was believed to kill lambs and sheep (hence the German name Lämmergeier or Lamergeyer) and sometimes falsely, kill even children (Everett 2008).

More recently, the Bearded Vulture is recorded as rare and locally threatened, despite its extremely wide range and its natural tendency to sparse, wide ranging populations. It is still commoner in Ethiopia, with populations in the Himalayas. Despite near extinction in Europe by the early 20th century, it is now considered endangered in Europe, 'with fewer than 150 territories in the European Union in 2007' (Margalida 2009: 236). Reintroductions have been undertaken in southern Spain (Andalusia), the Alps (France, Austria, Switzerland and Italy) and Sardinia in the Mediterranean Sea, with possible future projects in the Balkans (Margalida 2009). These reintroductions have had moderate success, especially in the Pyrenees mountains of Spain, the Swiss and Italian Alps, and parts of France. The largest population is in the Pyrenees, but here there are two major problems that compromise the birds' recovery: a decline in productivity (Carrete et al. 2006) and increased illegal poisoning (Margalida et al. 2008c). To counter these problems, the population in the Pyrenees is supported using feeding stations for enhanced breeding success and mortality reduction, especially in the pre-adult population (Heredia 1991; Margalida et al. 2009). Problems include poisons for carnivore-baiting, habitat modification, human disturbance of nesting birds and electrocution and collisions with powerlines (Ferguson-Lees and Christie 2001).

2.8 Cinereous Vulture, *Aegypius monachus* (Linnaeus 1766)

Physical appearance

The Cinereous vulture, has also been called the Black vulture or the Eurasian Black vulture, due to its all dark appearance. These latter names are still frequently used in publications in Europe, but there is confusion with the unrelated and smaller American Black vulture (*Coragyps atratus*). It has also been called the Monk vulture; this is a translation of the German name Mönchsgeier, which refers to the similarity between the vulture's bald head, feathered neck ruff and dark plumage and the cowl and clothing of a monk (Sibley and Monroe 1991). The name Cinereous is Latin for cineraceus, which means grey or ash-colored, and allows it to be distinguished from the American Black Vulture (Sibley and Monroe 1991).

This is a huge vulture, generally considered the largest Old World vulture, the largest member of the Family Accipitridae (and sometimes also the Order Falconiformes, if the slightly larger New World Condors are excluded) and therefore one of the world's largest flying birds (Sibley and Monroe 1991; Del Hoyo et al. 1994; Snow and Perrins 1998; BirdLife International 2014). Apart from the condors, its main rival in size is the Himalayan Griffon vulture; however the largest Cinereous vultures are recorded as exceeding the wingspan and weight of the largest Himalayan Griffon vultures, even if the long neck of the Griffon vulture gives it a slightly greater length (Brown and Amadon 1986; Ferguson-Lees and Christie 2001). The Cinereous vulture may be 98–120 cm (3 ft 3 in–3 ft 11 in) long from bill to tail, with females weighing 7.5 to 14 kg (17 to 31 lb), and the smaller males 6.3 to 11.5 kg (14 to 25 lb). The wingspan is about 2.5–3.1 m (8 ft 2 in–10 ft 2 in). Ferguson-Lees and Christie (2001) describe a cline of increased body size from west to east, as the birds from central Asia (Manchuria and Mongolia, southwards into northern China) are generally about 10% larger than those from southwest Europe (mostly Spain and southern France).

The bare skin of the head and neck is bluish grey and the head is covered with blackish down. The black to blue-grey bill is massive, similar to those of the other very large vultures (the Lappet-faced vulture, Bearded vulture and Himalayan Griffon vulture) (Fig. 2.8a,b,c). The legs are pale blue-grey. The adult plumage is all dark brown to blackish brown, with a short, slightly wedge-shaped tail. The juvenile plumage is brown above, and paler underside than in adults, with grey rather than blue-grey legs (Brown and Amadon 1986; Ferguson-Lees and Christie 2001). This dark bird may be distinguished from the other large vulture in the southern Middle East, the Lappet-faced vulture, as the latter bird has reddish to pinkish skin on the head and whitish feathers on the thighs and belly. The Gyps vultures are distinguished from the dark Cinereous vulture by their paler plumage and longer, down-covered necks.

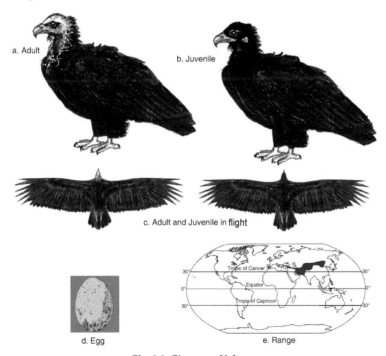

Fig. 2.8. Cinereous Vulture.

Classification

The Cinereous vulture, as noted in the description of the Genera Gyps, Sacrogyps, Torgos, and Trigonoceps, is classified by the nucleotide sequences of the mitochondrial cytochrome b gene as belonging to Aegypiinae, the larger of the two clades of Old World vultures (Wink 1995; Stresemann and Amadon 1979; Amadon and Bull 1988; Ferguson-Lees and Christie 2001). According to Wink (1995), the Cinereous vulture and the Lappet-faced vulture (Torgos tracheliotus) are sister species, which differ by 3.8% nucleotide substitutions.

Foraging

The Cinereous vulture forages in forests, and more open terrain such as steppe and upland grasslands (Heredia 1996). Flint et al. 1984 write that favored landcover includes arid hilly and montane habitat, with both semi-desert and wooded areas, mountainous areas above the treeline, and mixed agricultural and forested patches. Hiraldo and Donazar (1990: 131) note that the 'the foraging areas the Cinereous vultures use are preferentially plains and hills of gentle relief' (see also Hiraldo 1976; Amores 1979). In China it is observed in prairies and open farmlands (Weizhi 2006). In Israel, it is recorded from mountains, plateaus, desert wadis, and also grazed highlands and plains (Shirihai 1996). In Armenia, the favored habitat appears to be eroded cliffs with cavities and mountainous slopes with clay-gypsum, gravel and/or scree (Adamian and Klem 1999). In Kazakhstan it is recorded in semi-deserts, deserts and low mountains (Wassink and Oreel 2007). Juveniles, especially from the northern ranges move across large distances in open-dry habitats due to snowfall or hot summer temperature in local areas (Brown and Amadon 1986; Gavashelishvili et al. 2012).

In feeding at carcasses, the Cinereous vulture is dominant over other large vultures such as the Gyps and Bearded vultures and also over smaller mammals such as foxes (Brown and Amadon 1986). It is well equipped to tear open thick skins and ribs (Gavashelishvili et al. 2006). Moreno-Opo et al. (2010: 25) studied the eating preferences of the Cinereous vulture in Spain and found that 'the number of cinereous vultures that come to feed on the carcasses is related to the quantity of biomass present and to the types of pieces of the provided food.' They prefer 'individual, medium-sized muscular pieces and small peripheral scraps of meat and tendon.' Therefore, the time before the vultures began consumption depended on the 'biomass delivered, the number of pieces into which it is divided, and the type categories of the provided food.'

The carrion diet of the Cinereous vulture varies according to the region. For example, in Western Europe, carcasses consumed are usually those of smaller mammals, as there are fewer large non-domesticated mammals in this region than in Turkey and Asia. In the Iberian Peninsula the common food source was the European Rabbit (*Oryctolagus cuniculus*, Linnaeus 1758), but the decline of the rabbit numbers, due to the disease viral hemorrhagic pneumonia (VHP) contributed to vultures feeding on dead domestic sheep (Ovis aries, Linnaeus 1758), supplemented by pigs (*Sus scrofa domesticus* Erxleben 1777), red deer (*Cervus elaphus*, Linnaeus 1758) and fallow deer (*Dama dama*, Linnaeus 1758) (González 1994; Costillo et al. 2007a). In Turkey, the larger animals eaten included Argali (*Ovis ammon*, Linnaeus 1758), Wild Boar (*Sus scrofa*, Linnaeus 1758) and Gray Wolf (*Canis lupus*, Linnaeus 1758). Smaller animals eaten were the Red Fox (*Vulpes vulpes*, Linnaeus 1758) and domestic chickens (*Gallus gallus domesticus*, Linnaeus 1758); in some cases there was evidence of the ingestion of pine cones (Yamaç and Günyel 2010).

In Tibet, there are records of vultures eating the carcasses of wild and domestic Yaks (*Bos grunniens*, Linnaeus 1766), Bharal or Himalayan blue sheep (*Pseudois nayaur*, Hodgson 1833), Kiangs (*Equus kiang* Moorcroft, 1841), Woolly Hares (*Lepus oiostolus* Hodgson 1840), Himalayan Marmots (*Marmota himalayana* Hodgson 1841) and domestic sheep (*Ovis aries*, Linnaeus 1758) (Xiao-Ti 1991). In Mongolia, their main food was Tarbagan Marmots (*Marmota sibirica*, Radde 1862); also eaten were Argali (*Ovis ammon*, Linnaeus 1758) (Del Hoyo et al. 1994).

Living prey was also taken in China; these included 'calves of yak and cattle, domestic lambs and puppies, pig, fox, lambs of wild sheep, together with large birds such as goose, swan and pheasant, various rodents and rarely amphibia and reptiles' (Xiao-Ti 1991). Other Asian ungulates taken include neonatal or sickly lambs or calves, and also healthy lambs (Richford 1976). Other mammals killed are Argali, Saiga Antelope (*Saiga tatarica*, Linnaeus 1766), Mongolian gazelle (*Procapra gutturosa*, Pallas 1777) and Tibetan Antelope (*Pantholops hodgsonii*, Abel 1826) (Olson et al. 2005; Buuveibaata et al. 2012).

Breeding

Breeding areas also vary according to the region; however the main classification is of large nests in large trees and occasionally rocky cliffs and mountain faces with crevices and cavities (Brown and Amadon 1986; Ferguson-Lees and Christie 2001; Kirazlı and Yamaç 2013). Rock nesting is rare in Europe but more common in Asia (Batbayar et al. 2006; BirdLife International 2014). Breeding in Europe requires quiet slopes with forests, including low sierras and open valleys, sometimes also subalpine forests of *Pinus* spp. up to an altitude of 2,000 m and open parkland (Heredia 1996). In Spain, common habitats are forested hills and mountains 300–1,400 m high. In Asia, it inhabits higher altitudes up to 4,500 m, including scrub and arid and semi-arid alpine steppe and grasslands (Thiollay 1994). Xiao-Ti (1991) reports that 'the breeding habitat of the Cinereous vulture in China falls into two main types: mountainous forest and scrub between 780 m and 3800 m; and arid and semi-arid Alpine meadow and grassland, between 3800 m and 4500 m'.

In China, various authors have described the breeding habitats of Cinereous vultures (Xiao Ti 1991). For example, Futong Sheng et al. (1984) write that the favored habitat is mixed forest above 800 m in the Changbai mountains. For Cheng Tso-Hsin et al. (1963) the preferred habitat was above 2000 m at Xinjiang in the Altay mountains. Also noted were rocky, grass-covered mountain areas above 2000 in Sichuan (Li Gei Yan et al. 1985); regions of alpine meadow and scrub about 3000 m altitude in northwest Sichuan (Cheng Tso-Hsin 1965); above 1000 m in secondary mixed forest and scrub on the loess plateau of northern Shangxi (Yao Jianchu 1985); between 500 m and 3200 m in areas of mixed forest/scrub and farmland in South Shangxi (Cheng Gangmei 1976) and between 780 and 3400 m and mixed conifer and broadleaf forest (*Betula* spp.), with Alpine meadowland and scrub on the slopes of the Qinling Mountains (Cheng Tso-Hsin 1973).

Also recorded for China were semi-arid regions with coniferous forest, scrub and meadowland in Ningxia (Wang Xian Tin et al. 1977); 770 m altitude in an area of swamp, river, lake and desert around Lop Lake in Xinjiang (Gao Xingni et al. 1985); around 1000 m altitude in desert areas of South Xinjiang (Qian Yan Wen et al. 1965); between 1000 m and 4000 m in semi-arid, arid and mountainous regions of coniferous forest, Alpine scrub and pasture in Gansu (Wang Xiantin 1981); and between 3000 m and 4500 m in coniferous forest, Alpine scrub and pasture or semi-arid grassland in Qinghai (Ye Xiao Ti et al. 1988, 1989). Xiao Ti (1991) summarizes the different reports, noting that these fall into two different types; mountainous forest and scrub between 780 m and 3800 m, and arid and semi-arid Alpine meadow and grassland, between 3800 m and 4500 m altitude.

The Cinereous vulture usually breeds either in solitary nests or in loose colonies (Kirazlı and Yamaç 2013). The nests are usually situated in trees 1.5 to 12 m high (4.9 to 39.4 ft), often along cliffs. Common trees utilized for nesting are oak (*Quercus* L. spp.), juniper (*Juniperus* L. spp.), almond (*Prunus dulcis* (Mill.)) D.A. Webbor and pine trees (*Pinus* L.) (Brown and Amadon 1986). Birds start breeding at about 5–6 years. One white or pale buff egg with red to brown marks is laid (Fig. 2.8d). The egg laying period is usually from the beginning of February to the end of April, clustering from the end of February and to the beginning of March. Incubation lasts 50–54 days, and fledging more than 100 days (Hiraldo 1983, 1996; Reading et al. 2005; Gavashelishvili et al. 2012).

Population status

The distribution of the Cinereous vulture is shown in Fig. 2.8e. Several threats have reduced the population of the Cinereous vulture (Heredia 1996). Most of the declines in the past decades have occurred in the western part of the birds' range (i.e., France, Italy, Austria, Poland, Slovakia, Albania, Moldovia, Romania, and Morocco and Algeria) (del Hoyo et al. 1994; Snow and Perrins 1998; Ferguson-Lees and Christie 2001; Gavashelishvili et al. 2006).

Forestry disturbs breeding and removes trees used for nesting. Forest fires develop during the dry season and destroy nests and trees, especially in the Mediterranean region. For example, a fire in 1992 in Andalucía burned down eight nests with young birds, and 21 empty nests (Andalus 1993). Cinereous vultures are regularly killed after ingesting meat from carcasses laced with poisons (including strychnine and luminal) and pesticides in contravention of the Berne Convention. Recent measures have also reduced livetsock numbers in the field, removed carcasses and accommodated cattle and sheep indoors during winter. A few vultures are also shot for sport. Vultures are captured illegally for zoos in the post-USSR countries. Trapping and hunting of Cinereous Vultures is particularly common in China and Russia (del Hoyo et al. 1994). Mountain tourism, also disturbs nesting vultures especially in the Caucasus. Records show that human disturbance during incubation can result in egg losses due to crow predation (Heredia 1996).

Cinereous vulture populations nevetheless increased by over 30 percent between 1990 and 2000 (BirdLife International 2014). Overall, there are reported to be 7,200–10,000 pairs, of which 1,700–1,900 pairs are in Europe and 5,500–8,000 in Asia (BirdLife International 2004). In Europe, populations are reportedly increasing in Spain (probably 1,500 pairs) Portugal and France, and are stable in Greece and Macedonia, but they are decreasing in Armenia, Azerbaijan, Georgia, Russia, Turkey and the Ukraine (De la Puente et al. 2007; Eliotout et al. 2007; BirdLife International 2014). Asian birds are less studied, despite reported declines in populations (Baral 2005; Batbayar 2005; Fremuth 2005; Katzner 2005; Batbayar et al. 2006; Lee et al. 2006; Barov and Derhé 2011; BirdLife International 2014).

2.9 Conclusions

The vultures in this chapter are clearly very variable. Some have feathered necks, some are the smallest Old World vultures, others are the largest, some are urban denizens while most stay out of urban areas. The huge, eagle-like tearers contrast with the small peckers, and while they are more predatory than the Griffons, one is mainly a plant eater and another eats mainly bone marrow. In fact, they resemble the New World vultures, which also have three large tearers (Condors and King vultures) and four small peckers (the Genus Cathartes and the Black vulture). See Fig. 2.9 for a comparison of some of the similar species. In the Old World vultures the two peckers are the Egyptian and Hooded vultures, and the rest of the non-Griffons may be tearers. With such great variation in the Old World vultures, it is apparent that the habitat, population and conservation ecology will be a complex task. The next chapter will examine the New World vultures, after which all the species will be compared for their ecology and relations with people.

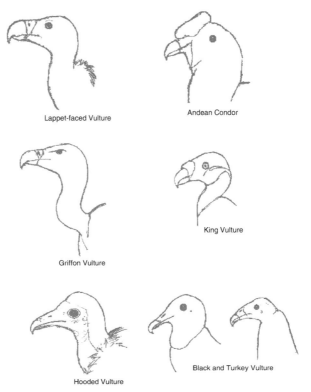

Lappet-faced Vulture

Andean Condor

Griffon Vulture

King Vulture

Hooded Vulture

Black and Turkey Vulture

Note the Old World tearers at the top, followed by the pullers and then the pickers resemble the Cathartids on the right, especially the pickers.

Fig. 2.9. Bills of Vultures.

3

Systematic List of New World Vultures

3 INTRODUCTION

This chapter examines the form, foraging and feeding habits, breeding patterns and status of the New World or Cathartid vultures. These are the Turkey vulture (*Cathartes aura*), Yellow-headed vulture (*Cathartes melambrotus*), Greater Yellow-headed vulture (*C. burrovianus*) Black vulture (*Coragyps atratus*), King vulture (*Sarcoramphus papa*), Andean condor (*Vultur gryphus*) and California condor (*Gymnogyps californianus*). New World vultures are given a separate chapter because they have disputed status; classified either in their own Family Cathartidae or their own Order Cathartiformes (Ericson et al. 2006; Remsen et al. 2008). As noted earlier, some scientists have even placed them with the storks in the Order Ciconiiformes or even as a subfamily of the storks (Sibley and Monroe 1990), although this latter grouping was criticized as an oversimplification (Avise et al. 1994; Griffiths 1994; Fain and Houde 2004; Cracraft et al. 2004; Gibb et al. 2007; Brown 2009). Some multi-locus DNA studies place these vultures within the Falconiformes (Hackett et al. 2008). In 2007, the American Ornithologists' Union's North American checklist also placed the Cathartidae within the Falconiformes (American Ornithologists' Union 2009), but later placed the family in the Ciconiiformes in the AOU's North American and South American Checklists (Remsen et al. 2008; American Ornithologists' Union 2010).

This detailed species by species account provides a background for later discussion of the evolution of these species, their contrasts and similarities with the Old World Accipitrid vultures and ancestral and related groups such as the storks. The list starts with the Black vulture and then the three related species of the Genus Cathartes: the Turkey vulture, the Greater

Yellow-headed vulture and the Yellow-headed vulture. This is followed by the description of the King vulture, usually held by some authors as intermediate between the smaller Cathartids and the two condors. The California condor and the Andean condor are then covered. For each species, current knowledge varies. The most important topics for current research concern ancestry, comparing each species with birds from the prehistoric era; foraging abilities, namely the balance between olfactory and visual ability; the current intra-family classifications; and their future trajectories either towards extinction through changed landscapes and food sources or expansion through adaptation and/or resurgent food sources. These points are crucial for Cathartids, given the small number of species, their specialized feeding habits and the rapidly changing landscapes of the Americas. Each of the anatomical and behavioral features will be compared with those of the Old World vultures, and of other raptors and storks in Part 2 of this book.

3.1 The Black Vulture *Coragyps atratus* (Bechstein 1793)

Physical appearance

The Black vulture is also known as the American Black vulture, to distinguish it from the much larger and unrelated Eurasian Black vulture. Its length from bill to tail measures 56–74 cm (22–29 in). The wingspan is 1.33–1.67 m (52–66 in). The weight varies from smaller figures for vultures from the tropical lowlands of South America [(1.18–1.94 kg (2.6–4.3 lbs)] to those from North America and the Andes which range from 1.6–2.75 kg (3.5–6.1 lbs) (Ferguson-Lees and Christie 2001). The plumage is glossy black. The wrinkled skin of the featherless head and neck is dark grey (Terres 1980). The legs are greyish white (Peterson 2001). Unlike all the Old World vultures, the nostrils are perforate (undivided by a septum, so one may see through the bill from one side to the other) (Allaby 1992).

The wings and tail are broad and comparatively short, compared to those of the Turkey vulture (Fig. 3.1a,b,c). In flight white underside wing patches are visible, due to the white bases of the primary feathers (Terres 1980). Occasionally, white individuals may occur; one such individual was seen in Piñas, Ecuador. It was a leucistic *Coragyps atratus brasiliensis* (Leucism is characterized by reduction in all types of pigmentation, unlike albinism which is caused by a reduction of melanin). This bird was described as white, but was not a full albino, as the tarsus and tail as well as some undertail feathers were black and the skin was the normal dark grey color.

The common bird that may be confused with the Black vulture is the Turkey vulture, with which it occurs throughout most of its range. Taylor and Vorhies (1933) gave an incisive way to distinguish the two species: the

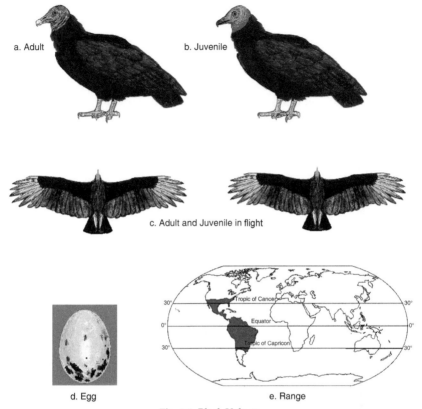

a. Adult

b. Juvenile

c. Adult and Juvenile in flight

d. Egg

e. Range

Fig. 3.1. Black Vulture.

Black vulture has a blackish rather than reddish head color; in flight, it is stockier than the Turkey vulture, with shorter, broader wings and a shorter, square tail; also the leading edge of the Black vulture's wing is straighter and less curved than the wing of the Turkey vulture. The primaries do not show separate feathers as in the Turkey vulture; there are white patches on the terminal third of the Black vulture's wings, easily visible in flight, while the turkey vultures wing primaries and secondaries are silvery grey. The Black vulture flaps its wings more, sometimes flapping six to nine times before a glide, while the Turkey vulture soars, with only occasional wing beats.

Classification

Three subspecies of the Black vulture have been recognized, with size increasing from south to north (Wetmore 1962). The North American Black vulture (*Coragyps atratus atratus*), the nominate subspecies, occurs from the eastern United States to northern Mexico. It was named by the German ornithologist Johann Matthäus Bechstein in 1793. The Andean Black vulture (*Coragyps atratus foetens*) is similar in size to the North American bird, but has a slightly darker plumage, as the white wing markings are smaller and the underwing coverts are darker. It occurs in western South America and the Andes mountains. It was named by Martin Lichtenstein in 1817. The South American Black vulture (*Coragyps atratus brasiliensis*) is smaller, with lighter underwing color than the other two subspecies and occurs from southern Mexico southwards into Central America, Peru, Bolivia, southern Brazil and eastwards into Trinidad (Wetmore 1962, 1965). It was named by Charles Lucien Jules Laurent Bonaparte in 1850. This species is generally regarded as monotypic (i.e., having a common ancestor, despite the differences among the subspecies (Monroe 1968).

Foraging

The Black vulture is common, foraging in most areas, including urban and cattle grazing areas, except heavily forested areas (American Ornithologists' Union 2010). It is generally more abundant around towns and refuse dumps than in natural habitats. It also occurs in lowland and degraded and moist forests, shrub/grasslands and swamps, pastures, and heavily degraded former forests and is rare in mountainous areas (Peterson 2001). By its feeding habits, its nearest equivalent among the Old World vultures would be the Hooded vulture, or in the past some behavorial characteristics of the formerly common White-rumped vulture in urban areas.

It is both a solitary and communal forager, sometimes occurring in flocks of hundreds, especially near urban rubbish dumps. Carrate et al. (2009) in a study of a mixed case study in Argentina, note that the Black vulture is not sensitive to habitat fragmentation, less so than the Turkey vulture. In Guyana, the main habitats are scrub or brush habitats, including white sand scrub, bush islands, and dense, low second growth; also habitats altered by humans, such as gardens, towns, roadsides, agricultural lands, disturbed forests and forest edge (Braun et al. 2007: 10).

The Black vulture locates food by sight as their sense of smell is either minimal or absent. It also follows other New World vultures (Genus Cathartes—Turkey Vulture, the Lesser Yellow-headed vulture, and the Greater Yellow-headed vulture) as these species are able to forage by scent, detecting ethyl mercaptan, the gas emanating from decayed flesh, and hence detect carrion below the forest canopy. Aggressive Black vultures may dominate the Turkey vultures (Gomez 1994; Muller-Schwarze 2006; Snyder and Synder 2006). It has been described as variably dominant over the Turkey vulture and the two Yellow-headed vulture species when feeding at a carcass. Some authors describe the Black vulture as dominant only when in large flocks of over 50 birds, otherwise the relationship is approximately even in one to one encounters (Wallace and Temple 1987).

The Black vulture is described as 'an example of a winning species positively responding to human transformations' (Carrete et al. 2010: 390; see also Carrete et al. 2009). These authors admit 'detailed studies on its large-scale geographic expansion are lacking' but the 'scarce information available shows how the species, once limited to highly productive tropical habitats, has progressively advanced until its current occupation of broad regions of North and nearly all of South America.'

Evidence of this range expansion may be seen in the literature with a date trend (Darwin 1839; Houston 1985, 1988; Tonni and Noriega 1988; del Hoyo, Elliot and Sargatal 1994; Buckley 1997; Schlee 2000; BirdLife International 2014). A determining factor is adaptation to food sources associated with human development, such as rubbish dumps, human discards, livestock and road kills (del Hoyo et al. 1994; Carrete et al. 2010). The consequence is that the Black vulture shares its range with all the other Cathartid vultures (Olden and Poff 2003). In the United States, the Black vulture is increasingly entering cities for foraging and roosting, creating problems for people (United States Department of Agriculture USDA 2003). It also roosts communally year-round, usually in trees, as a prelude to foraging (Rabenold 1983). Communal roosts are also recorded as enabling energy savings through thermoregulation, opportunities for social interactions, and a reduced risk of predation (Buckley 1998; Kirk and Mossman 1998; Devault et al. 2004).

This species has a wide range of food sources (Buckley 1999; Hilty 2003). Food sources include the carcasses of monkeys, coyotes and newborn calves (Lowery and Dalquest 1951; Sick 1993). Other sources include turtle eggs and fruits such as bananas, avocados, oil palm fruit, coconut flesh and even salt (Coleman et al. 1985). By location these include: oil palm fruit and coconut flesh in Surinam and Dutch Guiana (Haverschmidt 1947), sweet potatoes on Avery Island, Louisiana (McIlhenny 1945), salt from a cattle field in Pennsylvania (Coleman et al. 1985), avocados and oil palm (Brown and Amadon 1968), coconut husks in Trinidad (Junge and Mees 1961); avocados, the soft meat of coconuts and the oily pulp covering certain palm seeds in Panama (Sick 1993); coyole palm fruit in Veracruz, Mexico (Lowery and Dalquest 1951); cattle, coyote and human excrement (McIlhenny 1939; Coleman et al. 1985; Maslow 1986; Buckley 1996); and fresh vegetables and the excrement from several captive Galapagos giant tortoises (*Geochelone elephantopus*) at the zoological park in central Florida (Stolen 2003).

Other records include dead fish in Panama (Wetmore 1965); palm fruits in Brazil (Pinto 1965); the carcasses of cattle and horses, and refuse around human habitations in Argentina (di Giacomo 2005); competition with seabird colonies, for example on islands such as the Moleques do Sul, Santa Catarina (Sick 1993); and the successful competition against Greater Yellow-headed vultures for dead fish on at least eight occasions at Cocha Cashu, Peru (Robinson 1994). Some records also document the killing of live prey (Avery and Cummings 2004), such as birds (Baynard 1909), skunks and opossums (McIlhenny 1939; Dickerson 1983), turtle hatchlings (Mrosovsky 1971) and fish (Jackson et al. 1978) and livestock (Lowney 1999).

Breeding

Black vultures usually build no nests, but lay eggs on hard surfaces; for example on the surface in caves, tree or log hollows, in rock piles, in dense vegetation or even on high roof tops (e.g., on buildings in Sao Paulo, Brazil) (Sick 1993). Other examples are: a small cave in an escarpment in Oaxaca, Mexico (Rowley 1984); old churches in Antigua, Guatemala (Sclater and Salvin 1859); rock crevices in volcanic remnants, clumps of large rocks, cliff caves and cavities in Nispero (Manilkara chicle (Pittier) Gilly) trees in El Salvador (Dickey and van Rossem 1938; West 1988); house roofs in Costa Rica and bare rocks in a cave in Colombia (Todd and Carriker 1922); hollow trees and roots, and caves and cliffs in Trinidad (Williams 1922); between the buttresses of trees such as silk cotton tree (*Ceiba pentandra* L. Gaertn.) in Guyana; and in pita trees in La Jagua, Panama (Wetmore 1965).

In Brazil, reports show variable locations as well. While Oniki and Willis (1983) record four nests in tree hollows and a treefall in Belém, Sick (1993) records nests on building roofs, holes in tree roots and dead buriti palms. In Argentina, Di Giacomo (2005) recorded tree nests up to 6 to 7 metres above ground, located in holes in dead and living trees, living species including (*Prosopis* L. spp., *Caesalpinia paraguariensis* Parodi Burkart), *Enterolobium contortisiliquum* ((Vell.) Morong), *Aspidosperma quebracho-blanco* (Schltr.), *Sideroxylon obtusifolium* ((Roem. & Schult.) T.D. Penn), *Schinopsis balansae* (Engl.) and *Albizia inundata* ((Mart.) Barneby and J.W. Grimes).

The egg laying period varies regionally. Ferguson-Lees and Christie (2001) describe the breeding season as related to the latitude. In Florida, breeding starts as early as January, while in Ohio it could start in March. Argentinian and Chilean birds start egg-laying in September, but in northern South America, October is more common, and in Trinidad November is usual. In the United States, Brock (2007) recorded eggs in April and May and young chicks between May and July. In the Yucatan, Stone (1890) recorded a nest with eggs on 15 February. In El Salvador, egg laying occurs from December to March (Dickey and van Rossem 1938; West 1988). In Panama, a similar breeding season has been recorded (Wetmore 1965). Similar dates were recorded by Willis and Eisenmann (1979) but they also recorded eggs in a nest between 28 October and 6 November 1976. In Colombia, two eggs were collected from a nest at Tabanga in the Santa Marta region on 12 July (Todd and Carriker 1922). In Trinidad, Williams (1922) reported eggs in February 1918, and young birds on 22 May 1918; and 24 February 1919. In Guyana, Young (1929) recorded an egg and a half-grown nestlings in October 1924. In Brazil, Oniki and Willis (1983) describe two eggs which hatched on about 29 May 1968 in Belém, and two downy young on 13 August 1972. In Argentina, nesting records exist from Reserva El Bagual, Formosa Province, from 26 July to 15 December (Di Giacomo 2005).

Black vulture clutches usually comprise two pale greenish-white, dark brown blotched eggs (Fig. 3.1d), this number being recorded in Panama (Willis and Eisenmann 1979), Guyana (Young 1929) and Argentina (in a few cases one or three eggs in this country) (Di Giacomo 2005). The color may vary; in the United States, eggs have been recorded as having a very pale bluish tinge (Wolfe 1938). In South America however, eggs from several South American countries, including Chile, Trinidad, Brazil and Paraguay did not have the bluish tinge seen in North America (Wolfe 1938). In Argentina, eggs are described as variably creamy white, whitish, pale greenish, pale bluish and either unmarked or heavily marked with reddish, brownish or blackish markings. Eggs in the clutch have different markings (Di Giacomo 2005). In Ecuador, the incubation period varies, e.g., in Argentina, it was 39–41 days (Di Giacomo 2005); in Tennessee and Ecuador it was 35 days (Crook 1935; Marchant 1960); and in Brazil it was 32 to 39 days (Sick 1993). Nesting periods also varied; for example in Brazil it would be 70 days (Sick 1993), while in Argentina it would be 74–81 days (Di Giacomo 2005). Fergus (2003) notes that young remain in the nest for two months, and can fly after 75 to 80 days.

Population status

The distribution of the Black vulture is shown in Fig. 3.1e. The Black vulture is protected in the United States under the Migratory Bird Treaty Act of 1918 (US Fish and Wildlife Service 2013). In Canada it is protected by the Convention for the Protection of Migratory Birds in Canada (1917) and the Migratory Birds Convention Act, 1994. In Mexico, it is protected by the Convention for the Protection of Migratory Birds and Game Mammals (1937).

The Black vulture has been described as the most common bird of prey in the western hemisphere (Brown and Amadon 1968). Although in the past it was thought to be declining (Rabenold and Decker 1990; Buckley 1999), it 'thrives today' (Blackwell et al. 2006: 1976). Blackwell et al. (2006) note that despite opinions supporting the decline of vultures in the southeastern United States (see Stewart 1984; Rabenold and Decker 1990; Buckley 1999), possibly due to nest site losses and eggshell thinning (Jackson 1983; Kiff et al. 1983; Rabenold and Decker 1990), evidence shows that adult survival (in addition to the secondary factor of fertility) is contributing to stronger populations. This is possibly because among other factors, birds may be breeding earlier in life than assumed (Blackwell et al. 2006).

The population is estimated to be 20 million birds (Rich et al. 2004). Furthermore, its population is increasing. Seamans (2004) and Sauer et al. (2001) write that Black vulture populations have increased at annual rates of 2.3% in eastern North America, 1966–2000. Trend records from the Breeding Bird Survey (BBS) and the Christmas Bird Count show population increases in the United States (Sauer et al. 1996, 2005; Avery 2004) and range expanses northeastward into southern New England and northwards into Ohio (Greider and Wagner 1960; Buckley 1999; Blackwell et al. 2006).

Unlike the other species mentioned so far in this book, which are mostly endangered or declining and require supports in their relationships with people; the interactions between the Black vulture and people is largely concerned with increasing conflicts, centred on collisions with aircraft (DeVault et al. 2005; Blackwell and Wright 2006), attacks on livestock (Lowney 1999; Avery and Cummings 2004) and damage to property (Lowney 1999). The United States Department of Agriculture (2003) notes that 'More damage is attributed to black vultures, although turkey vultures have been implicated in some situations.' Danger to livestock includes 'plucking the eyes and eating the tongues of newborn, down, or sick livestock, disemboweling young livestock, killing and feeding on domestic fowl, and general flesh wounds from bites.' Both vultures have been implicated in roosting problems which include fecal matter droppings, water pollution and electricity outages.

The solutions to these problems are divided into the non-lethal methods (Avery et al. 2002; Seamans 2004) and those that require the killing of vultures; the latter is illegal due to the vulture's protected status (e.g., Holt 1998). Nevertheless, non-lethal methods, which may involve displacement of the offending birds may not be successful, as the vulture is very mobile and may return or relocate, causing problems elsewhere (Blackwell et al. 2006). Population reduction methods are possible but problematic, as more information on the birds' demographics and life history may be needed, such as age-specific survival, age-at-first-breeding, fecundity, and age distribution (Parker et al. 1995; Buckley 1999; Humphrey et al. 2004; Blackwell et al. 2006).

3.2 Turkey Vulture (*Cathartes aura*, Linnaeus 1758)

Physical appearance

The Turkey vulture, is also named the turkey buzzard or buzzard in some parts of the United States, and the John Crow or Carrion Crow in the Caribbean. It is one of the three species of the Genus *Cathartes* Illiger, 1811. The others are the Yellow-headed and Greater Yellow-headed vultures which are also sometimes named as Turkey vultures. The Turkey vulture is sometimes called the Red-headed Turkey vulture to distinguish it from these other Cathartes vultures.

The Turkey vulture is considered a large bird in the Americas, but it is small compared with the large Old World vultures and slightly smaller than the Hooded and Egyptian vultures. The head, body and tail measures 62–81 cm (24–32 in) and the wingspan is about 160–183 cm (63–72 in). The bird weighs about 0.8 to 2.3 kg (1.8 to 5.1 lbs). This species is another example of Bergmann's Rule, with size increasing northwards. Central and South American birds are usually smaller, about 1.45 kg (3.2 lbs), while those from northern North America may weigh about 2 kg (4.4 lbs) (Hilty 1977; Kirk and Mossman 1998; Ferguson-Lees and Christie 2001; Meiri and Dayan 2003).

In coloration, the body feathers are brownish-black (Fig. 3.2a,b,c). The flight feathers on the underwings are silvery-gray, which contrasts with the blackish-brown to black wing linings. The head of the adult is red and featherless, although some individuals may have a small amount of blackish down on the top of the neck. The bill is white and the feet are pink, but may be stained white due to the Cathartid habit of defecating on the legs for cooling purposes (Terres 1980; Feguson-Lees and Christie 2001). As the nostrils are perforate, not divided by a septum, it is possible to see through the bill (Allaby 1994). Although it is of similar size to the Black Vulture, it has a longer tail and wings, and flies with less flapping. In flight, it may be distinguished from the Bald and Golden Eagles, and large Buteo buzzards by the much smaller head, and the wings held in a broad 'V' shape. Also, its wings wobble from side to side when in soaring flight, unlike the flat wing positions of the eagles and buzzards. Similar to the Black vulture there are leucistic (sometimes mistakenly called 'albino') Turkey vultures (Kirk and Mossman 1998).

Classification

The number of subspecies of Turkey vulture is disputed. Some publications report five, others six subspecies. Where five subspecies are listed, they are; *C. a. aura* (L. 1758), *C. a. jota* (Molina, 1782), *C. a. meridionalis* (Swann, 1921) or *C. a. teter* (Friedmann, 1933), *C. a. septentrionalis* (zu Wied, 1839) and *C. a. ruficollis* (von Spix, 1824). *C. a. aura* the nominate and smallest

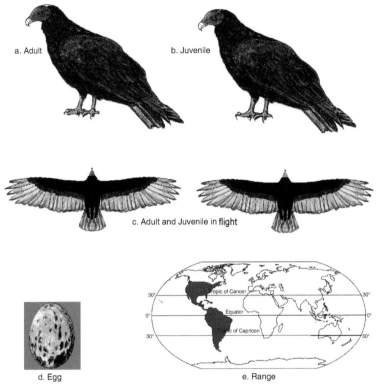

Fig. 3.2. Turkey Vulture.

subspecies, occurs from Florida and Mexico south to Costa Rica, and the Greater Antilles into South America and in coloration is very similar to *C. a. meridionalis* (Amadon 1977).

C. a. meridionalis and *C. a. teter*, named the Western Turkey vulture are synonymous. The name meridionalis was created by Whetmore (1964) for the western birds of *C. a. teter*, which was classified as a subspecies by Friedman in 1933. This subspecies breeds from southern Canada to the southwest of the United States and northern Mexico (Peters et al. 1979). It also migrates to South America, wintering from California to Ecuador and Paraguay and there occurring with the smaller *C. a. aura*.

C. a. ruficollis is found on the island of Trinidad, southern Costa Rica and Panama, and south to southern Brazil, Uruguay and Argentina (Brown and Amadon 1968). It is a darker bird than *C. a. aura*, and the brown wing edgings are usually absent or narrower (Brown and Amadon 1968). In addition, the dull red head may have yellowish- or greenish-white markings, and a yellow crown patch in adults (Blake 1953).

C. a. jota also called the Chilean Turkey vulture, is a larger, browner bird than *C.a. ruficollis* and may also be distinguished from *C. a. ruficollis* in being slightly paler, within grey margins to the secondary feathers and wing coverts (Blake 1853; Hellmayr and Conover 1949). It occurs from southern Colombia to southern Chile and Patagonia.

C. a. septentrionalis, the Eastern Turkey vulture, which occurs in southeastern Canada south through the eastern United States, may be distinguished from *C. a. teter*, as in the latter the edges of the lesser wing coverts are darker and narrower (Amadon 1977). These two subspecies also differ in wing and tail proportions and the eastern bird rarely migrates south of the United States (Amadon 1977). This species winters in the southern and eastern United States.

C. a. falklandicus (or *C. a. falklandica*, named by Sharpe 1873, Bang 1972 and Breen and Bildstein 2008) is a sixth subspecies not always mentioned. It occurs on the coast of western South America from Ecuador in the north to Tierra Del Fuego and the Falklands (Malvinas) in the south. However, Dywer and Cockwell (2011) and Van Buren (2012) record the Turkey vultures on the Falklands as *C. a. jota*. Hellmayr and Conover (1949) considered the separation of the birds of the Falklands Islands from *C. a. jota* as 'impracticable.'

Other subspecies have been mooted. These included *C. a. insularis* and *C. a. magellanicus*. These were not recognized by many authors including Stresemann and Amadon (1979). *C. a. teter* was seen as a synonym of *C. a. meridionalis* and *C. a. insularum* was classified as a synonym of *C.a. aura*. Some authors, including Houston (1994) have merged *C. a. meridionalis* and *falklandica* into *C. a. aura* and *C. a. jota*, respectively. Also, molecular studies may result in two subspecies, *C. a. ruficollis* and *C. a. jota*, being upgraded into species (see for example Jaramillo 2003).

Foraging

Unlike the Black vulture which is described as nonmigratory, 'or nearly so' throughout its range, the Turkey vulture is migratory in most of its range (Jackson 1983: 245). It frequents many landcover types, including pastures, grasslands, and sub-tropical and tropical forests (Hilty 1977; Kaufman 1996). In Argentina, Carette et al. (2009) record it as more affected by habitat fragmentation than the Black vulture. In Guyana, recorded habitat includes savanna grasslands, scrub or brush habitat, white sand scrub, bush islands and habitats altered by humans, such as gardens, towns, roadsides, agricultural lands, disturbed forests and forest edge (Braun et al. 2007: 9). In the United States, it frequently forages and roosts in urban areas (United States Department of Agriculture USDA 2003).

Turkey vultures and other members of the Genus Cathartes have a strong sense of smell (Stager 1964; Houston 1986, 1988; Graves 1992). For this reason, they can detect food in close-canopied forests. The olfactory lobe (which processes smells) is larger than that of most birds (Snyder and Snyder 2007). Lowery and Dalquest (1951) report that Turkey vultures routinely located meat invisible from the air above, under logs or tree base hollows. Due to its sense of smell it leads the Black vulture and King vulture to carcasses (neither have a sense of smell) but may be displaced by them in competition on the ground (the social Black vulture may prevail due to its greater numbers, the King vulture may prevail due to its size). Wallace and Temple (1987) noted that in one-to-one encounters at carcasses, Turkey vultures prevailed over Black vultures in slightly more than half of the encounters, although the Turkey vultures could be overwhelmed by superior numbers of Black vultures. In some cases, the stronger King vulture assists the Turkey vulture by tearing open carcasses (Gomez et al. 1994; Muller-Schwarze 2006; Snyder and Snyder 2006).

It feeds on all types of carrion, including garbage, coconuts, oil-palm fruit and even salt (Pinto 1965; Coleman et al. 1985; ffrench 1991). In some cases, they have been known to attack live fish, young ibises and herons and tethered or netted birds (Mueller and Berger 1967; Brown and Amadon 1968). Campbell (2007: 275) writes that the Turkey vulture 'also feeds on a wide variety of small and large wild and domestic birds as well as dead fishes, amphibians, reptiles, stranded mussels, snails, grasshoppers, crickets, mayflies, and shrimp when they are available' (see also Brock 1896; Keyes and Williams 1888; Pearson 1919; Tyler 1937; Rivers 1941; Rapp 1943; James and Neale 1986; Buckley 1996). Campbell et al. (2005: 108) also listed several food sources in British Colombia, Canada; these included invertebrates (sea cucumbers, marine worms, sea stars, sea urchins, octopus), fishes (several salmon species and dogfish), toads, reptiles (garter snakes, rattlesnakes), birds (mostly waterbirds and domestic fowl), and mammals (wild and domestic species, ranging in size from moles to black bears).

In the United States, they have also been recorded eating fish (Cringan 2007). Other records are iguana in Mexico (Lowery and Dalquest 1951); snakes, rats and other animals killed in burnt cane fields in Mexico (Young 1929); road kills, feces and fruits, including macaúba palm (*Acrocomia sclerocarpa* (Jacq.) Lodd. ex Mart.) and oil palm or dendê (*Elaeis guineensis*, Jacq.) in Brazil; large birds, such as storks, mid-sized mammals such as armadillos, crab-eating foxes, coatis, crab-eating raccoons, nutrias, and tayras, and reptiles such as turtles and snakes in Argentina (Di Giacomo 2005); and pickings from sea lion colonies and sheep ranches in the Falklands (Woods and Woods 2006). In the case of sheep, it has been blamed for killing sheep during the lambing period, and eating the tongues and eyes of sheep

(Brooks 1917) or attacking weak sheep (Woods 1988). The bird defends itself by regurgitating semi-digested meat, which may repel an attacker or nest raider (Fergus 2003).

Turkey vultures roost communally year-round. In one study, the possibility of roosting sites being used for the spread of foraging information was examined (Rabenold 1983). Communal roosts are also recorded as offering energy savings through thermoregulation, opportunities for social interactions, and a reduced risk of predation (Buckley 1998; Kirk and Mossman 1998; Devault et al. 2004). The Turkey vulture is gregarious and roosts in large community groups, breaking away to forage independently during the day. Several hundred vultures may roost communally in groups which sometimes even include Black Vultures. It roosts on dead, leafless trees, and will also roost on man-made structures such as water or microwave towers.

Breeding

Turkey vultures, similar to Black vultures, normally do not build a nest; the eggs are laid in a crevice in rocks or on a cave floor, a hollow stump or even in sugarcane fields (Cuba), or old hawk nests (Brazil) (Brown and Amadon 1968). In Saskatchewan, nests were recorded in caves or on the ground in dense cover, brush piles or vacant buildings (Houston 2006; Houston et al. 2007). Nests in the United States have been recorded under rocks or logs, in caves, or on the ground, rarely in tree cavities (Maslowski 1934). Sick (1993) recorded nests in dead buriti palms in northwestern Bahia, Brazil. In the Falkland Islands, nesting usually takes place in deserted buildings, in caves or on the ground under tussac grass (Woods and Woods 1997).

The egg-laying period starts at different times in different periods, but the incubation period is 30–41 days. In Trinidad the recorded period is from November to March (Belcher and Smooker 1934). In British Columbia and Saskatchewan, Canada, egg-laying was recorded from mid-May to early June, and the eggs hatched from late June to early July, with fledgling in September (Campbell et al. 2005; Houston 2006). In the Falkland Islands, the egg-laying mostly occurred between mid-September and late October (Woods 1988; Woods and Woods 1997), with fledgling in January (Woods and Woods 2006). The clutch size is usually two creamy-white, reddish or brown spotted eggs (Fergus 2003) (Fig. 3.2d).

Population status

The Turkey vulture has the widest range of the New World vultures from southern Canada to southern South America. The total population is estimated at about 4,500,000 individuals (BirdLife International 2012). Seamans (2004) and Sauer et al. (2001) write that Turkey vulture populations have increased at an annual rate of 3.4% in eastern North America, 1966–2000. Roost and nest sites isolated from humans have become limited due to increased urbanization (Rabenold and Decker 1989). The United States Department of Agriculture USDA (2003: 1) states that 'Turkey vultures have become increasingly abundant throughout the northeast' parts of the United States. Along with the Black vulture, the Turkey vulture has been linked to 'property damage and nuisance problems' and 'both species jeopardize aircraft safety when in or around airport environments and are involved in wildlife-aircraft collisions (birdstrikes).' Also, similar to the Black vulture in the United States, Turkey vultures roost in urban areas such as 'backyard trees, in town, or even on suburban rooftops' (see also Tyler 1961; Rabenold 1983; Mossman 1989; Davis 1998; Lowney 1999; Lovell and Dolbeer 1999).

The USDA defines nuisance behavior as 'unwanted congregations of these birds around areas of human activity (homes, schools, churches, shopping areas) resulting in accumulations of feces on trees and lawns, residential and commercial buildings, electrical and radio transmission towers, and other structures.' The results of their presence can include 'unpleasant odors emanating from roost sites' and in some cases 'accumulations on electrical transmission towers have resulted in arcing and localized power outages... and public water supplies have been contaminated with fecal coliform bacteria as a result of droppings entering water towers, springs, or other sources.' Other property damage attributed to vultures includes the tearing and possible consumption of asphalt shingles and rubber roofing material, rubber, vinyl, or leather upholstery from cars, boats, tractors, and other vehicles, latex window caulking, and cemetery plastic flowers.

3.3 Yellow-headed Vulture (*Cathartes burrovianus* Cassin 1845)

Physical appearance

The Yellow-headed vulture, also known as the Lesser Yellow-headed vulture, Lesser Yellow-headed Turkey vulture or Savanna vulture, is the smallest member of the Genus Cathartes, generally smaller than the Turkey vulture or the Greater Yellow-headed vulture. The head to tail length is about 53–66 cm (21–26 in), with a wingspan of about 150–165 cm (59–65 in). The weight range is about 0.95 to 1.55 kg (2.1 to 3.4 lb) (Ferguson-Lees and Christie 2001). The plumage is black, with a greenish or bluish gloss, and the under parts are usually duller and more brownish (Fig. 3.3a,b,c). Blake (1977) describes the head as yellowish to orange, 'varied by prominent blue markings bordered with green on crown'. The

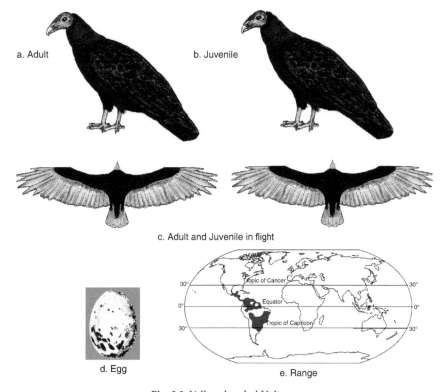

a. Adult　　b. Juvenile

c. Adult and Juvenile in flight

d. Egg　　e. Range

Fig. 3.3. Yellow-headed Vulture.

forehead and nape are usually reddish and the crown is greyish blue to bluish grey. The bill is buff colored to dull ivory white and the legs are dull white or off-white. The rounded tail is short, shorter than that of the Turkey vulture, but like the latter it has a perforated nose. The juvenile has browner plumage a whitish rather than reddish nape and darker colored skin on the head, and more yellowish legs (Blake 1977; Hilty 1977). This species may be distinguished from the Greater Yellow-headed vulture by its smaller size and lighter build; shorter, thinner tail; browner, less glossy black plumage; lighter colored legs; more orange, less yellow skin on the head; less steady flight; and primarily savanna rather than forest habitat (Amadon 1977; Ferguson-Lees and Christie 2001).

Classification

The Lesser Yellow-headed vulture was first recorded by John Cassin (Cassin 1845). Two subspecies are generally recognized, similar except in size

(Blake 1977). The nominate subspecies *Cathartes burrovianus burrovianus* Cassin, 1845, is the smaller of the two, and occurs from Mexico southwards through Central America to northwestern South America (Amadon 1977). The second subspecies, *Cathartes burrovianus urubitinga* recorded by August von Pelzeln in 1861, is larger and ranges from Colombia southwards to Argentina (Blake 1977). Despite the existence of two subspecies, Dickinson (2003) described the species as monotypic.

This species was in the past regarded as a superspecies with the Greater Yellow-headed Vulture (Amadon and Bull 1988) despite the fact they are sympatric in many contexts, and despite the habitat differences of savanna and forest (the latter bird is sometimes called the Forest vulture) (Houston 1994). In the past, it was wrongly named *C. urubitinga* (Pinto 1938; Hellmayr and Conover 1949; Wetmore 1950), but later the name burrovianus was prioritized (Stresemann and Amadon 1979).

Foraging

The Yellow-headed vulture flies singly and in flocks, and has been described as flying at lower altitudes than the Turkey vulture (Belton 1984). Like the Turkey vulture, this species has a strong sense of smell (Stager 1964; Houston 1986, 1988; Graves 1992). Reports from different countries illustrate adaptive behavior in foraging, which are enabled by its olfactory powers.

It generally occurs in open terrain (hence its other name, Savanna vulture) including freshwater or brackish and marshes, moist savannas, mangrove swamps, scrubland and scrub savanna, open and moderately wooded margins of rivers, and farmland and ranches in some areas (Ferguson-Lees and Christie 2001). In a case study in Brazil, it occurred only in savanna with lagoons (Zilio et al. 2013). In Guyana, Braun et al. (2007) gave its habitat as savanna grasslands, scrub or brush habitats, including white sand scrub, bush islands; also fresh water habitats, including lakes, impoundments, ponds, oxbows, marshes, and canals. In Nicaragua, habitat types included forest edge and secondary pine dominated savanna (Martínez-Sánchez and Will 2010). In Brazil, habitats included uncultivated, mixed riverine and marsh vegetation (Sick 1993). In Mexico, some individuals were even recorded behind harvesting equipment (Pyle and Howell 1993).

Fishes and reptiles are mentioned as common favorite foods (Wetmore 1965; Pyle and Howell 1993; Sick 1993). For example, some individuals killed and ate pool-stranded fish in Costa Rica (Stiles and Skutch 1989) and Panama to the neglect of mammal carcasses (Wetmore 1965). However, Di Giacomo (2005) records this species feeding on road-killed cats and dogs, and also dead wild animals such as anteaters (*Tamandua tetradactyla*), crab-eating foxes (*Cerdocyon thous*), nutrias (*Lontra longicaudus*), coatis (*Nasua nasua*), and capybaras (*Hydrochaeris hydrochaeris*). It was also observed eating carcasses of reptiles such as boas (*Eunectes notatus*), other snakes (e.g., *Hydrodynastes gigas*) and lizards; also toads (Rhandia cv. quelen) (*Tupinambis merianae*), and fish (Hoplias cf. malabaricus) and eels (*Synbranchus marmoratus*).

Breeding

No nest is built, similar to other members of the Genus Cathartes. The eggs are laid in a tree hollow (Sick 1993) or on the ground in dense grass (Yanosky 1987). For example, Di Giacomo (2005) describes 13 nests in Argentina, all on the ground, in dense grass patches called 'chajapé' (*Imperata brasiliensis*), in 'paja colorada' (*Andropogon lateralis*), dense stands of large bromeliads (*Aechmea distichantha*) and grass 'pajonal de carrizo' (*Panicum pernambucense*) (Di Giacomo 2005).

The average clutch size is of two creamy-white (with grey to reddish splotches) eggs (Wolfe 1938; Di Giacomo 2005) (Fig. 3.3d). The incubation period at an Argentine nest was 40 days, and the nestling period was between 70 and 75 days (Di Giacomo 2005).

Population status

The distribution of the Yellow-headed vulture is shown in Fig. 3.3e. The range is very large, and the population is recorded as stable. It is described as common to fairly common throughout most of its range. It is fairly common in Mexico and Central America (Howell and Webb (1995). This was also the more recent assessment for southern and eastern Mexico (southern Tamaulipas, Veracruz, Tabasco, Chiapas, Yucatan Peninsula, and on both slopes of Oaxaca) southward into Central America (AOU 1998).

It has variable status in Central America. In Belize it was fairly common in open country and mangroves (Russell 1964; Jones 2003). In Guatemala it has been extensively recorded in lowlands, coastal areas and lagoon fringes (Dickerman 1975; Thurber et al. 1987; Dickerman 2007; Eisermann and Avendaño 2007). In El Salvador it is described as a visitor or a resident (Thurber et al. 1987; Komar and Dominguez 2001; Jones 2004). In Honduras, it is common in the pine savanna of Mosquitia and marshy habitat throughout, outnumbering the Turkey vulture. Opinions differ on its breeding status (Monroe 1968; Jones 2003; Anderson et al. 2004). It is regarded as rare in Nicaragua, but possibly more common in the Caribbean Region (Martínez-Sánchez and Will 2010). Sightings have been confirmed across the country, especially on the Caribbean coast, but also on the Pacific side (Howell 1972; Jones and Komar 2007, 2008; McCrary et al. 2009). In Costa Rica it is a variably common resident, especially in the Guanacaste

Province and Río Frío region (Slud 1964; Orians and Paulson 1969; Stiles and Skutch 1989; Jones and Komar 2010). In Panama, it is fairly common in the savannas, grasslands and open marshes, especially the Pacific coast, but some argue it might be migratory as most sightings are during the rainy season (Wetmore 1965; Ridgely and Gwynne 1989).

This species also has a variable status in South America. In Colombia, it is common in open marshy landcovers and moist grasslands (Hilty and Brown 1986; Márquez et al. 2005). In Ecuador, there are disputed and unconfirmed sightings, at least one believed to be *C. melambrotus* (Salvadori and Festa 1900; Albuja and de Vries 1977; Pearson et al. 1977; Tallman and Tallman 1977; Ridgely and Greenfield 2001). It is rare in Peru, but occurs along grassy, riverine marshes east of the Andes (Clements and Shany 2001). It is fairly common in Venezuela (Hilty 2003) and Guyana (Braun et al. 2000). In Brazil, it occurs mostly in the Northeast and the Amazon region (Sick 1993) and parts of the Rio Grande do Sul in the south (Belton 1984). In Paraguay it is common throughout, but rare in the humid forested Alto Paraná (Hayes 1995; del Castillo and Clay 2004). It is common in Argentina, especially at Reserva El Bagual, Formosa Province (Di Giacomo 2005). It is resident but not common in Uruguay, usually in lowlands near Laguna Merín and uncommon in the rest of the country (Arballo and Cravino 1999). On the Caribbean coast of South America, it is commonest in Guyana (Braun et al. 2007), and less common in French Guiana (Thiollay and Bednarz 2007) and Suriname (Haverschmidt and Mees 1994). In Trinidad, its occurence is disputed (Belcher and Smooker 1934; Junge and Mees 1958; Herklots 1961; ffrench 1991). One possible sighting was recorded, but this could actually have been a Turkey vulture, as the local Turkey vulture usually has a yellow nape (Murphy 2004).

3.4 Greater Yellow-headed Vulture (*Cathartes melambrotus*, Wetmore 1964)

Physical appearance

The Greater Yellow-headed Turkey vulture is also known as the Forest vulture, thus distinguishing its preferred habitat from that of its close relative the Yellow-headed or Savanna vulture. It is larger than the Yellow-headed vulture; the length is 64–75 cm (25–30 in), and the wingspan about 166–178 cm (65–70 in). The weight is about 1.65 kg (3.6 lb). The body plumage is blacker, with a glossier, greenish or purplish sheen than that of the Yellow-headed vulture (Fig. 3.4a,b,c). The bare skin of the head is orangish-yellow to yellowish-orange. The nape and the skin close to the nostrils are pink. In flight, the upper wing surface and back are glossy black, while on the underside the flight feathers are slightly lighter grey. The rest of the wing feathers are

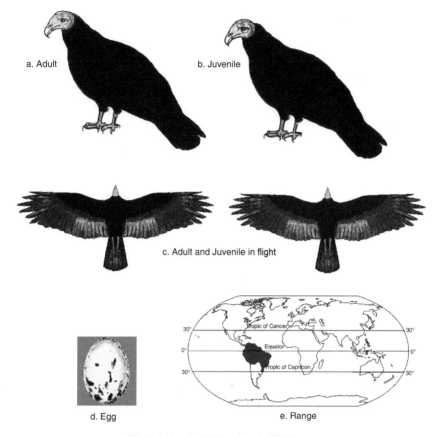

a. Adult

b. Juvenile

c. Adult and Juvenile in flight

d. Egg

e. Range

Fig. 3.4. Greater Yellow-headed Vulture.

black. Towards the primary wing feathers, there is a darker band of feathers from the front edge to the back of the wing, similar in color to the tail. This contrasts with the lighter colored flight feathers, and is almost as black as the feathers of the underside of the body. Similar to the Turkey vulture, the tail is long and rounded. The legs are blackish, but may be lighter due to the habit of defecating on the legs. The plumage of the juvenile is similar to adult, but has greyish skin on the head (Brown and Amadon 1968; Hilty 1977).

This species is commonly contrasted with the Yellow-headed vulture, therefore identification must be based on a comparison. It may be distinguished by its larger size and bulkier build; broader, longer tail and broader wings; darker, glossier plumage; darker colored legs; yellower skin on the head, with less orange and pink; steadier flight movement; and darker inner primaries, contrasting with the lighter colored outer primaries and secondaries (Ferguson-Lees and Christie 2001). It also prefers forest, unlike the savanna preference of the Yellow-headed vulture (Amadon 1977). Compared with the similarly sized Turkey vulture, the Greater Yellow-headed vulture is distinguished by the yellow rather than red head, and the band of darker feathers vertically crossing the wings, and the slightly darker primaries and secondaries (Ridgely and Greenfield 2001).

Classification

The Greater Yellow-headed vulture was considered as the same species as the Yellow-headed vulture, until 1964 (Wetmore 1964). Although Amadon and Bull (1988) described the two species as merely one super species, they are largely sympatric, but occur in different habitats, forest and savanna.

Foraging

This species forages in the moist, dense, lowland forest, while generally avoiding high-altitudes and non-forested areas (Hilty 1977; BirdLife International 2014). For example, in a study in Guyana, Braun et al. (2007: 10) describe its habitat as lowland forest, including both seasonally flooded and non-seasonally flooded forest. It generally roosts on tall trees, allowing landscape surveys. A generally solitary bird, it only rarely flies in groups (Hilty 1977). Schulenberg (2010) notes that the Greater Yellow-headed vulture is 'ecologically separated from the other members of the genus, occurring exclusively over large tracts of undisturbed lowland forest in Amazonia and the Guyanas.'

As it commonly forages over dense forests, possibly more than any other New World vulture, it necessarily uses smell to locate carcasses, usually those primates, sloths and opossums (Stager 1964; Houston 1986, 1988; Graves 1992). Graves (1992: 38) wrote that 'although the olfactory capacities of the Greater Yellow-headed vulture (*C. melambrotus*) are unknown, they are thought to be similar to those of its congeners', namely the other members of Cathartes (see also Houston 1988). It has also been recorded as preferring fresh meat, which may be more difficult to detect at the earlier stages of putrefaction (Snyder and Synder 2006; Von Dooren 2011; Wilson and Wolkovich 2011).

Graves (1992) gives evidence of the olfactory senses of the Greater Yellow-headed vulture in a case study of dense forest on the east bank of the Rio Xingu, 52 km SSW of Altamira, Patti, Brazil. The occurrence of the different vulture species was segregated by the vegetation; in the pristine forest, the Greater Yellow-headed vulture was by far the most numerous.

The Black vultures, with their poor sense of smell, only foraged for fish over sandbars and never over unbroken forest. King vultures were observed soaring high over the river.

However, due to its comparatively weak bill, it is often dependent on larger vultures, such as the King vulture, to open the hides of larger animal carcasses. The King vulture's olfactory senses are too weak to track carrion in the forest; hence, it follows the Greater Yellow-headed vultures to carcasses, where the King vulture tears open the skin of the dead animal, allowing the smaller vultures to feed. In some instances, the Greater Yellow-headed vulture is driven from food by the larger vulture (Gomez et al. 1994). In relations with the Turkey vulture, some authors conclude that the Greater Yellow-headed vulture is dominant (Schulenberg 2010), while others state the reverse (Hilty 1977).

Breeding

Schulenberg (2010) states that 'perhaps due to its preference for un-disturbed lowland rainforest and the general inaccessibility of this habitat, no nest site has ever been found for this species.' However, numerous accounts describe the nesting features and habits. This species, similar to others in the Genus Cathartes does not build a nest, but lays its eggs on cave floors, in crevices or in tree hollows. The eggs are cream-colored with brown blotches (Fig. 3.4d). It lays one to three eggs, but the common number is two. Fledging takes place about two to three months later (Hilty 1977; Terres 1980; Howell and Webb 1995).

Population status

The distribution of the Greater Yellow-headed vulture is shown in Fig. 3.4e. Like the other members of the Genus Cathartes, the Greater Yellow-headed vulture has a large range, and is therefore not considered a threatened species, though it may be declining in numbers (del Hoyo et al. 1994; Ogada et al. 2011; BirdLife International 2014). It is found in the Amazon Basin of tropical South America; specifically in south-eastern Colombia, southern and eastern Venezuela, Guyana, French Guiana, Suriname, northern and western Brazil, northern Bolivia, eastern Peru and eastern Ecuador. It has not been recorded in the Andes, in the lowlands to the west or north of these mountains, in more open landcover in northern South America, eastern South America, or in the subtropical regions to the south of the continent (Ferguson-Lees and Christie 2001).

3.5 King Vulture (*Sarcoramphus papa*, Linnaeus 1758)

Physical appearance

The King vulture is the only member of the Genus Sarcoramphus (Duméril 1805). After the condors, the King vulture is the largest New World vulture. From bill to tail, it averages 67–81 centimeters (27–32 in), with a wingspan of 1.2–2 meters (4–6.6 ft) and a weight of 2.7–4.5 kg (6–10 lb) (Ferguson-Lees, James; Christie, David A 2001). The plumage is mostly white, usually with a perceptible rose yellow tint, while the wing coverts, flight feathers and tail are contrasting dark grey to black. The wings are broad and the tail is also broad and square (Howell and Webb 1995). The head is complicated, the most multi-colored among the New World vultures. The bald skin, includes yellow, orange, blue, purple, and red, with a fleshy, yellow caruncle on its beak (Fig. 3.5a,b,c). There is also a feathery, grey-white ruff (Gurney 1864;

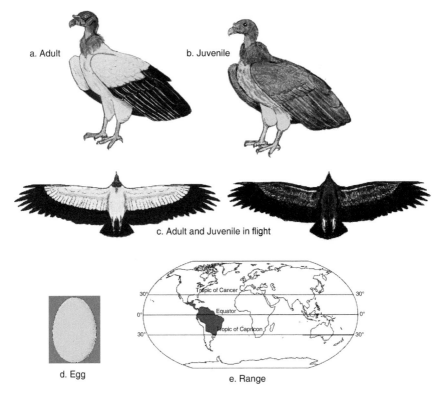

a. Adult   b. Juvenile

c. Adult and Juvenile in flight

d. Egg   e. Range

Fig. 3.5. King Vulture.

Terres 1980; Houston 1994). This species has the largest skull and strongest bill of the New World vultures (Fisher 1944; Likoff 2007).

In the juvenile, the crest or caruncle is absent until the fourth year (Gurney 1864). The juvenile also has a downy, gray neck and a dark bill and eyes, slate gray plumage, that gradually changes to the adult colors by the fifth or sixth year (Howell and Webb 1995; Eitniea 1996). The juveniles resemble Turkey vultures, but may be distinguished by their flat-winged flight (the latter carries its wings in a broad "V" (Hilty 1986).

Classification

The classification of the King vulture has been changed over the years. It was first classified in the Genus *Vultur* (where the Andean Condor is now) as *Vultur papa* by Carl Linnaeus in 1758 in the tenth edition of his Systema Naturae, using a specimen from Suriname. In 1805 the French zoologist André Marie Constant Duméril, changed its classification to the Genus Sarcoramphus. In 1841, Constantin Wilhelm Lambert Gloger placed it in the Genus Gyparchus, but this was not universally recognized (Peterson 2007). Despite the removal of the King vulture from Vultur, the Andean Condor (*Vulture gryphus* Linnaeus, 1758) is recognized as its closest living relative (Amadon 1977). Some past authors placed the Andean Condor and the King vulture in a separate subfamily from the rest of the New World vultures, but this classification was not universally upheld (Coues 1903: 721). Although Amadon (1977) described this classification as invalid, the basis was to group the New World vultures into two groups. Baird et al. (1874: 336–337) defined the two groups as follows:

a) 'Crop naked. Male with a fleshy crest or lobe attached to the top of the cere. Bill very robust and strong, its outlines very convex; cere much shorter than the head (Vultur, Sarcoramphus).'

b) 'Crop feathered. Male without a fleshy crest or other appendages on the head. Bill less robust, variable as to strength, its outlines only moderately convex, cere nearly equal to the head in length (Coragyps, Cathartes, Gymnogyps).'

This classification was also followed by Sharpe (1874: 20–29). Friedmann (950: 9–10) classified them according to cervical vertebrae: Vultur and Sarcoramphus 17, Gymnogyps 15, Cathartes 13, Coragyps 14. Amadon (1977: 414) however argued that the five genera of Cathartidae are more accurate, 'at least until further studies have been made' and he arranged the genera as: Coragyps, Cathartes, Gymnogyps, Vultur, Sarcoramphus, 'thereby placing Gymnogyps between Cathartes and Vultur.'

Fisher (1944) also used the skulls of the New World vultures to determine that the King vulture had the strongest bill, with the most raptor-like hook to the upper mandible and the widest gape. This feature was noticed in its feeding, as it has a greater ability to tear through the skin of carcasses than the smaller Cathartid vultures (Schlee 2005). In this respect, the King vulture performs a role similar to the Old World tearers (Lappet-faced, White-headed and Red-headed vultures), while the smaller Cathartids have the role of the pickers (the Hooded and Egyptian vultures).

Foraging

The King vulture, similar to the Greater Yellow-headed vulture, is primarily a denizen of the tropical lowland forests from southern Mexico to northern Argentina, mostly below 1500 m (5000 ft). It is also found in the nearby savannas and grasslands, and swamps or marshes (Wood 1862; Brown 1976; Houston 1994; Ferguson-Lees and Christie 2001). The King vulture is described as being outnumbered by the Greater Yellow-headed vulture in the Amazon rainforest, while it is also outnumbered by the Lesser Yellow-headed, Turkey and American Black vultures in more open grassland and savanna (Houston 1988, 1994; Graves 1992; Restall et al. 2006). It is generally a solitary forager, but may also bathe and feed in groups where the food is abundant. Despite its status as a carrion eater, it is also recorded taking injured small animals, newborn calves and small animals such as lizards (Baker et al. 1996; Ferguson-Lees and Christie 2001).

Lemon (1991: 700) admits that 'Little is known about the foraging behavior or physiology of King Vultures', but notes that it is a forest specialist (see also Schlee 1995). There is debate about the strength of the sense of smell of the King vulture. Currently, the debate continues; there are records that it follows the Cathartes vultures with strong olfactory senses, to carcasses (Houston 1984, 1994; Beason 2003). However, Lemon (1991) provided evidence that King vultures located carrion in dense forest without the presence Cathartes vultures. At carcasses, the King vulture displaces the smaller, weaker billed Cathartes and Black vultures, but defers to the Andean Condor (Wallace and Temple 1987; Lemon 1991; Houston 1994). It uses its rasp-like tongue to remove flesh from bones and eats mainly the skin and harder tissue. It also occasionally eats fruit such as that of the Moriche Palm (*Mauritia flexuosa* L.f.) (Schlee 2005).

Breeding

The King vulture is non-migratory and is usually a solitary rooster and breeder. Unlike the Turkey, Lesser Yellow-headed and American Black vulture, it generally lives alone or in small family groups. Similar to these other vultures, no nest is built, but eggs are laid on bare surfaces in caves, crevices in cliffs, tree stumps or in the base of spiny palms. Schlee (1995: 269) notes that 'Several nest records indicate that king vultures are ground-nesters.'

Examples include: two nests in Panama, one in a tree stump cavity 30 cm high, in dense moist forest near a river; and another near the base of a spiny palm on the forest floor, composed of leaves and dirt scraped together (Smith 1970). A nest near the Rio Candelaría in the Yucatan Peninsula, Mexico was in a cavity 6–7 m high in a Pucté tree (Gómez de Silva 2004). A nest in El Salvador was in a cavity in a buttressed volador (Terminalia oblonga) tree and another was on a decayed tree stump (Thurber et al. 1987; West 1988). In Brazil a nest at Matozinhos, Minas Gerais was located in a crevice 70 m high in a limestone wall (Carvalho Filho et al. 2004) although others were in trees (Sick 1993).

The clutch size is usually one creamy or dirty white egg (Wolfe 1951; Smith 1970; West 1988; Sick 1993) (Fig. 3.5d). The incubation period is usually 50–58 days (Heck 1963; Cuneo 1968; Sick 1993). Fledging is about 130 days (Carvalho Filho et al. 2004). It is mooted that the King vulture breeds once every two years (Carvalho Filho et al. 2004).

Population status

The distribution of the King vulture is shown in Fig. 3.5e. It has a wide range, occurring in southern Mexico and northern Argentina, but although not endangered its population is believed to be declining due to habitat (forest) destruction (Ferguson-Lees and Christie 2001; BirdLife Inernational 2014). Schlee (1995: 271) noted that 'Unlike the smaller cathartids, the king vulture is reported as not adapting well to human presence and as being more frequently seen in heavily forested areas that do not have permanent human occupation (Clinton-Eitniear 1986; Whitacre et al. 1991; Berlanga and Wood 1992). Schlee (1995: 272) notes that deforestation can 'severely strain and possibly eliminate' king vultures. More recent assessments of the population also hint at gradual decline (BirdLife International 2014).

In the older literature it was already described as either declining or moderately stable. In Mexico, it was recorded as extinct in former habitat in Los Tuxtlas, Veracruz (Winker 1987). It was also recorded as endangered in El Salvador (Komar and Dominguez 2001) and in Ecuador where it declined greatly over the 20th century, due to deforestation and agricultural development (Ridgely and Greenfield 2001). It was described as near threatened in Paraguay (del Castillo and Clay 2005) and rare in Brazil, due to trophy hunting and medicinal use (Sick 1993). On the other hand, in Panama there was less evidence of a decline (Ridgely and Gwynne 1989).

3.6 Californian Condor (*Gymnogyps californianus*, Shaw 1797)

Physical appearance

The California condor is the only living member of the Genus Gymnogyps (Lesson 1842). Other members or possible members of this genus are extinct; for example *G. californianus amplus*, slightly larger than the modern condor, from the Late Pleistocene. The California condor is one of the largest flying birds in North America, nearly the length and weight of the Trumpeter swan (*Cygnus buccinator*, Richardson 1832), Mute Swan (*Cygnus olor*, Gmelin 1789), American White Pelican (*Pelecanus erythrorhynchos*, Gmelin 1789) and Whooping crane (*Grus americana*, Linnaeus 1758).

The California condor is usually slightly smaller than the Andean condor, with slightly shorter wings and a slightly longer body. The length is usually from 109 to 140 cm (43 to 55 in) and the wingspan from 2.49 to 3 m (8.2 to 9.8 ft), despite unverified reports up to 3.4 m (11 ft) (Wood 1983). The weight is generally 7 to 14.1 kg (15 to 31 lb), with estimations of average weight ranging from 8 to 9 kg (18 to 20 lb) (Ferguson-Lees and Christie 2001). In coloration, the plumage is black with triangular white patches on the underside of wings (Fig. 3.6a,b,c). The bill is white and the bare skin of the head varies from yellow to reddish orange (but can flush more reddish according to emotional state). There is a ruff of black feathers around the base of the neck and the legs and feet are grey. The juvenile plumage is mottled dark brown, with mottled grey replacing the white wing patches (BirdLife International 2014).

Classification

The California condor was described by the English naturalist George Shaw in 1797 as *Vultur californianus*, in the same Genus as the Andean Condor (*V. gryphus*), but diferences resulted in the separation of this species from the Genus *Vultur* (Nielsen 2006). Amadon (1977) also suggested merging Gymnogyps with Vultur, but this idea was later abandoned (Amadon and Bull 1988).

Foraging

The California condor has a very large foraging range, up to 250 km (150 mi). The favored landcover is rocky shrubland, coniferous forest, and oak savanna in flatland and more mountainous areas. An important former foraging habitat was the littoral zone of the Pacific Coast, which is gradually being recolonized by birds released from captivity. Currently, there are two condor sanctuaries for birds released from captivity, the Sespe

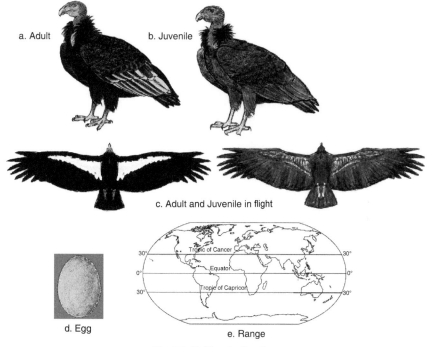

a. Adult

b. Juvenile

c. Adult and Juvenile in flight

d. Egg

e. Range

Fig. 3.6. California Condor.

condor Sanctuary in the Los Padres National Forest and the Sisquoc Condor Sanctuary in the San Rafael Wilderness.

Before North America was colonized, the main sources were small and large wild animals, from ground squirrels to large ungulates and marine mammals. Later, cattle and sheep carcasses on ranches predominated. Condors tend to feed mostly on muscle and viscera when at larger carcasses. Unlike the Cathartes vultures, the California condor does not have a sense of smell. Therefore, condors follow these species to food sources (Nielsen 2006). Condors are usually dominant over Turkey and Black vultures and ravens at carcasses, but are usually driven away by Golden eagles (Koford 1953).

Breeding

Cliff caves or crevices, rocky outcrops or large trees are used as nest sites (United States Fish and Wildlife Service 1996). The egg-laying period is January to April. One bluish-white egg is laid every other year (Fig. 3.6d). In most cases, when an egg or chick is lost, another egg is laid as a replacement. Eggs are laid from January to April (Snyder and Snyder

2000). The incubation period is about 53 to 60 days and the fledging period is about five to six months.

Population status

The distribution of the California condor is shown in Fig. 3.6e. Condors ranged widely across the North American continent at the commencement of human habitation until a few hundred years ago (Miller 1931, 1960; Wetmore 1931, 1932, 1938; Majors 1975). There was a sharp decline in the population of the California Condor during the 20th century; the main factors were poaching, lead poisoning, and habitat destruction (Graham 2006; Cade 2007; Parish 2007). Other factors were collisions with electric power lines (Kiff et al. 1979), DDT poisoning (Church et al. 2006), shooting (Finkelstein et al. 2012) and poor food (Rideout et al. 2012). Thacker et al. (2006) note that lead bullets pose a problem for condors, due to their strong digestive juices, which is less of a problem for smaller scavengers such as ravens and Turkey vultures.

In 1987, all the 22 remaining wild condors were captured, and placed for protection and breeding in the Los Angeles Zoo and the San Diego Zoo Safari Park. In 1991, Condors were reintroduced into the wild. By May 2012, the total population was recorded as 405, of which 226 were in the wild and 179 were captive (Hunt et al. 2007; California Condor Recovery Program 2012; BirdLife International 2014).

Austin et al. (2012: 13) note that 'Lead poisoning continues to be the number one diagnosed cause of mortality' for the condors reintroduced into the wild. Over 90% of condors released in Arizona had lead in their system (BirdLife International 2014) and some birds have died (Austin et al. 2012) while others have been treated for different ailments (Parish et al. 2007; Walters et al. 2010).

3.7 Andean Condor, *Vultur gryphus* (Linnaeus 1758)

Physical appearance

The Andean condor is the only member of the Genus Vultur (Linnaeus 1758), as other species such as the King vulture and the California condor have been removed from this genus. The Andean condor is about seven to eight cm shorter in length than California condor, about 100 to 130 cm (3 ft 3 in to 4 ft 3 in) (Del Hoyo et al. 1996; Ferguson-Lees and Christie 2001). However, it usually has a larger wingspan of 270 to 320 cm (8 ft 10 in to 10 ft 6 in) and generally weighs more, about 11 to 15 kg (24 to 33 lb) for males and 8 to 11 kg (18 to 24 lb) for females.

The Andean condor is sometimes cited as the world's largest flying bird; however it has some rivals. In terms of the heaviest average weight for a flying bird measured by weight and wingspan, it is exceeded by the Dalmatian Pelican (male bustards may weigh more, but are not considered full flying birds, rather ground dwellers) (Del Hoyo et al. 1996; BirdLife International 2014). In terms of the wingspan, although the largest albatrosses and pelicans have longer wings (the Wandering Albatross 3.6 m/12 ft, the Southern Royal Albatross, the Dalmatian and the Great White Pelican 3.5 m/11.6 ft), the Andean condor lays claim to the largest wing surface area of any living bird (Wood 1983; Harrison 1991; Ferguson-Lees and Christie 2001).

The bare skin of head and neck are dull red, which may change to a stronger red color when the bird is excited, for example during the mating season (Whitson and Whitson 1968). The male has a dark red comb or carbuncle on the top of the flattened head (Fig. 3.7a,b,c). There is a visible ruff of white feathers at the base of the neck. In the adult, the plumage is

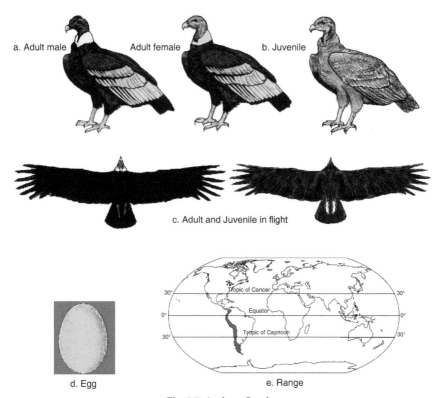

Fig. 3.7. Andean Condor.

black with large white patches on the wings, which do not appear until the completion of the bird's first moulting (Hilty 1977). Juveniles have blackish skin on the head and neck, brown collar ruffs and grey-brown plumage (Blake 1953).

Classification

The Andean condor is the only accepted living species of its genus, Vultur. As noted by (Griffiths 1994), it was formerly included in the Genus Sarcoramphus, with the Kibg vulture, but using syringeal morphology it was returned to Vultur (Griffiths 1994). There is little prehistoric fossil evidence for the Andean condor, compared with the California condor, which is well documented in fossil remains. Some fossil remains from the Pliocene-Pleistocene were revealed to be the modern species, rather than an ancestral species, and remains found in a Pliocene deposit of Tarija Department, Bolivia, possibly could be a smaller subspecies, named *V. gryphus patruus* (Fisher 1944).

Foraging

The condor, primarily a scavenger of large carcasses, prefers high, open terrain, such as grasslands, alpine meadows, rocky mountains up to 5,000 m (16,000 ft); occasionally it descends to lowlands in eastern Bolivia and southwestern Braziland and lowland desert areas in Chile and Peru, and is found over southern-beech forests in Patagonia (Sibley and Monroe 1990; Houston 1994; Parker et al. 1996; BirdLife International 2014). Similar to the California condor, and the Old World giants the Cinereous and Himalayan Griffon vultures it forages over very large areas, traveling more than 200 km (120 mi) a day in search of suitable food (Lutz 2002). Due to its weight and large size, it generally roosts in elevated areas such as rock cliffs which allow take-off without much wing-flapping effort, and where thermals are easily available (Benson and Hellander 2007).

Food sources include carcasses of large, wild animals such as llamas (*Lama glama*, Linnaeus 1758), alpacas (*Vicugna pacos*, Linnaeus 1758), guanacos (*Lama guanicoe*, Müller 1776), rheas (*Rhea americana*, Linnaeus 1758) and armadillos (*Dasypodidae* spp., Gray 1821). More recently, food has included the carcasses of domestic livestock (cattle, horses, donkeys, mules, sheep, pigs, goats and dogs) and introduced species such as boars (*Sus scrofa*, Linnaeus 1758) and red deer (*Cervus elaphus*, Linnaeus 1758) (Newton 1990; del Hoyo et al. 1994; Swaringen et al. 1995). Commoner remains are those of domestic animals, such as cattle, horses, sheep, goats and dogs. In coastal areas, beached carcasses of marine cetaceans are also eaten. Andean condors may also hunt small, live animals, such as rodents and birds and rabbits. In feeding, male condors dominate females and displace them to lower quality areas (Donazar et al. 1999; Carrete et al. 2010).

It is usually dominant over other avian scavengers, such as Turkey, Yellow-headed and Black vultures (Wallace and Temple 1987), although some studies have shown otherwise (Carrete et al. 2010). Recently, the increased range of the Black vulture has increased the presence of the smaller bird at carcasses. Although dominance hierarchies are usually related to size (Wallace and Temple 1987; see also the chapters in this book on Old World vultures, where the larger vultures such as the Cinereous, Lappet-faced and large griffons, are dominant over smaller vultures), between the Andean Condor and the Black vulture 'carcass consumption seemed to be determined by species abundance' rather than only size (Carrete et al. 2010: 385; see also Mikami and Kawata 2004). Although during this study, Andean condors arrived first to 76% of carcasses in mountains, and Black vultures arrived first to 72% of carcasses in plains, the numbers of male and female Andean condors feeding at carcasses were negatively related to the abundance of Black vultures (both in plains and mountains, but more so in the plains). Therefore, Black vultures may 'represent a serious obstacle' to Andean Condor feeding (Carrete et al. 2010: 385).

Breeding

The Andean condor is recorded as principally a cliff or rock ledge nester. Some nests have been recorded at elevations of up to 5,000 m (16,000 ft) (Fjeldså and Krabbe 1990). In places with few cliffs, such as coastal Peru, a nest may be created in cavities scraped among boulders (Haemig 2007). The nest, composed of a few sticks on the bare ground, contains one or two bluish-white eggs (Fig. 3.7d). The egg-laying period is from February to March every second year, and incubation lasts about 54 to 58 days. The period before fledging is about six months (Cisneros-Heredia 2006).

The Andean condor has been described as communal rooster at least on some occasions. Donazard and Feijoo (2002) noted that within large groups of birds there are hierarchies. Birds segregated by sex and age in summer and autumn communal roosts in the Patagonian Andes. It was observed that condors preferred roosting places that received earlier sun at sunrise (summer) and also later sun at sunset (autumn). There was also selection of sheltered crevices. It is hypothesized that sunny areas enabled occupying birds to increase foraging, plumage care and maintenance times, and also to avoid the colder temperatures.

In competition for favored places, adults dominated juveniles and males dominated females within each age class. In consequence, 'fighting and subsequent relocating led to a defined social structure at the roost', with mature males clustering as the most dominant, and immature males as the least dominant; it is inferable that 'irregularity in the spatial distribution and aggregation patterns of Andean Condors may be the result of requirements for roosting' (Donazard and Feijoo 2002: 832).

Population status

The distribution of the Andean condor is shown in Fig. 3.7e. The breeding range in the 19th century stretched from western Venezuela to Tierra del Fuego, i.e., along the whole of the Andes mountain range. Currently, the impacts of human activity have contributed to a contraction of this range (Haemig 2007). Negative effects may also stem from reductions in predatory mammals and livestock (Lambertucci et al. 2009) and interspecific competition with Black vultures (Carrete et al. 2010). It still occurs very rarely in Venezuela and Colombia, and also southwards to Argentina and Chile (BirdLife International 2014). The greatest declines appear to be in the northern area of its range, namely Venezuela and Colombia (Beletsky Les 2006). Recently, it has been recorded as declining in Ecuador (Williams 2002), Peru, Bolivia, Colombia (BirdLife International 2014), and Venezuela (Cuesta and Sulbaran 2000; Sharpe et al. 2008). By contrast, it is more stable in Argentina (Pearman 2003), especially in the largest known population, which is in north-west Patagonia (Lambertucci 2010).

   The Andean condor is considered near threatened by the International Union for the Conservation of Nature. The factors for its rarity include foraging habitat loss, secondary poisoning from animals killed by hunters and direct persecution, the last factor due to the belief that it attacks livestock (Reading and Miller 2000; Roach 2004; Tait 2006; Ríos-Uzeda and Wallace 2007).

   Similar to the policy for the California condor, from 1989 there have been reintroduction programs releasing captive-bred Andean condors from North American zoos into the wild in Argentina, Venezuela, and Colombia (Hilty and Brown 1986; Houston 1994; Chebez 1999; Roach 2004). Improvements may also be due to increased tourism in Chile and Argentina which showed the ecotourism value of the species and reduced persecution (Imberti 2003).

3.8 Conclusions

The New World vultures are clearly superficially similar to the Old World vultures, and the differences between the two groups are an entire subject in its own right. The Old World vultures themselves are a complex group, and some members superficially resemble the New World vultures more than others.

For example, the small Old World vultures, the Hooded and Egyptian vultures, classified as pickers, resemble the ecological roles of the Genus Cathartes and the Black vulture, all these having small bills for picking small morsels from caracasses. Possibly the role of the New World King vulture would be more similar to that of the African White-headed vulture, another medium-sized tearer, despite the far more eagle-like appearance and behavior of the latter. The two condors would be similar to the giant tearers, namely the Cinereous and Lappet-faced vultures, due to their large bills and role in opening tough skinned carcasses. Missing from among the New World vultures would be the equivalents of the pullers, the Griffons with their large size, very long, down covered necks and ability to reach deep into the carcasses.

The next chapters will examine in more detail the similarities and differences of both groups of vultures in terms of their ecological relations with other scavengers, predators and each other, and with natural human-modified landcover. Vultures use similar foraging techniques, but they face different challenges. For example, large competitors such as hyenas and Marabou storks in Eurasia and Africa are absent from the New World, and the large cats and eagles are fewer in number. The reader is invited to visualize how the facts presented in these chapters create different ecological settings for the two groups of vultures, and how similarities may show interesting adaptations from both groups.

PART 2

Vulture Ecology and Evolution

This part looks at the evolution of vultures. This introduction to the next three chapters provides a background, by describing a few of the larger animals, mainly pachyderms, armored mammals and large cats, that provided food for the vultures. It seeks to answer the question; where did the vultures, both Old and New World, come from, and what was the context within which they evolved. Vulture evolution followed the evolution of birds, which parallel to mammals followed that of reptiles and other terrestrial and marine life forms.

The birds we now call New and Old World Vultures, which are an example of convergent evolution, developed in a very different world from the current context. Palaeontogical evidence shows that the vultures developed during the peak of the Age of Mammals, similar to a more developed version of the modern African savanna with vast herds of large mammals, but found in North America and Europe as well. The megafauna consisted of huge herbivores that provided bountiful food for avian scavengers (probably a factor for the huge size of some vultures) and the huge carnivores that killed their huge prey, opening the carcasses of thick skinned giants and enabling the vultures to feed. The result was that vultures developed in a much more plentiful world than the current scenario, where they are reduced to feeding on small livestock, road kills and garbage. Such ecosystems possibly were a factor for the convergent evolution of avian scavengers, as different Orders, Families and Genera of birds may have capitalized on the abundant meat to evolve similar forms: large, carrion eating, soaring, bare headed birds, that did not need to hunt smaller prey in the midst of carrion abundance, but increasingly found it difficult to adapt as the Age of Mammals shifted to smaller forms of life.

The evolution of prehistoric ecosystems is the key to understanding the development of vultures. The time sequences, Eons, Eras and Periods are fundamental to such understanding (in order of duration, Eons are the longest, followed by Eras and then Periods). To evaluate the ecosystems of the time, it is first necessary to create time-series scenarios involving the main carnivores, herbivores and scavengers of the various periods. In this section, the first focus is on the geologic time scale (GTS), during which the different animals roamed the earth.

The geologic time scale (GTS) links stratigraphy to time. Stratigraphy is the study of sedimentary and volcanic rock layers and layering (stratification), and has two main branches; *lithologic stratigraphy* or *lithostratigraphy*, which looks at physical and chemical differences in rock layers, and *biostratigraphy* or *paleontological stratigraphy*, which addresses the fossil record revealed in the rock layers. The main concern in this book is biostratigraphy, and how this reveals the animals that lived during the various geologic time periods. The first period of the Earth's existence was the Precambrian, which is believed to have lasted from the formation of the

Earth approximately 4600 million years ago, to the start of the Cambrian Period, about 540 million ago. During the Precambrian, life forms appear to have evolved from unicellular to multicellular forms. The Cambrian Period was the first period of the current Phanerzoic Eon. The Phanerzoic Eon, from about 540 million years to the present is the main period for the evolution of species. It is divided into the Paleozoic, Mesozoic and Cenozoic Eras.

The Paleozoic lasted from 541 to 252.2 million years ago (ICS 2004), and was the longest of the eras within the Phanerozoic period. The Paleozoic was divided into six geologic periods, from the Cambrian (541.0 ± 1.0 to 485.4 ± 1.9 million years BP), Ordovician (485.4 ± 1.9 to 443.4 ± 1.5 million years BP), Silurian (443.4 ± 1.5 million years BP), Devonian (419.2 ± 3.2 Mya (million years BP), Carboniferous (358.9 million years ago, to the beginning of the Permian Period, about 298.9 ± 0.15), and Permian (298.9 ± 0.2 to 252.2 ± 0.5 million years BP).

The Cambrian Period saw the extensive development of animal life, also termed the 'Cambrian Explosion.' Life changed from mostly unicellular forms to mostly aquatic, multicellular forms of molluscs and arthropods in the sea, while the land was possibly still dominated by unicellular animals and plants. During the next period, the Ordovician, life was more sophisticated, mostly sea dwelling mollusks and arthropods and a few fish ancestors. During the Silurian, bony fish appeared in the seas, as did terrestrial arthropods and small moss-like land plants. During the Devonian, there was further development of vascular plants, with leaves and roots in forest communities, with spores later giving way to seed bearing plants. Fish developed further, and tetrapods began the colonization of land, with the fins changing into legs. During the succeeding Carboniferous Period, terrestrial life was dominated by amphibians, from which reptiles later evolved, and larger arthropods (for example the Meganeura), were also common in the forest dominated landscapes. The succeeding Permian period was the end of the Paleozoic, and saw more development of large reptiles and also flying reptiles and the ancestors of mammals, and insects.

The next Era, the Mesozoic lasted from 252 to 66 million years BP, and has been termed the Age of Reptiles, and included three geological periods: the Triassic (252.2 to 201.3 million years BP); the Jurassic (201.3 to 145 million years BP) and the Cretaceous (145 to 66 million years BP). The end of the Permian and the beginning of the Trassic, which was also the boundary of the Paleozoic and Mesozoic Eras, was marked by the Permian–Triassic (P–Tr) extinction event, sometimes termed the Great Dying, which happened 252.28 million years BP and was the largest known extinction event. It is estimated that 70 percent of terrestrial vertebrates and 96 percent of all marine species were extinct after this event, and that the recovery of life on Earth took up to 10 million years after this event. During this period, and possibly a little earlier, the archosaurs appeared. These are recognized as the ancestors of birds and crocodilians, and include the dinosaurs, the two main clades being Avemetatarsalia, including birds and their extinct relations; and Pseudosuchia, which includes crocodilians and their extinct links. The Avemetatarsalia, named by the palaeontologist Benton (1999) for those archosaurs more related to birds than crocodiles, included the subgroup, Ornithodira, which included the Pterosauromorpha, among which were the first vertebrates considered capable of flight. Also appearing during this period were the first true mammals, a specialized subgroup of Therapsids which possessed many of the traits that later developed in more advanced mammals.

During the next period, the Jurassic, the first real birds or Avialans, like Archaeopteryx, evolved from the forms of the Triassic. The large dinosaurs, the herbivorous sauropods and the largely carnivorous Theropods dominated the land. The succeeding Cretaceous period saw further development of birds and mammals (firstly marsupials and then placentals), and the extinction of the very large reptiles. As noted by Olson and Parris (1987) 'fossils of Cretaceous birds are scarce and usually difficult to interpret.' The Cretaceous–Paleogene extinction event began during this period (about 66 million years BP), possibly due to a massive comet/ asteroid impact during which the dinosaurs, and large numbers of birds (including all non-neornithean birds, such as the toothed enantiornithines and hesperornithiforms), mammals and marine life became extinct (Alvarez et al. 1980; Martin et al. 1996). This period may be regarded as the main background to current life forms, and forms the transition to the Cenozoic Era, which continues today.

The Cenozoic Era, the current geological era covers the period following the extinction event to the present, and is also known as the Age of Mammals (possibly due to the extinction event that ended the Age of Reptiles of the Mesozoic). It is divided into three periods: the Paleogene (66 to 23.03 million years BP, including the Paleocene, Eocene and Oligocene Epochs), the Neogene (23.03 to 2.588 million years BP, including the Miocene and Pliocene Epochs) and the Quaternary (2.588 to years BP to the present, including the Pleistocene and Holocene epochs). Birds and mammals proliferated, and increased in size possibly because of the extinction of the large reptiles. Most of these developments happened during the Paleogene, and by the Neogene bird and mammal species were almost similar to the modern species. Compared with the previous epoch, mammals and birds in some cases increased in size. Hominids, believed to be ancestral to humans, appeared at this time. During the Pleistocene Epoch of the Quaternary Period (2,588,000 to 11,700 years BP), many mammals and birds were much larger than their modern equivalents, and human/hominids forms were more similar to modern *Homo sapiens*.

The Cenozoic Era is the primary concern in this book, as it includes the evolution of avian scavengers, their changes in size and ecology, and the evolution and extinction of their avian and mammalian competitors, which also served as their predators and food sources. The main issues are the provision of food by the carcasses of the megafauna, the carnivores that killed the megafauna and hence provided food for the vultures, and also competed with them, and scavenging competitors, including birds, mammals and reptiles. Prehistoric megafauna refers to the large species of mammals that dominated some places during this period. Most of these animals became extinct in may be termed a Quaternary extinction event, this altering the food sources of scavengers and carnivores, both aerial and terrestrial. The main extinctions, which have been blamed on increased human hunting, climatic change, disease and/or an asteroid or comet (Scott 2009), occurred in the Americas, northern Eurasia and Australia, while subSaharan Africa and southeast Asia retained some larger mammals. The actual extents of change have been challenged.

For example, in North America, 35 genera of primarily large mammals became extinct by the end of the Pleistocene (Grayson 2006). The common assumption is that most of these extinctions happened between 10,000 and 12,000 BP. However, some writers have noted that only 16 of these extinct genera have conclusively been dated to this period, and there is the possibility that the extinctions took place over a longer period (Grayson and Meltzer 2002). The first birds appeared in the late Jurassic/early Cretaceous (150 million years BP). There is evidence of massive bird extinctions during the late Mesozoic. The beginning of Cretaceous/Tertiary (about 65 million years BP) also saw extensive diversification of bird forms (Feduccia 1995; Wink 1995). Fossil evidence from the Eocene and Oligocene (35 million years BP) reveal the presence of all orders of birds (including raptors, except Passerines). The expansion of bird species to most of the orders, 'an extraordinary explosive evolution', may have occurred in only 5 to 10 million years (Feduccia (1995).

Mammal groups that became extinct during this period included the Genera Pampatheriuma and Holmesina (family Pampatheriidae), Glyptotherium (family *Glyptodontidae*), Megalonyx (family *Megalonychidae*), Eremotherium and Nothrotheriops (family *Megatheriidae*), and Paramylodon (family *Mylodontidae*). These all of the Order *Cingulata* (SuperOrder Xenarthra), were large, armored, placental, vegetarian mammals related to modern armadillos Dasypodidae Gray, 1821. Glyptodonts, among the largest, weighed as much as two tonnes and measured up to 3.5 m, with a covering protective shell of hundreds of 2.5 cm-thick bony plates or osteoderms or scutes. Other large mammals were of the Order *Artiodactyla*. These included Mylohyus and Platygonus (family *Tayassuidae*, similar to modern peccaries, only larger); Camelops and Hemiauchenia (family *Camelidae*, similar to modern camels, but some species larger); Navahoceros and Cervalces (family *Cervidae*, related to modern deer); Capromeryx, Tetrameryx and Stockoceros (family Antilocapridae, similar to the modern pronghorn (*Antilocapra americana* Ord, 1815); and Saiga, Euceratherium, Bootherium, *Oreamnos harringtoni* (family Bovidae, similar to the modern musk ox (*Ovibos moschatus*, Zimmermann 1780) and mountain goat (*Oreamnos americanus*, Blainville 1816). Also present were the American mastodon (*Mammut*, Blumenbach, 1799; family Mammutidae) and the mammoth (*Mammuthus primigenius*, Blumenbach, 1799; family Elephantidae).

These herbivores, generally similar or larger than their modern relatives, coexisted with large carnivores, which were capable of killing them and leaving remains for scavengers. Carnivore species that became extinct in Pleistocene North America included: Brachyprotoma, a skunk species (family Mustelidae); and the dire wolf Canis dirus (family Canidae, similar to the modern smaller dhole Cuon alpinus, Pallas 1811), the only surviving member of the Genus Cuon. The dire wolf was larger and fiercer than the modern wolf (*Canis lupus*, Linnaeus 1758). The bears of this period were the huge short faced bear (Arctodus simus, Cope 1897 and the Florida spectacled bear (Tremarctos floridanus, Gidley 1928; family Ursidae). The large cats of the family *Felidae*, comprised the subfamily Machairodontinae. These were an extinct subfamily of the family Felidae (true cats). They were endemic to Asia, Africa, North America, South America, and Europe from the Miocene to Pleistocene living from about 23 million until about 11,000 years ago.

The Machairodontinae line started in the early or middle Miocene of Africa, possibly from the early felid *Pseudaelurus quadridentatus*, which developed large, elongated upper canines (Augusti 2002). The earliest recorded genus of machairodont was Miomachairodus from middle Miocene Africa and Turkey (Ostende et al. 2006). During the earlier Miocene, other large carnivores with long upper canines, called barbourofelids coexisted with the machairodontines (Ostende et al. 2006). Three groups of machairdontines are recognised: the massively built Smilodontini, with the largest upper canines (sometimes referred to dirk toothed, as a dirk is long dagger about twelve inches long) including Megantereon and Smilodon; the Machairodontini or Homotherini, including the Machairodus or Homotherium, with smaller upper canines (sometimes referred to as scimitar toothed); and the Metailurini, with yet smaller teeth including Dinofelis and Metailurus (also sometimes classified as scimitar toothed) (Martin 1989; Barnett et al. 2005; Van Valkenburgh 2007; Slater and Van Valkenburgh 2008; Meloro and Slater 2012). Some recent writers put the Metailurini within the Felinae (the subfamily of modern cats) while others dispute this classification (Wesley-Hunt and Flynn 2005; Ostende et al. 2006; Meloro and Slater 2012; Salesa et al. 2012). The last machairodontine genera were Smilodon and Homotherium (Fabrini, 1890), which existed until the late Pleistocene in the Americas, about 10,000 years BP (Berta 1985; Kurten and Werdelinb 1990; Rawn-Schatzinger 1992; Turner and Antón 1997; Antón et al. 2005; Ascanio and Rincón 2006; Andersson et al. 2011).

Megantereon, with a size range from 60 to 150 kg (130–330 lb) which is commonly seen as the ancestor of the most well known sabertoothed cat, the Smilodon, lived during this period in Africa, North America and Eurasia, with the oldest fossil evidence dated to 4.5 million years from the Pliocene of North America (Turner and Antón 1997). The African records date to about 3–3.5 million BP, and Asian records date to 2.5 to 2 Million years BP. The oldest European records are from Les Etouaries (France), about 2–2.5 million years BP (Martin 1989; Turner and Antón; Barnett et al. 2005; Slater and Van Valkenburgh 2008). At the climax of the Pliocene, Megantereon evolved into the larger Smilodon in North America, but in the Old World it remained until the Middle Pleistocene (Turner and Antón 1997; Hemmer 2002).

The three species of Smilodon were *S. populator* (Lund 1842); S. fatalis (Leidy 1869); and S. gracilis (Cope 1880). *Smilodon gracilis* was the first Smilodon, living from 2.5 million–500,000 BP and weighing about 55 to 100 kg (120 to 220 lb) (Christiansen and Harris 2005). It is considered to have evolved from the North American Megantereon in North America. *Smilodon fatalis*, which lived 1.6 million–10,000 years ago, replaced S. gracilis in North America and also occupied western South America during the Great American Interchange (Kurten and Werdelinb 1990; Rincón et al. 2011). It weighed 160 to 280 kg (350 to 620 lb) (Christiansen and Harris 2005; Turner 1997; Christiansen 2007; Jefferson 2001; Shaw and Cox 2006; Van Valkenburgh and Sacco 2002). The largest, *Smilodon populator*, lived 1 million–10,000 BP, in eastern South America (Kurten and Werdelinb 1990; de Castro and Langer 2008). It is considered possibly the largest known felid, weighing 220 to 400 kg (490 to 880 lb) (Christiansen and Harris 2005; Sorkin 2008). The Smilodons survived in Africa until 1.5 million years BP, Eurasia until 30,000 years BP and North America until 10,000 years BP (Turner 1997; Turner and Antón 1997).

The larger cats, by opening the megafauna carcasses with very thick skin, provided avian scavengers with access to the internal meat. Kurten and Werdelinb (1990) note that an association of Homotherium species with proboscidean (elephant and mastodon) and rhinoceros remains, suggests that Homotherium preyed selectively on these tough-skinned animals. Smilodon, a massive apex predator, hunted large bison, camels, ground sloths, horses and mastodons, and pig-like Platygonus and the llama-like Hemiauchenia, possibly in competition with the American Lion (Panthera leo atrox) and the Dire Wolf (Vanvalkenburgh and Hertel 1993; Coltrain et al. 2004; Fennec 2005; Christiansen and Harris (2005).

The constructed scenario of megafauna provides a justification for vulture evolution. These evolved into very large birds, the largest and strongest probably capable to tearing through the skin of the huge carcasses left from the carnivores. Each massive meal could feed large flocks of large vultures much more easily than could an alternative predatory life style. There is little doubt that such a proliferation of large carcasses could fuel the evolution of large avian scavengers, possibly from different sources within the avian kingdom. The main question is when vultures appeared, or when birds may be classified as the ancestors of either the present New World or Old World vultures. Is the ancient bird discovered a vulture, a vulture ancestor or a different species altogether? Furthermore, what is a vulture? The questions will be dealt with in Chapter 4.

4

The Biological Evolution of Vultures

4 INTRODUCTION

This chapter examines the history of both groups of vultures, from the earliest, disputed forms, to the more recognizable species that spawned the Old and New World vultures. The origins of the New World vultures are more disputed than those of the Old World vultures. The New World vultures have been classified within the Acipitridae, Ciconidae, and also independently, in their own Order. The Old World vultures, by contrast, are seen as evolving within the Accipitiformes.

At the beginning of vulture evolution, five main issues stand out. First, the earliest 'vultures' are not necessarily recognized as true 'vultures'. Here, a vulture is defined as a largely carrion eating bird, with feet not as clearly designed for killing as those of hawks and eagles, with bare or semi-bare heads, broad wings for soaring flight and bills designed for either pulling, tearing or picking. Second, what is termed Old World or New World vulture is less definable by location, as Cathartid ancestral groups have been discovered in the Old World and the Accipitrid vulture ancestors are fossilized in the Americas. As noted by Zhang et al. (2012b: 7) 'birds fly, and their ancestral distributions may not correlate with their modern distributions.'

Third, the classification of fossils, based on skeletal form compared with modern forms, does not conclusively determine the diet of such birds. For example, an extinct counterpart of the Black vulture could be a vegetarian, and as such its 'vulture' status would be disputed. Fourth, much of the evidence is circumstantial; merely because one bird form was dated to a period before another similar form, does not conclusively prove that the latter evolved from the former. Fifth, vultures may not necessarily have reached their final form; the new ecosystems they inhabit, with less food and human modified landscapes may encourage further convergence or divergence.

With these ideas in mind, this chapter first looks at the prehistoric scavengers that are considered ancestral to modern vultures. It follows up with a detailed consideration of the New World and Old World vultures.

4.1 Convergent evolution; What is it?

New and Old World vultures are generally seen as an example of convergent evolution. Critical examination of the biochemistry, morphology and anatomy of Old and New World vultures shows that the two groups 'represent a phylogenetically inhomogenous, i.e., polyphyletic group, whose shared characters are based on convergence' (Wink 1995; as also noted by Mundy et al. 1992; Sibley and Ahlquist 1990), which some have called a 'textbook example of convergent evolution' (Del Hoyo et al. 1994).

Both groups eat carrion, but Old World vultures are in the eagle and hawk family (Accipitridae) and use mainly eyesight for discovering food; the New World vultures are of obscure ancestry, and some (Genus Cathartes) use the sense of smell as well as sight in hunting. Birds of both families are amongst the largest flying birds in the world, have unfeathered heads and necks, search for food by soaring, the larger species live in mountains, all circle over a sighted carrion, nest in tall trees and high cliffs, and may roost in flocks in trees. Other shared characteristics are long intestines, strong stomach acids to digest rotten meat and bones and kill microorganisms in carrion, feet for movement on the ground, and lack of the raptors' pronounced sexual dimorphism (Newton 1990; Brown and Amadon 1968; Mundy et al. 1992; Glutz von Blotzheim et al. 1971; Cramp and Simmons 1980; del Hoyo et al. 1994).

Such similarities are apparently possible, if species from different Orders and Families slowly adapt to similar opportunities and through time, resemble each other in both appearance and behavior. For example, all avian carrion eaters will need soaring flight, as carrion exists in widely spaced areas, and soaring birds outcompete terrestrial bird or mammal scavengers. They may need bare heads and necks (even the Marabou Stork (*Leptoptilos crumenifer*, Lesson 1831), a habitual carrion-eater, has a bare head to avoid bloodied feathers); they may be large, as speed is less important than the ability to rip open carcasses, and to store large quantities of food in their crops against lean seasons; and they may not need the strong feet of predatory eagles and hawks, for terrestrial locomotion. Any Families, such as hornbills, gulls, storks or raptors that gradually adapt to this lifestyle, possibly due to the abundance of carrion (such as during the Age of Mammals) and the relative ease of carrion consumption compared with catching live prey, might gradually acquire similar appearance and habits.

All vultures were once thought to have evolved 'only once among extant diurnal birds of prey' (Sielbold and Helbig 1995: 163). But in the late 19th century, scientists queried the status of the Cathartidae as vultures (Garrod 1873; Sibley and Ahlquist 1990). Rea (1983: 26) argues that 'from the very beginning the relationship of the cathartid vultures to the Falconiformes has been controversial.' Further, 'so far in the fossil records of Cathartids, Accipitrids and Falconids, there is no time when Cathartid vultures were more raptorial or the raptors less raptorial, there is no fossil candidate for a common ancestor' (*ibid.* 34). Vultures were initially hypothesised to be a monophyletic group, but with the increasing awareness of distinctions between New and Old World vultures, vultures were increasingly seen as polyphyletic. In this sense, a monophyletic group is a taxon (group of organisms) forming a clade, i.e., consisting of one ancestral species and all its descendants. A polyphyletic group has convergent features not inherited from one common ancestor, but from several such species.

Based on studies using anatomy, morphology and biochemistry, there is evidence that vultures must represent a phylogenetically diverse group (i.e., they share different ancestors; a polyphyletic group) and the shared characteristics are due to convergence (Mundy et al. 1992; Sibley and Ahlquist 1990). Del Hoyo et al. (1994) even called them a 'textbook example of convergent evolution'. A study using biochemical nucleotide sequences of the mitochondrial cytochrome b gene (which holds information on the biological relations among animals) found evidence that the Cathartidae (New World Vulture family) was not closely related to the other raptors in Acipitridae (including the Old World Vultures). 'Since also the fossil record provides no evidence that the different families within the Falconiformes had a common ancestor, 'raptors' should be a result of an evolutionary convergence between bird groups of polyphyletic origin'; the similarities 'found in morphology are indeed based on convergence and not on close genetic relatedness' (Wink 1995: 880). The evidence for convergence, found by morphological (the study of the appearance and shape of the skeleton and body), karyological (the study of chromosomes) and DNA-based studies show vultures are in at least three monophyletic clades (this means that each group possibly has a different common ancestor) showing polyphyly (multiple ancestors) (see also Sibley and Ahlquist 1990; Mundy et al. 1992; Del Hoyo et al. 1994).

Convergent evolution is the independent evolution of similar or identical features in species of different Genera, Families or Orders. In convergent evolution, similar structures with similar form and/or function that were non-existent in the ancestral groups, may develop in the descendant groups. These similarities are termed *analogous structures*, which differ from *homologous structures* (which may share a common origin but have different functions). Convergent evolution is the opposite of divergent evolution, in which similar, related species develop different traits. The most similar phenomenon to convergent evolution is parallel evolution. The latter occurs when two independent but similar species change in similar ways. The difference lies with the ancestors; if the ancestors were similar then the evolution is parallel, if they were different, then the evolution is convergent.

Problems emerge when the ancestral forms are not definitively known; in this case parallel and convergent evolution are less distinguishable. Further complications emerge when different structures perform a similar function. It is also possible that where lineages do not evolve together and then do so, the result may be convergent evolution at some later point. For example, if all birds share a common ancestor, then evolved into different species in different places (for example, into raptors and storks), later convergent evolution could create the two groups of vultures, but there may have been parallel evolution at some earlier point in time.

It must be noted that the definitions of traits may also determine whether a form of change may be considered divergent, parallel or convergent. For example, if we consider only the heads of vultures, there is convergent evolution of bald heads. Likewise, the broad wings for soaring flight indicate convergent evolution. As also shown in Fig. 2.9, there are peckers and tearers among both the New and Old World vultures, with few differences between the Egyptian and Hooded vultures on one side and the Cathartes and Black vultures on the other. The condors and King vulture may be seen as tearers, although their bills are proportionately slightly smaller than the giant bills of the Lappet-faced, Cinereous and Red-headed vultures. However, other traits show divergence, such as the weaker feet of the New World vultures (see Fig. 4.1 for a comparison of vulture, eagle and stork feet), their habits of defecating on their legs and feet, their perforate nostrils and in the Cathartes vultures, the sense of smell. The pronounced brow ridges above the eyes of the Old World vultures are more similar to raptors like eagles, while the small brow ridges of the New World vultures (the eyes appear to pop out of the skull) are more similar to the storks (Fig. 4.2). These would count as divergence if they shared a common ancestor with the Old World vultures, as the latter do not have these traits. Taking all the traits of both groups together, the result is termed convergent because it is believed that there are distinct ancestral groups.

Rea (1983) notes that the New World vultures differ from the Old World vultures and other raptors in the following: mycology of thigh and breast; pteryography; absence of patagium connecting the secondary flight

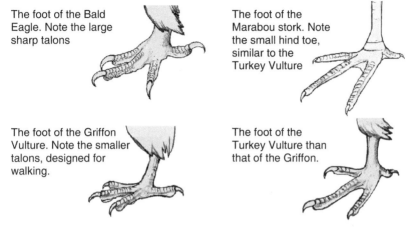

The foot of the Bald Eagle. Note the large sharp talons

The foot of the Marabou stork. Note the small hind toe, similar to the Turkey Vulture

The foot of the Griffon Vulture. Note the smaller talons, designed for walking.

The foot of the Turkey Vulture than that of the Griffon.

Fig. 4.1. The Feet of Vultures, Eagles and Storks.

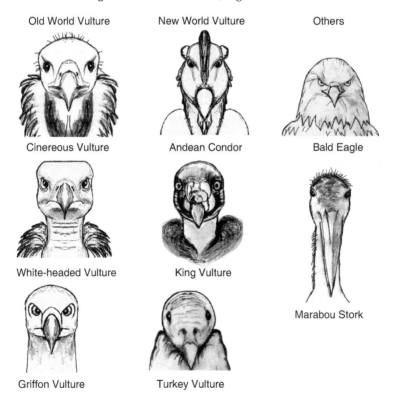

Old World Vulture

New World Vulture

Others

Cinereous Vulture

Andean Condor

Bald Eagle

White-headed Vulture

King Vulture

Marabou Stork

Griffon Vulture

Turkey Vulture

Note the larger, eagle-like brow ridge above the eyes in the Old World Vultures compared with the stork-like head and face of the New World vultures.

Fig. 4.2. The Heads of Vultures, Eagles and Storks.

feathers; arrangement of wing coverts; lack of syrinx and intrinsic syringeal muscles; in the configuration of the laryngeal apparatus; the shape of wing claw; relative sizes of liver lobes; thermoregulation and nest defence; and chromosomes, hematology, egg white protein and physical structure of eggshells. This author (p. 45) states 'there is nothing in the fossil history to unite the groups or to suggest that one might be derived from the other.' Furthermore, the Catheridae must be placed with Ciconiiformes, as 'the most inescapable conclusion from an examination of the Catheridae is that they have virtually no trait, whether anatomical, physiological, cytological or behavioral, that suggests that they are related to the Falconiformes.'

An important issue in the study of vulture biology concerns the lack of sexual dimorphism in both the New and Old World vultures. In raptors, the female is much larger than the male, but in both condors the male is slightly larger, and in many other vultures the distinction is minimal (although size may favor the male in New World vultures, and the female in the Old World vultures). This makes it difficult in some cases to assess the population structure and ecological behavior (Fry 1983). Methods used to distinguish between male and female birds include: direct gonadal (sex organs, testes and ovaries) examination by laparotomy (incision) or laparoscopy (incision with a camera); analysis of blood or urinary steroids; karyotype analysis through identification of sex chromosomes; and behavioral differences between males and females. Some of these methods are more easily used than others. Urinary steroids can be measured from fecal samples, but the other methods require physical handling of the bird to obtain blood samples, and judging sex by behavior is unreliable (Fry 1983).

4.2 The Evolution and Classification of New World Vultures

The history of vultures dates back to at least 50 million years BP, i.e., well into the Cenozoic Era (Rich 1983). Crucially, the grouping into New World and Old World vultures is challenged by fossil evidence, as some New World Vulture ancestral fossils are located in the Old World and some Old World Vulture ancestral fossils are located in the New World. Olsen (1985: 191) notes that 'if the available fossil record is any guide, the so called New World Vultures are definitely misnamed, as their early history is almost completely confined to the Old World'. For example, fossils of *Lithornis vulturinus* (a possible candidate for Cathartid ancestry) have been dated to 60 million years BP in English sediments. The Genus Lithornis (Lithornis means stone bird in ancient Greek), comprised extinct paleognathous (reptilian) birds from the Upper Paleocene to the Middle Ecoene, which the evidence suggests were good flyers, but some have classified as related to modern tinamous and ratites (the former are mediocre flyers, the latter are flightless).

Houde (1988) describes six species of Lithornis, but the evidence for their validity is disputed. These species are: *Lithornis vulturinus* (described by Owen (1840) from an Early Eocene fossil in London Clay deposits on the Isle of Sheppey, Kent, and from Denmark (Leona et al. 2005); *Lithornis hookeri*, described by Harrison (1984) (Mayr 2008); *Lithornis nasi* described by Harrison (1984); *Lithornis celetius* from the Bangtail Quarry, Sedan Quadrangle, Park County, Montana, USA, described by Houde (1988); *Lithornis promiscuus*, from the Early Ecocene Willwood Formation, Clark Quandrangle, Wyoming, USA; and the locationally similar *Lithornis celetius* and *Lithornis plebius* described by Houde (1988) from the Bangtail Quarry, Sedan Quadrangle, Park County, Montana, USA.

However, some authors question whether these species are actually Cathartids or Cathartid ancestors. Rich (1983) points out that the evidence for classification is not strong enough. Problems include specimen losses during World War 2, and uncertain classification based on questionable evidence (for example Cracraft and Rich 1972). One classification, by Harrison and Walker (1977) even concluded that Lithornis was neither a Cathartid or a member of Falconiformes, but in the family Threskiornithidae (large wading birds, such as the ibises and the spoonbills).

Apart from the Lithornis group, a number of other discoveries purport to be Cenazoic ancestors of the Cathartids (or Vulturids), but these also are disputed. For example, a species named *Teracus littoralis* (Milne-Edwards 1867-71) from the early Oligocene of France, was believed to be a Cathartid, but was classified with Incertae Sedis (Olsen 1978a). This is Latin for 'of uncertain placement', for any taxonomic group with unknown or uncertain relationships. Also, bones of a species named *Eocathartes robustus* were found in Saxony, Germany, but one author stated 'it is almost certainly not a vulturid', but possibly close to the hornbill (*Geiseloceros robustus*, Lambrecht Olsen 1978a).

Other, possibly stronger examples are *Diatropornis ellioti* (Milne-Edwards 1892) and *Plesiocarthes europaeus* (Gaillard), from late Eocene to Oligocene Phosphorites du Quercy in France. Several writers classified these as Vulturids (Cracraft and Rich 1972; Mourer-Chauvire 1982). Another example is *Plesiocathartes gallardi* of early Miocene (Burdigalian) Spain and another vulture discovered in the early Oligocene of Mongolia (Kurochkin, in Olsen 1985). This fossil evidence may lead to the conclusion that 'New World' vultures had a presence in the Old World. Olsen notes that although the Cathartids were evidently in the Old World as far back as the middle Paleogene, there is no evidence of their presence in the New World until the late Neogene (p. 192).

Other possible Cathartid fossils have been found in the Americas. These were mostly from the Oligocene, Eocene, Pliocene, Pleistocene and Quaternary periods. The pre-Pleistocene fossils are mostly disputed as Cathartid links. Condor-like fossils were dated to the Pliocene (5 million BP) or Pleistocene deposits (2 million BP) (Wink 1995). Olson argued that only *Sarcoramphus kernense* (L. Miller) from the Pliocene of California and *Pliogyps fisheri* (Tordoff) from the Pliocene of Kansas are pre-Pleistocene Cathartids in the Americas. Examples of the pre-Pleistocene fossils are *Palaeogyps prodomus* (Wetmore) and *Neocathartes grallator* (Wetmore) from Wyoming, but both of these have been described as related to the Bathornithidae, which is now believed to be linked to the modern seriemas of the Order Cariamiformes. There were also *Phasmagyps patritus* (Wetmore) from Colorado, disputed as the earliest Cathartid (Olson 1985; Stucchi and Emslie 2005). *Paracathartes howardae* from Wyoming was also argued to be as earliest known Cathartid (Harrison 1978), but this position was challenged by Rich (1983). It was later included in the order Lithornithiformes and family Lithornithidae by Houde (1988).

There were two other extinct, possible Cathartids from the Pleistocene which were unearthed in the Americas. *Vultur patruus* (Loonberg) of the Pleistocene deposits in Bolivia, has been compared with the current species Andean Condor (*Vultur gryphus*). It is considered to have a stronger link to the Cathartids than those mentioned above (Fisher 1944; Campbell 1979). Another fossil from the Pleistocene of the New World, which appears more strongly linked to the Cathartids is *Breagyps clarki* L. Miller from Rancho La Brea, California (Howard 1974; Emslie 1988).

The Golden Age for the Cathartids appears to be from just before to during the Pleistocene and to a lesser extent the Pliocene. These include: *Diatropornis* (European Vulture), Late Eocene/Early to Middle Oligocene of France; *Phasmagyps*, Chadronian of Colorado; Cathartidae gen. et sp. indet. (similar in size to the modern Black Vulture) from the Late Oligocene of Mongolia and Middle Pliocene of Argentina and Cuba; *Brasilogyps*, Late Oligocene–Early Miocene of Brazil; *Hadrogyps* (American Dwarf Vulture), Middle Miocene of SW North America; Cathartidae gen. et sp. indet. Late Miocene/Early Pliocene of Lee Creek Mine, USA; *Pliogyps* (Miocene Vulture), Late Miocene–Late Pliocene of S North America; *Perugyps* (the 'Peruvian Vulture') Pisco, Late Miocene/Early Pliocene of SC Peru; *Dryornis* (Argentinian Vulture), Early–Late Pliocene of Argentina, which was similar to the modern Genus Vultur; *Aizenogyps* (South American Vulture), Late Pliocene of SE North America; *Breagyps* (Long-legged Vulture), Late Pleistocene of SW North America; *Geronogyps*, Late Pleistocene of Argentina and Peru; *Gymnogyps varonai*, late Quaternary of Cuba; *Wingegyps*

(Amazonian Vulture), Late Pleistocene of Brazil (Wetmore 1927; Emslie 1988; Suarez 2003, 2004; Alvarenga and Olson 2004; Stucchi 2005).

An important example of vulturine ancestors were the giant teratorns (Greek Τερατορνις Teratornis, monster bird), the extinct species of the family Teratornithidae (Miller 1925). All the remains for these birds date from the early or late Pleistocene, although some are from the Pliocene (Campbell and Tonni 1983). These giant birds are believed to be linked to the giant megafauna of the period (for example the huge mastodons) and to have died out with their associated mammals (Tambussi and Noriega 1999). There were four genera: Teratornis (species *Teratornis merriami,* Miller 1909) Aiolornis (species *Aiolornis incredibilis,* Howard 1952), at least 43 percent larger than *Teratornis merriami* (Howard 1952); Cathartornis (species *Cathartornis gracilis,* Miller 1910); and Argentavis (species *Argentavis magnificens,* dated to the Miocene, Campbell and Toni 1980). Teratorns were the largest flying birds known, with the largest *Argentavis magnificens* reaching a wingspan of 6–8 m and a weight of 72–79 kg (Campbell and Toni 1980; Campbell and Marcus 1992). *Teratornis merriami* is the most known of the teratorns, with over one hundred specimens discovered in the La Brea Tar Pits, California and also in southern Nevada, Arizona, and Florida. It was estimated to stand about 75 cm, with a wingspan of around 3.5 to 3.8 m and a weight of 15 kg. The taxonomic history of the Teratorns is highly disputed. The Teratorns are currently placed with the storks (Ciconiidae) and in the order Ciconiiformes (Jollie 1976, 1977; Rea 1983; Olson 1985; Emslie 1988; Chatterjee et al. 2007).

They were first classified with the family Cathartidae, as there appeared to be strong skeletal resemblance between *Teratornis merriami* (Miller 1909) and the California condor. This was disputed, and arguments arose that classified it as a predator with both raptor and stork attributes (Campbell and Toni 1980, 1982, 1983; Campbell 1995). The taxonomy of *T. incredibilis* was changed to the Genus Ailornis (Campbell et al. 1999). Miller and Howard (1938: 169) acknowledged that *Cathartornis gracilis* was 'markedly similar to *Teratornis merriami,* though it is undoubtedly a distinct species'. This is disputed by Campbell et al. (1999) who argue that the maintenance of Cathartornis as a separate genus is weak.

Several differences between teratorns and modern condors have been gleaned from the fossil remains of *Teratornis merriami.* The legs of the extinct bird are described as stouter than those of the Andean condor. It is also described as less adapted to the running takeoff, possibly more likely to jump and beat its wings, as the legs were smaller in proportion to the body, with a proportionately smaller stride than in the modern condors (Fisher 1945). Also, it possibly hunted animals including fish (Hertel 1995), as they appear to possess larger stronger bills than the modern condors (Campbell and Tonni 1983).

Large numbers of these birds have been found fossilized in the La Brea Tar Pits. One possible factor for this frequent, curious entrapment was their attraction to dead and dying mammals trapped in the tar (Hertel 1995). The huge birds possibly helped the smaller species (smaller vultures, ravens and eagles) by opening up the carcasses, and would have an advantage over the mammalian scavengers which would be mired in asphalt if they approached. The teratorns may also have attempted fishing from water on the surface of the asphalt.

There are many debates as to the reasons for the extinction of the teratorns. Commonly accepted factors are the climatic shifts at the end of the Ice Age (late Pleistocene/early Holocene). This led to ecological change, extinction of the large mammals, reduced aquatic vertebrate populations, competition from predators for small prey, and from more adaptable smaller vultures (including the condor) for smaller carrion.

Other vulture species also evolved during the Pleistocene. There was a prehistoric species of the Black vulture, *Coragyps occidentalis*, known as the Pleistocene Black vulture in North and South America, possibly differentiated from the smaller modern Black vulture (*Coragyps atratus*, Bechstein 1793). Some have argued that it evolved into the modern species by reducing in size after the Ice Age (Fisher 1944; Howard 1962; Steadman 1994; Hertel 1995). If this were the case, the final stage of the evolution may have taken place during the period of human habitation; this is because a fossil bone of the extinct species was discovered in a native midden (or dump for domestic waste) at Five Mile Rapids near Dalles, Oregon, dated about 9000–8000 years BCE (Miller 1957).

There have been disputes in classification of the prehistoric variants, subspecies or ancestral species of the Black vulture. For example, Howard (1968) classified pre-historic birds found in Mexico as *Coragyps occidentalis mexicanus*, in contrast to the northern sub-species which was classified as *C. o. occidentalis* (Howard 1968). There was little difference in size between the extinct Mexican birds and modern Black vultures, except that the former had slightly wider and flatter bills, and stouter tarsometatarsus (Arroyo-Cabrales and Johnson 2003). Some writers hold that the prehistoric and modern variants form a continuum, and include *C. occidentalis* in *C. atratus* (Steadman 1994).

Concerning the current New World vulture species, there are seven species and five genera currently recognised: Coragyps, Cathartes, Gymnogyps, Sarcoramphus, and Vultur. The characsistics of New World vultures are featherless heads and necks (Zim et al. 2001); long, broad wings and a stiff tail suitable for soaring (Ryser and Ryser 1985); clawed but weak feet (Krabbe and Fjeldså 1990), long front, slightly webbed toes (Feduccia 1999); the lack of a syrinx or voice box, limiting their vocalizations to hisses or grunts (Kemp and Newton 2003; Howell and Webb (1995); a slightly hooked, comparatively weak bill, compared to other raptors (Ryser and Ryser 1985; Krabbe and Fjeldså 1990); oval, perforate nostrils not divided by a septum, i.e., one can see through the bill by the nostrils (Terres 1991; Allaby 1992); the absence of the bony brow of raptors (Terres 1991) (see Fig. 4.1); and the storks' habit of urohidrosis, defecating on their legs for evaporative cooling. Cathartes is regarded as the only genera that is not montypic; where monotypic refers to the possession of only one immediately subordinate taxon (for example a subgenera). Therefore, the Genus Cathartes has three species, while the rest have only one species each (Myers et al. 2008).

Regarding the classification of the New World vultures, there have been disparate views since the late 19th century. Initially, these were placed in their own family within the Falconiformes (Sibley and Ahlquist (1991). The conflicting views may be classified into three: (1) those that hold that the New World vultures are related to the storks and belong in the Ciconiiformes; (2) those that hold that they should be in the Falconiformes; and (3) those that hold that they should be in their own Order. Earlier studies were based on morphology, but in the late twentieth century DNA and other biochemical based studies were also published.

Brown and Amadon (1968: 17) recognised four main groups of vultures, 'singling out the New World vultures as the most likely to be unrelated.' A close phylogenetic (multi-ancestral) relationship with storks was suggested (Garrod 1873; Ligon 1967; Konig 1982; Rea 1983). Garrod (1873) placed Cathartids between storks and herons, due to thigh muscle form. Goodchild (1886) used a study of secondary wing coverts (small wing feathers) to determine that the Cathartids were similar to Ciconiiformes, Procellariformes and four families of Pelicaniformes, but different from the Accipitriformes (see also Beddard (1889, 1898). Sharpe (1891) used the details of intestinal coil patterns to place the Cathartids with the Ciconiiformes, Procellariformes and Pelicaniformes. Hudson (1948) used the pelvic musculature of the Falconiformes, and concluded that the Cathartids, Accipitrids and Sagittarids evolved as different lineages. Friedmann (1950) stated that 'the exact relationship of the Cathartidae, is however somewhat complex. It has been fairly clearly demonstrated that they are not very distantly related to the Ciconiiformes, Pelicaniformes and Procellariformes.' Ligon (1967) used evidence from osteology, natal plumage, myology, patagial tendons and syringeal structure, to combine the Cathartids and Ciconiiformes; placing the herons by themselves in the Order Ardeiformes.

Jollie (1976) used pterylography (the study of feather arrangement), osteology (the study of bone structure) and myology (the study of muscle structure) to conclude that the Falconiformes included four phylogenetically unrelated groups; an important factor for this decision was the physical distinctions of the New and Old World vultures. The distinctions between these groups concerned the lack of the raptor grasping foot and a syrinx in the New World vultures, that distanced these birds from raptors. Factors that related the Cathartids to the storks were the defecation on the legs, wing spreading, bill clapping and bill interlocking before and during copulation, the inflation of neck sacs and gender—neutral incubation (see also Konig 1982; Rea 1983; Sielbold and Helbig 1995).

Fisher (1955) used only morphological evidence to distinguish the New World vultures from the Old World vultures. Basically, the skulls of the cathartid vultures 'may be distinguished from the skulls of the other Falconiformes' in that: 'the external nares are perforated; the rostral area of the skull is elongated (except in Sarcoramphus); an imperfect frontonasal hinge is present; the lachrymals are completely fused to the frontals and are directed downward; the premaxilla is highly vaulted; the opisthotic processes are extremely long; the articular process of the squamosal is weak or absent; the sphenoidal rostrum is excavated in front of the basitemporal plate; the bones within the olfactory chamber are more completely ossified; the zygomatic arch is split anteriorly; and the skull is indirectly desmognathous' (ibid.).

For an explanation of these terms, the nares are nostrils. In New World vultures, these are perforated; it is as if a hole has been drilled through the bill at right angles to the direction of the beak. In the Accipitrid vultures, the nasal passage is divided by a septum, composed of cartilage and bone. The rostral area of the skull refers to the part of the skull connected to the bill, as the rostrum refers to the bill itself. New World vultures (with the exception of the king vulture), have more elongated heads, compared with the Old World vultures. The former also differ from the long heads of the Hooded and Egyptian vultures, as in these two small Old World vultures the skull is not long, only the bill is long (see Fig. 2.9). The frontonasal hinge refers to the connection between the bill and the frontal and nasal bones, a bony plate composed of extensions of the nasal, frontal and premaxillary bones. The frontal bones comprise the front of the cranial cavity and forms the forehead. The nasal bones refer to the bones that together form the bridge of the nose. The lachrymals refer to the glands that produce tears, which are fused to the frontal bones in the New World vultures. The opisthotic bone, along with the prootic and the epiotic, forms the petrosal or periotic bone of the head, behind the ear (Coues 1883). The articular process of the squamosal refers to the articulation with the quadrate bone of the lower jaw. The sphenoidal rostrum and the basitemporal plate are both parts of the skull. The olfactory is the cavity inside the nose, used for respiration and in some cases smell. The zygomatic arch is the cheekbone on the side of the skull (Olsen and Joseph 2011).

Fisher (1944) also pointed out differences between the skull of the King vulture and some of the other Cathartids. In the former, the angle of the bill forms an acute angle of about 30°, while in Cathartes it is 12 to 18°. This gives the skull and bill of the King vulture a 'definite predatory aspect in contrast to the weak-billed, flat-topped condition found in the other cathartids' (*ibid.* 278). This is also argued to indicate a stronger bill in the King vulture, justifying its status as a tearer. The lower mandible is also strongest in the King vulture, and progressively weaker in the Andean condor, the Cathartes vultures and the Black vulture. See Fig. 4.3 for the profiles of some of the Cathartid heads. The King vulture can open its bill more widely than can the other Cathartids. The Black vulture, having a longer bill than the rest, also has a wider gape. It is noted that

> '*ability to open the bill more widely may be an adaptation in two ways. In grasping live, struggling prey a widely opened bill is an aid to securing and maintaining a strong hold. In the case of Sarcoramphus it probably aids in this manner as well as in enabling larger chunks of carrion to be swallowed. The latter adaptation is perhaps the more important in Coragyps, but here again the aid in grasping prey may be significant since McIlenney (1939) has shown that Black Vultures attacking in groups kill skunks, opossums and other small mammals*' (ibid. 285).

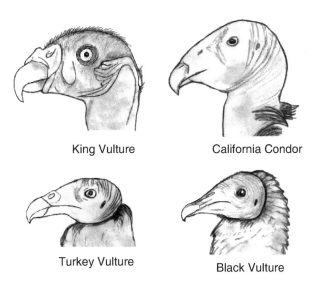

King Vulture   California Condor

Turkey Vulture

Black Vulture

Fig. 4.3. The Heads and Bills of some Cathartids.

The Cathartids were nevertheless included within the Falconiformes, or Acciptriformes until the 1990s. Studies supportive of this view include those of Wetmore (1960), Brown and Amadon (1968), Stresemann and Amadon (1979), Cracraft (1981) and Sielbold and Helbig (1995). Later writers during the 1990s however continued to acknowledge the possiblity of a Cathartid and stork link and convergent evolution (Konig 1982; Sibley and Ahlquist 1990).

The raptor placement was challenged in favor of a closer relation with storks during the late 20th century, referring to behavioral, karyotype, molecular and morphological information (Ligon 1967; de Boer 1975; König 1982; Rea 1983; Seibold et al. 1993, 1995; Wink 1994; Wink et al. 1993a,b, 1996). The karyotypes studies are based on the number and appearance of chromosomes in the nucleus of a eukaryotic cell, and/or the set of chromosomes in a species. Behavioral traits include defecation on the legs for cooling purposes, this being shared with storks. Morphological issues concern the composition of the uropygial gland secretions, leg and pelvic muscle anatomy, and the distribution of feather lanes (Wink 1995).

During this period, the placement with or near the storks was also challenged using both molecular and morphological evidence (Griffiths 1994; Cracraft et al. 2004; Ericson et al. 2006; Gibb et al. 2007; Slack et al. 2007; Livezey and Zusi 2007; Hackett et al. 2008). These new studies used marker genes, the better to describe a 'more precise phylogenetic reconstruction of relationships within and between genera, subfamilies and families of birds' (Wink 1995: 868; see also Avise 1994; Sibley 1994) because marker genes can be easily amplified by PCR and sequenced (Kocher et al. 1989; Hillis and Moritz 1990; Taberlet et al. 1992; Edwards and Cracraft 1993; Kornegay et al. 1993; Hedges and Sibley 1994; Meyer 1994). More recent studies used deoxyribonucleic acid (DNA) (Sibley and Ahlquist 1990; Avise 1994; Sibley 1994). DNA is a molecule that contains the codes of the genetic instructions for the development and functioning of living organisms. These studies use the nucleotide sequence of the mitochondrial cytochrome b (cyt b) and are able to record phylogenetic events (evolutionary changes in organisms) as far as the last 20 million years (Seibold et al. 1993, 1995; Wink 1994; Wink et al. 1993a,b, 1996). This technique has been justified in some studies from the 1990s (Edwards et al. 1991; Richman and Price 1992; Helm-Bychowsky and Cracraft 1993; Kocher et al. 1989; Meyer 1994; Ta berlet et al. 1992; Heidrich and Wink 1994; Heidrich et al. 1995; Wink 1994; Wink et al. 1993a,b, 1994, 1996; Seibold et al. 1993, 1995).

The results of these new studies were not conclusive, as some found that New World vultures were a subfamily of storks (Ligon 1967; Sibley and

Monroe 1990), others found they were more closely related to storks than to raptors, while other results showed they were not related to either storks or raptors (Rea 1983; Jacob 1983).The placement as a subfamily of storks was criticized as an oversimplification (Griffiths 1994; Fain and Houde 2004). An early DNA sequence study in this direction was based on erroneous data and was subsequently retracted (Avise 1994; Cracraft et al. 2004; Gibb et al. 2007; Brown and Midell 2009).

Other studies did not place them with storks, but nevertheless found a close relationship, possibly closer than with raptors. Avise et al. (1994) looked at the nucleotide sequences of cyt b from New World vultures in relation to storks and other members of the Ciconiiformes, and found these vultures appear to be more related to storks than to Old World vultures. The relation to storks was supported by many studies (Konig 1982; Rea 1983; Sibley and Ahlquist 1990; Avise et al. 1994; Seibold 1994; Seibold et al. 1993, 1995). However, some studies found only weak links with storks and Accipitrids. Siebold and Helbig (1995) tested the phylogeny of 11 species of Old World vultures, three species of New World vultures and related species, using 1026 nucleotides of the mitrochondrial cytochrome b gene. New World vultures were found to be unrelated to the Falconiformes, but the evidence of a closer link to storks than to Falconiformes was insufficient.

Wink (1995: 880) showed an interesting result; the Cathartidae are not closely related to the storks, but also not related to the Accipitridae. This author noted that despite morphological data and DNA-DNA hybridisation linking New World vultures and storks ancestrally (Garrod 1873; Ligon 1967; Konig 1982; Rea 1983; Sibley and Ahlquist 1990), the analysis showed separation between storks (*Ciconia*, *Leptoptilos* and *Mycteria*) and New World vultures, and similar genetic distances were found between the New World vultures and Old World vultures, condors and storks, and storks and Ac cipitridae. The close relationship of cathartid vultures with storks was thus unlikely. The results also did not support a close relation between the stork *Jabiru mycteria* and the Black vulture as had been indicated by Avise et al. (1994).

Wink's study was described as more comprehensive as it combined both sets of sequence data for 6 of the 7 New World vultures and 11 of 15 Old World vultures, and looked at the phylogenetic relations within Old World vultures, between New and Old World vultures and also the relations between both groups and other Accipitridae raptors, and also with the Ciconiiformes. The results also justified the classification of carthartid vultures into several monotypic genera (except for Cathartes with 3 species) (Wink 1995). Findings included a closer link between the California condor and Black vulture than between the latter and the Andean condor, and links between the Turkey vulture and the Yellow-headed vulture (Wink 1995). The first result is less obvious than the second, using morphological data.

A new, independent Order of New World vultures, Cathartiformes was also mooted (Ericson et al. 2006). This was challenged in 2007 by the American Ornithologists' Union Committee on Classification and Nomenclature and North American checklist, which placed the Cathartidae back into the Falconiformes (Banks et al. 2007), but left open the possibility of error: i.e., it is 'probably misplaced in the current phylogenetic listing but for which data indicating proper placement are not yet available' (American Ornithological Union 2009). Nevertheless, the AOU's draft South American checklist put the Cathartidae in their own order, Cathartiformes (Remsen et al. 2007, 2008). In the AOU's 2010 North American checklist, the Cathartidae family was placed in the Order Ciconiiformes. Further challenges arose, as a recent multi-locus DNA study indicates that New World vultures are related to the raptors, with the exception of the true falcons Falconidae (Hackett et al. 2008). In this analysis, the New World vultures should rather be part of a new order Accipitriformes (American Ornithologists' Union 2010).

4.3 The Evolution and Classification of Old World Vultures

As the location of Old World vultures in the Falconiformes is less disputed, and the classification controversies are less intense than those for the New World vultures, this section dealing with the Old vultures is shorter than the previous section covering the New World vultures. Here we look at the origins and relations of Old World vultures within the larger Order Falconiformes and Family Accipitridae. Fundamentally, are Old World vultures one group of birds with one ancestor, or several groups with different ancestors, and which raptors are their closest relatives—eagles, sea eagles, buzzards or kites?

As was mentioned earlier, Old World vultures contrast with the New World vultures, as they have stronger, grasping, feet and talons, a voice box and rounded nasal openings (Feduccia 1999). Their ancestors are held to be eagle-like, in fact two species, the Bearded vulture and the Palm-nut vulture, are commonly seen as intermediate between vultures and eagles (Feduccia 1999). The White-head vulture is also seen as having some eagle-like attributes. The fossil record of Old World vultures occurs in both the Old and the New Worlds from about the Late Oligocene, with increased numbers up to the Pleistocene (Zhang et al. 2012a). Old World vultures were widespread in both the Old World and North America, during the Neogene (Wink 1995), the Early Miocene of North America (Palaeoborus) and the Late Oligocene of Europe (Palaeohierax) (Miller and Compton 1939; Brodkorb 1964).

Zhang et al. (2012b: 1) argue that 'among the great surprises of avian paleontology was that compared with the Old World, the New World has an unexpectedly diverse and rich fossil component of Old World Vultures.' According to Feduccia (1999) in 1916, Miller described a fossil specimen of *Neophrontops americanus* and *Neogyps errans*, from the Pleistocene Rancho La Brea tar pits of southern California. The former was closely related to the modern Egyptian vulture *Neophron percnopterus* (Savigny, 1809) (Miller and Demay 1942). The status of these Old World vultures in the New World is disputed (Rich 1980); however, the balance of opinion seems to favor Old World species in the New World, especially in the case of *Neophrontops*, which is 'markedly like that of the Recent Old World vulture *Neophron*' (Howard 1966b: 3), especially as the differences between these two species 'are of less note than those which exist between *Neophron* and its contemporaries among the vultures today' (Howard 1932, 70).

Concerning the current Old World vulture species, two subfamilies are usually recognized; (1) the *Gypaetinae*, including Genus Gypaetus (Lammergeier or Bearded Vulture *Gypaetus barbatus*); the Genus Gypohierax (Palm-nut Vulture *Gypohierax angolensis*); and Genus Neophron (Egyptian Vulture *Neophron percnopterus*); and (2) the much larger *Aegyptiinae*, containing the Genus Aegypius (Cinereous vulture, *Aegypius monachus*); Genus Gyps (Griffon Vulture *Gyps fulvus*, White-rumped vulture *Gyps bengalensis*, Rüppell's vulture, *Gyps rueppelli*, Indian vulture *Gyps indicus*, Slender-billed vulture, *Gyps tenuirostris*, Himalayan vulture *Gyps himalayensis*, White-backed Vulture *Gyps africanus*, Cape Vulture, *Gyps coprotheres*; the Genus Necrosyrtes (the Hooded Vulture, *Necrosyrtes monachus*) Genus Sarcogyps (Red-headed vulture *Sarcogyps calvus*); Genus Torgos (Lappet-faced vulture *Torgos tracheliotus*); and the Genus Trigonoceps White-headed Vulture *Trigonoceps occipitalis*.

As noted above, in the past, vultures were once thought to have evolved 'only once among extant diurnal birds of prey' (Sielbold and Helbig 1995: 163). The Old World vultures have generally always been considered as within the raptors. Therefore disputes on their classification usually concern the closeness of the relation between particular vultures and other raptors (especially some eagles, kites and buzzards), and the possibility of some species (e.g., the feather-headed Bearded vulture and Palm-nut vulture) being not fully vultures, but between vultures and other raptors.

Old World vultures (Aegypiinae and Gypaetinae) have been proposed to be monophyletic (Brown and Amadon 1968; Thiollay 1994) or polyphyletic with Gyophierax, Neophron, and Gypaetus forming one or more groups separate from the others (Jollie 1977b; Mundy et al. 1992; Seibold and Helbig 1995). In this sense as noted above, a monophyletic group is a taxon (group of organisms) forming a clade, i.e., consisting of an ancestral species and all its descendants. A polyphyletic group has convergent features not inherited from a common ancestor, but from several ancestral species. Griffiths et al. (2007) point out that the subfamilies of the Accipitridae have usually been based on differences in behavior and feeding ecology, and physical similarities and differences. As such distinctions may not have a genetic/ biochemical basis, the subfamily and Genera monophyly is possibly subjective and inaccurate. They argue that phylogeny for Accipitridae based on morphology is difficult to assess (e.g., Brown and Amadon 1968; Jollie 1976, 1977a,b; Thiollay 1994; Seibold and Helbig 1995; Griffiths et al. (2007). Recent phylogenetic analyses have used osteology, DNA sequences from a single mitochondrial gene and mitochondrial plus nuclear DNA sequences (Wink 1995).

An incisive study was that of Lerner and Mindell (2005) who presented a 'full or nearly complete taxonomic representation of five accipitrid subgroups (sea and fish eagles, harpy eagles, booted eagles, snake eagles, and Old World vultures'. Both mitochondrial and nuclear sequences were used for representative species of 51 out of 65 genera (78%) and almost half of the known Accipitridae species (n = 111). The three species in Gypaetinae were highly divergent from the remaining 11 species in Aegypinnae (see also Mundy et al. 1992). Reference was also made to the study by Seibold and Helbig (1995), who used cyt-b sequence from 11 Old World vulture species and found evidence of polyphyly for the Old World vultures. This relates to polarized opinions on Old World vultures, that they are either monophyletic (Brown and Amadon 1968; Thiollay 1994) or polyphyletic with Gyophierax, Neophron, and Gypaetus forming one or more groups separate from the others (Jollie 1977b; Mundy et al. 1992; Seibold and Helbig 1995).

Key issues concern not only the monophyletic or polyphyletic status, but also possible 'sister groups' within the Old World vultures, including other raptors that may be related to vultures (Lerner and Mindell 2005). The Palm-nut vulture may be a transition between vultures to sea eagles (Brown and Amadon 1968). The Gypaetinae (including Gypohierax, Gypaetus, Neophron and the serpent eagle Eutriorchis) would be of a different evolutionary origin, diverging earlier, possibly related to the Perninae hawks (Wink 1995). The Aegypiinae are a monophyletic group (possibly close to the Circaetinae snake eagles and Aquiline eagles (Wink 1995). The 'phylogenetic position of Necrosyrtes (the Hooded vulture) within the Aegypiinae remains uncertain' (Lerner and Mindell 2005). It is more closely related to (but divergent from) the Gyps than the other Aegypinae. Although the Genera Aegyptiinae, Torgos, Aegypius, Sarcogyps and Trigonoceps are classified separately, they may even be within the same genus (ibid.).

These findings were complemented by those of Wink (1995), which went further, and for the part concerning Old World vultures, sought to identify phylogenetic relationships within Old World vultures and between these and other raptors using molecular data. This study found that each of the Gypaetinae species (*Gypohierax angolensis, Eutriorchis astur, Neophron percnopterus, and Gypaetus barbatus*) was highly divergent from each other genetically, but more closely related to each other than to the other Accipitridae species. A strange result was the finding that *Eutriorchis astur*, the Madagascan Serpent Eagle is a member of the Gypaetinae. *Aegypius monachus* and *Torgos tracheliotus* clustered 'as sister species' (*ibid.* 875). Necrosyrtes was uncertain in relation to the Gyps or Aegypius-clades, despite morphological similarities with Neophron (del Hoya et al. 1994). The Gyps vultures were closely related, possibly from a common ancestor about 500000 years BP. *Gyps africanus* and *G. bengalensis* were a better fit in Gyps than in the separate genus Pseudogyps. and were found to be separate rather than same species (see also Dowsett and Dowsett-Lem aire 1980). Neophron and Gypaetus (the Egyptian and Bearded Vultures) were related, as had been found in the past using karyological (DeBoer and Sinoo 1984) morphological, anatomical (Jollie 1976, 1997a,b) and embryological data (Thaler et al. 1986). The Gypaetus/Neophron group was not close to other Old World Vultures.

There was also a close relation between Short-toed Snake Eagles (*Circaetus gallicus* Gmelin, 1788), and the Gyps/Aegypius clade was confirmed as it was earlier by morphological studies (Jollie 1976, 1977a,b; Mundy et al. 1992). The Honey buzzard (*Pernis apivorus* Linnaeus, 1758) was found to be near to Neophron/Gypaetus (which was previously at the base of the accipitrid tree) (Wink 1995).

4.4 Conclusions

This chapter has examined the evolution and biological classification of vultures, relative to each other and to birds of other Orders. The most conclusive results are that the New World and Old World vultures are different. Also it appears that while the New World vultures arguably originate from a common ancestor, the Old World vultures may have at least two ancestral groups. Thus convergent evolution may not only be between the New and Old World vultures, but also within these groups.

New biochemical analytical methods from the 1990s have complemented and also refuted the results of older physical studies. New conclusions may emerge with more sophisticated methods, more fossil discoveries and more comparisons with other species both living and extinct. Also possible are further studies of the origin and classificatory placement of vultures and their current lifestyles in comparison with other species. Therefore the next chapter looks at the relations between vultures, facultative (part time) scavengers and predators.

5

Vultures, Facultative Scavengers and Predators

5 INTRODUCTION

This chapter looks at the relationships between New and Old World vultures and other birds and mammals. These relations concern predation, competition, cooperation, and adaptation. Carrion removal is a broad activity and a vital service for ecosystems (Sekercioglu 2006). Most animal carcasses are consumed by other vertebrates (Brewer 1994; DeVault et al. 2011). These vertebrate scavengers include not only obligate scavengers such as vultures, but predators such as raptors and carnivorous mammals (termed facultative rather than obligate scavengers). Disturbed or highly altered habitats may have less efficient ecosystem services (including scavenging) than intact, diverse habitats (Perfecto et al. 2004; Sekercioglu 2010). This is because the former may require behavioral adaptation from the scavengers and some species may not adapt to ecosystem instability (Campbell 2009). The competitive relationship between vultures and non-obligate scavengers is based on a shared attraction to carrion and may intrude into other relationships across the ecosystem.

Carrion presence influences the presence of obligate scavengers, facultative scavengers, predators and decomposers such as microbes and insects, at different tropic levels of the ecosystem (Janzen 1976, 1977; Rose and Polis 1998; DeVault et al. 2003, 2004, 2011; Wilmers et al. 2003a,b; Adams et al. 2010). Some avian predators situate foraging or breeding habitats partly based, at least in part, on carrion availability (Watson et al. 1992; Marr et al. 1995). Cortés-Avizanda et al. (2009a) found that the abundance of facultative scavengers such as common ravens (*Corvus corax*), red foxes, and jays (*Garrulus glandarius*) increased in the vicinity of ungulate carcasses, while the abundance of common prey species decreased, possibly as a result of prey avoidance or increased predation by the facultative scavengers. Predation by facultative scavengers on ground-nesting birds may also increase near carcasses (Cortés-Avizanda et al. 2009b). Competition between obligate scavengers, and non-obligate scavengers occurs both in undisturbed and disturbed environments. In such cases the adaptability of either group increases in importance (Campbell 2009).

This chapter compares the prehistoric, historic and current scenarios for vultures in all the continents. Other species are avian scavengers (raptors, storks, corvids, Larus gulls) and mammals (hyaenas, Canids, Felids and rodents). The hypothesis that vultures are relics from a past era of larger mammals is also examined, as is the argument that they may be supplanted in modern times by smaller more adaptable scavengers, such as feral dogs and cats, small mammal predators, corvids and gulls, suited to a cleaner, cultural world with little carrion. The extinction of the past megafauna, the recent mammal extinctions (e.g. that of the bison in North America), competition from other more adaptable avian species under human induced landscape and food-source change, and the ability of vultures to adapt to these changes at the Species, Genus and Family level are important issues. The current trend is towards more disturbed environments, in which scavengers adapted to past environments may have adjustment problems. Much research has been conducted in undisturbed habitats (Selva et al. 2005; Selva and Fortuna 2007; Wilmers et al. 2011). However, also important for vultures are highly disturbed, human-dominated habitats (Campbell 2009). These include farmed lands and urbanized areas (Benton et al. 2003; Swihart et al. 2003; Prange and Gehrt 2004). In other words, can the majestic vultures that once fed on mastodons prepared by sabretoothed tigers now feed on rats, road kills and garbage?

5.1 New World Vultures and Other Avian Scavengers

This section looks at the relationship between New World vulture species and each other, and with Golden *Aquila chrysaetos* (Linnaeus, 1758) and Bald eagles *Haliaeetus leucocephalus* (Linnaeus, 1766), ravens, crows, and buzzard species in northern North America, and with tropical eagles and hawks in Central and South America.

Concerning the relationships between the vulture species, as summarized in Chapter 3, research has been conducted on the dominance hierarchies between Andean Condors and King vultures (where the condor is usually described as dominant); King vultures and the Cathartes and Black vultures, where the King vulture is dominant (except where crowds of Black vulture swamp the other species); and between Cathartes vultures and Black vultures (where there are different opinions on which group is dominant, but the Black vulture is usually numerically dominant). Size is important, but in the case of the Black vulture, so is numerical superiority.

Numbers at a food source, arrival time, size and aggressive behavior influence the relationships between the smaller vultures. The Cathartes vultures, with their sense of smell, are described as the first arrivals at carcasses, especially in forests where vegetation may obscure food sources (Stager 1964; Olrog 1985; Houston 1986, 1988; Graves 1992; Snyder and Snyder 2007). Black vultures lack a strong sense of smell, and tend to follow the Cathartes vultures to food, and have been described as dominant over them, while other studies find they are rather subordinate except when in large numbers (Wallace and Temple 1987). Turkey vultures prevail over Black vultures in slightly more than half of the one to one confrontations, but Turkey vultures can be dominated by large numbers of Black vultures (Stewart 1978; Wallace and Temple 1987; Buckley 1996). For the Greater Yellow-headed vulture, a forest scavenger with a strong sense of smell, the main relation is to lead other vultures to carcasses. In encounters with the Turkey vulture at carcasses, the Greater Yellow-headed vulture may be dominant (Schulenberg 2010), or the Turkey vulture may be dominant (Hilty 1977).

The King vulture is the common larger vulture of Central and South America, although it is out-numbered by the Greater Yellow-headed vulture in the Amazon rainforest, and by the Lesser Yellow-headed, Turkey and American Black vultures in more open habitats (Houston 1988, 1994; Graves 1992; Restall et al. 2006). It may find carrion in dense forest (Lemon 1991), but it may also follow the Cathartes vultures with strong olfactory senses, to carcasses (Houston 1984, 1994; Beason 2003). It is dominant over the smaller, weaker billed Cathartes and Black vultures, but may help them by tearing open carcasses, and is itself dominated the Andean Condor (Wallace and Temple 1987; Lemon 1991; Houston 1994; Gomez et al. 1994; Muller-Schwarze 2006; Snyder and Snyder 2006). Similar to the King vulture, the California condor does not have a sense of smell and often follows Turkey vultures to carcasses (Nielsen 2006). Condors are usually dominant over Turkey and Black Vultures (Koford 1953).

The Andean condor is usually dominant over other avian scavengers, such as Turkey, Yellow-headed and Black vultures (Wallace and Temple 1987). However, in some cases, smaller vultures may cause problems for the condor (Carrete et al. 2010). As described in Chapter 3, flocks of Black vultures were able to dominate the Andean condor even when Andean condors arrived first to carcasses (especially in mountains) and also when the Black vultures arrived first (mostly in the open plains). Therefore, the numbers of condors were negatively related to the abundance of black vultures (both in plains and mountains, but more so in the plains). 'Although black vultures do not completely prevent the arrival of Andean condors to carcasses, they represent serious obstacles for feeding. Thus, while dominance hierarchy at carcasses could be related to body size, carcass consumption was determined by species abundance' (Carrete et al. 2010: 390).

Although some studies recorded some relation between the two species (the condor being dominant, but helping smaller vultures by opening carcasses), it is also mentioned that they did not coexist until the recent expansion of the range of the Black vulture. The hypothesis was that both species are in competition for food, with the degree of competition and dominance dependent on human presence and the abundance of either species. The results showed that the superior size of the condor could be a factor in dominance, but the Black Vulture's superior numbers where enough for dominance over the condor. The Black vulture therefore was a serious hindrance for condor access to food supplies (Carrete et al. 2010).

A study by Wallace and Temple (1987) examined the relations between Andean condors, King vultures, Turkey vultures, Black vultures and a non-vulture, the Caracara *Polyborus plancus* in competition at carcasses. The results showed that Turkey vultures arrived first at the carcasses (92% of the carcasses), Black vultures second (72% of the carcasses) and Andean Condors third (70% of the carcasses). Aggressive inter-and intra-species interactions were ranked from most to least dominant, considering size, gender, age, for each species. These were: (1) adult male Andean condors; (2) older juvenile male Andean condors; (3) adult female Andean Condors; (4) older juvenile female Andean condors; (5) younger juvenile male Andean Condors; (6) younger juvenile female Andean condors; (7) adult King vultures; (8) older juvenile King vultures; (9) younger juvenile King vultures; (10) adult Caracaras; (11) juvenile Caracaras; (12) adult Turkey vultures; (13) juvenile Turkey vultures; (14) adult Black Vultures; and (15) juvenile Black Vultures. In conflicts, adult Turkey Vultures dominated juvenile Turkey vultures, but less so than adult Black vultures dominated their juveniles. Adult and juvenile Caracaras dominated Turkey vultures (91, 90% respectively) and Black Vultures (85, 72% respectively, but not when the Black vultures were numerous). Turkey vultures were slightly dominant over Black vultures one to one, but not when the Black Vultures were numerous (commonly over 50 birds). King vultures were always dominant over the smaller vultures, but lost all encounters with Condors. Large numbers of Black vultures could snatch food in Condor presence with frequent aggressive contact (Wallace and Temple 1987).

The evidence hints that numbers are at least as important as size, and among the New World vultures at least, the smaller birds are much more numerous. An important relation among the vultures therefore concerns the inability of larger vultures to obtain food. Adaptability is also an issue: Black vultures (and to a lesser extent Turkey vultures) are expanding their foraging habitats into urban and other modified environments, while still occupying natural environments. This competition between numerous, small generalists and large, rarer solitary foragers is an important problem and falls within the opinion that the larger vultures are past relics from a time of megafauna carrion.

Roosting is another important activity that influences inter-species relations. All Cathartids, with the exception of the King vulture are known to form communal roosts, as do nine Old World vulture species, including the seven Gyps vulture species. Communal roosts are hypothesized to serve as information centers were birds share foraging information (Ward and Zahavi 1973; Rabenold 1983). This hypothesis was tested for the relationship between communally roosting Turkey and Black vultures. The evidence of shared information would be revealed when their departures from communal roosts at the commencement of foraging would 'deviate from random both in direction taken and time' (Rabenold 1983: 303). A common characteristic of both Black vultures and Turkey vultures is to commence morning foraging with flapping, directional flight, where they are aware of the food location. When the food source is unknown, then the behavior is different; foraging commences later in mid-morning, synchronized with the thermals, with soaring until later directional flight towards a sighted food source.

Evidence shows that Turkey vultures leave roosts earlier than Black vultures, and most return within two hours of nightfall, sometimes to different roosting spots for each night (Rabenold 1983). There was evidence that departures were grouped in time, the departing birds were directionally related (not correlated with weather conditions), with these directions changing daily and the departing groups were more numerous than the returning groups. Possibly, these behaviors reveal a common purpose for foraging; departure is more important than later individual foraging, and group directional change indicates adaptation to different, daily food locations (Rabenold 1983). Such conclusions for vultures are supported by studies of other species (e.g., Emlen 1971; Krebs 1974; Erwin 1978; Jones 1978 on bank swallows *Riparia riparia*, great blue herons *Ardea herodias*, six species of terns and reed cormorants *Phalacrocorax africanus*), despite the fact that the latter species were in breeding mode, and possibly found safety from predators in group foraging. Turkey and Black vultures, despite being communal roosters, tend to be solitary breeders, with pairs siting nests in remote tree or cliff locations, with roosting in the old location continuing for the bird not on the nest with the young (Rabenold 1983).

Roosts as information centers is not conclusively proved (Ward and Zahavi 1973; Rabenold 1983). The group behavior could be to ensure more effective foraging and not necessarily include transfer of information at a roost. Despite the visual evidence, why would a bird lead others to a food source (e.g., Alexander 1974)? Possibly, due to their keen eyesight, this is unavoidable. The need for help in breaking into a carcass is one possible reason (Jackson 1975; McHargue 1977; Rabenold 1983). Information sharing might be demonstrated by observing roosting birds that sighted food on the first day and determining if other birds followed them to food sources on the second day, before these other birds sighted their own food sources. It is important to determine if the primary bird is followed before it sights food, as after it sights food it would be followed anyway, whether or not information had been shared at the roost. As the author (Rabenold 1983) acknowledges, more research must be done to establish the actual meanings behind such possibilities. To this we may add, there must be more research on both New and Old World vultures, as if communal roosting confers advantages in food access, there are important links with habitat changes that affect roosting places.

Concerning the relations between New World vultures and other avian facultative scavengers, the main species groups are eagles, corvids and smaller raptors such as caracaras. Some authors have written that Black and Turkey vultures have no 'major predators in North America away from their nests' (Stolen 1996: 43; see also Townsend 1937; Jackson 1988a, 1988b). These observations suggest that they have reason to be wary of large predatory birds. Stolen (1996: 43) acknowledges that there are 'few published accounts of interactions of either the Black Vulture or the Turkey vulture with large predatory birds' however, 'competitive interactions due to overlap in use of carrion might be expected.' Glazener (1964) gives examples of Turkey vulture interactions with the Crested Caracara *Caracara plancus* (the caracaras harassed the vultures at the carcasses until the latter disgorged morsels, which the caracaras then ate.). Regarding eagles, Oberholser (in Bent 1937) stated that 'Bald Eagles *Haliaeetus leucocephalus* do not . . . hesitate even to pursue the vultures and compel them to disgorge, when if it fails to catch the coveted morsels before they reach the ground it alights and devours them.' Audubon (in Bent 1937) noted that the arrival of eagles at a carcass caused vultures to '. . . retire and patiently wait until their betters are satisfied'. Coleman and Fraser (1986) recorded an incident where an immature Golden eagle (*Aquila chrysaetos*, Linnaeus 1758) disturbed a large, feeding flock of Turkey and Black vultures and attacked one Black vulture, causing it to regurgitate before it flew away. In one incident a Bald Eagle actually killed a vulture that 'for some reason was unable completely to disgorge' (Audubon 1937).

Stolen (1996) gives an interesting study in central Florida, where small mammal carcasses were placed to attract vultures, contributing to interactions between both Black and Turkey vultures and Bald eagles. In both incidents several vultures (seven Black and three Turkey vultures in the first incident, and nine Black vultures and one Turkey vulture in the second incident) were feeding or sitting near a raccoon carcass, and flew away at the approach of one Bald Eagle, which merely flew over the carcass. Another study by the same author concerns the impact of Bald Eagles on a communal roost of mostly Black vultures with a few Turkey vultures. In one incident, Black vultures 'appeared agitated' and '. . . flew between perches and looked around more' when Bald Eagles were harassing Ospreys within 100 meters (*ibid.* 43). In another incident, a Bald Eagle was sitting in the vultures' roost tree, and all the vultures were sitting in low trees about 200 m distant (possibly to avoid the eagle). However, in two other incidents, the presence of Bald Eagles at the roost site appeared to have no effect on the behavior of the vultures. Stolen (1996: 43) notes that 'These observations suggest that interactions with large predatory birds may constitute a selective pressure on Black and Turkey Vulture behavior.' This author argues that the eagle–vulture interactions seem to be more important at carcasses; the relevant evidence concerns how vultures regurgitate food in response to harassment or threats from predators which may distract potential attackers (for different opinions, see also Maynard 1881; Brown and Amadon 1968; Jackson 1988a, 1988b; Ritter 1983; Townsend in Bent 1937).

Concerning the Golden Eagle, there is evidence that this bird is dominant over all the New World vultures in North America, including the California condor. Snyder and Synder (2000) found that California condors avoided nesting where Golden Eagles were common. Another study concerned the relationship between California condors, Golden Eagles, Turkey vultures, Ravens and Coyotes (Koford 1953). Other animals discussed in this case study were Cougars (*Puma concolor* Linnaeus, 1771) and Black Bears (*Ursus americanus* Pallas, 1780). *S*pecies not discussed but possibly sharing the condor's foraging areas were American Crows, Red tailed Hawks, feral dogs and wolves. Golden Eagles, Turkey vultures and ravens usually arrived to start feeding before condors, and left later than them. Turkey vultures roosted much closer to carcasses than condors. Cougars only provided food for the scavengers by killing animals and were not competitors at kills. Black bears by contrast were described as possible competitors for carcasses, but were not recorded in the study.

Condors were dominant over ravens and Turkey vultures. Although ravens occasionally fed within one meter of condors, they jumped from feeding at the approach of condors, flew away when condors descended, and flew from alarm faster than condors. Ravens crowded at a carcass appeared to attract condors. Turkey vultures did not approach condors as closely as ravens did. They would fly from a carcass when a condor approached, and rarely tolerated condors within 1.5 meters. Normally, Turkey vultures waited several meters away, for the condor to finish feeding before approaching the carcass.

Two animals dominated condors: the golden eagle and the coyote. When one eagle was feeding and condors approached, the eagle drove the condors away and then resumed feeding, a feat repeated several times. One eagle chased 16, 14 and 12 condors in separate, successive incidents.

When the eagles left, the condors rushed the carcass. A common tendency was for condors to become bolder after being chased several times. In one case a condor was feeding on a sheep, on the opposite side from an eagle. Then a group of condors approached and crowded the eagle off the food. In another incident, an eagle and a condor stood seven meters apart, then the eagle approached and pecked the condor, which retreated and stood distant. The results of the study showed the dominance ranking to be golden eagles at the top, followed by condors, then Turkey vultures and finally ravens, with the strongest difference between eagles and condors, and the smallest between Turkey vultures and ravens.

Rodriguez-Estella and Rivera-Rodriguez (1992) also recorded Crested Caracaras being dominant over Turkey vultures at carcasses in the Cape Region, Baja California Sur, Mexico. The Crested Caracara (*Polyborus plancus*, now named *Caracara cheriway* Jacquin 1784) is described as an opportunistic but mostly carrion feeding raptor (see also Sherrod 1978; Glazener 1964) and is smaller than the Turkey vulture; length is 49–58 cm (19–23 in), wingspan of 107–130 cm (42–51 in), and weight 800–1,300 g (1.8–2.9 lb). Caracaras and some other raptors are generally dominant over Turkey and Black vultures (e.g., Striated Caracara *Phalcoboenus australis* Gmelin 1788; see Strange 1996; Catry et al. 2008).

In another study by Dwyer and Cockwell (2011) in the Falkland Islands (Malvinas), interactions were studied between the Variable Hawk (*Buteo polyosoma*, now classified as *Geranoaetus polyosoma* Quoy and Gaimard 1824), Striated Caracara, Southern Caracara (*Caracara plancus,* Miller 1777), and Turkey vulture (*Cathartes aura jota*). Variable hawks were aggressors in 98% (96–100%) of interactions with Striated Caracaras, 82% (69–95%) of interactions with Turkey vultures, and 80% (72–88%) of interactions with Southern Caracaras. Southern Caracaras were aggressors in 100% of interactions with Striated Caracaras, and 90% (80–100%) of interactions with Turkey vultures. Turkey vultures were aggressors in 71% (61–82%) of interactions with Striated Caracaras. These results put the Turkey vulture as more passive towards the Variable hawk and the Southern Caracara, and dominant over the Striated Caracara.

In North America, there are other scavengers that take the place of vultures during the winter. Studies of these scavengers, in the absence of vultures are important, because they illustrate the potential for competitive conflict for periods when vultures are present, the role of facultative scavengers as substitutes for obligate scavengers, and the importance of vultures if scavenging is incomplete or has different result from vulture presence.

For example, Wilmer et al. (2003) give the example of the Greater Yellowstone Ecosystem in the northern Rocky Mountains in winter. In this study, neither the Black vulture nor the Turkey vulture is mentioned. The primary winter scavengers were, in order of their dominance at carcasses, coyotes, Golden eagles, Bald eagles, ravens and magpies. Five species of mammal carnivores existed in the area: coyote, wolf (*Canis lupus*), cougar (*Puma concolor*), grizzly bear (*Ursus arctos*), and black bear (*U. americanus*). Elk were the main prey of wolves and human hunters (the main carcass providers in the area) and consequently were also the main food of the scavengers (see also Murie 1940; Gese et al. 1996; Lemke et al. 1998; Crabtree and Sheldon 1999a,b; Mech et al. 2001).

Bald eagles and ravens had the largest feeding radii, due to long distance flying capabilities (see also Heinrich 1988; Buehler 2000). As communal rosters, they might transfer resource location information among themselves (see also Marzluff et al. 1996; Buehler 2000; Dall 2002). Magpies had small feeding radii due to their lower mobility (see also Bekoff and Andrews 1978; Trost 1999). The Golden Eagle, although similar in flying ability to the Bald Eagle, was relatively sedentary and solitary, with smaller feeding radii, as 'they lack the degree of social interaction that enhances information transfer concerning resource locations' (Wilmer et al. 2003: 997, see also LeFrank Jr. and Clark 1983).

Where humans provided carcasses in localized areas, these were dominated by local scavengers with lower feeding radii (such as magpies). Where wolves provided the carcasses, these were widely dispersed and dominated by scavengers with wider feeding radii (such as Golden Eagles). Although vultures were absent, the other sources cited in this book allow speculation on their presence. Both Black vultures and Turkey vultures have wide feeding radii, but as noted in the study by Koford (1953) Golden Eagles are dominant over Turkey vultures at food sources. Turkey vultures and Black vultures are approximately even, but the Black vulture is numerically dominant. Turkey vultures have been found to be dominant over ravens, and presumably Black vultures, at least in large numbers would also be dominant.

5.2 New World Vultures and Mammal Scavengers

This section looks at the relationship between New World vultures and mammal species that indulge in scavenging, namely feral dogs, jaguars, cougars, smaller wild cats, black and brown bears, and small mammals such as the raccoon.

The commonest mammalian scavengers in North America are coyotes (*Canis latrans* Say 1823), raccoons (*Procyon lotor* Linnaeus 1758) and opposums (*Didelphis virginiana* Gray 1821) (Wilmers et al. 2003; DeVault et al. 2011). These are commonly termed mesopredators, or medium-sized predators. In a study by Koford (1953) on the interactions between condors and coyotes, coyotes were dominant over condors; when a coyote approached in three incidents, 3, 10 and 20 condors flew away. Occasionally, coyotes would also eat sick or young condors.

In another study, where vultures were absent or did not detect small carcasses in an intensively farmed region in Indiana, USA, DeVault et al. (2011) found that raccoons and Virginia opossums removed 93% of mouse carcasses. Possibly, this dominance by abundant mesopredators had negative consequences for other scavengers. In highly modified contexts, effective mesopredators are likely to exist (Huston 1997; Wardle 1999). Such mesopredators indulging in scavenging in highly modified contexts are also recorded in other studies (see Litvaitis and Villafuerte 1996; Oehler and Litvaitis 1996; Crooks and Soulé 1999; Gibbs and Stanton 2001; DeVault and Rhodes 2002; Sikes and Raithel 2002; DeVault et al. 2003, 2004; Prange and Gehrt 2004; Prugh et al. 2009; Ritchie and Johnson 2009).

An important factor for the large number of coyotes appears to be the decline and local extinction of wolves *Canis lupus* Linnaeus 1758 through much of North America (Berger and Gese 2007). Before and immediately after the European colonization of North America, wolves roamed over most of the continent, but with time they were exterminated by human action. Wolves were replaced by the more adaptive coyotes. Berger and Gese (2007) examined data on cause-specific mortality and survival rates of coyotes in wolf-free and wolf-abundant locations, and wolf and coyote presence in Grand Teton National Park (GTNP), Wyoming, USA, and the Greater Yellowstone Ecosystem (GYE) (located in Wyoming, Montana, and Idaho, USA). This study found that coyotes were numerically dominant over wolves, but their population densities varied with wolf abundance. Coyote population densities were actually 33% lower in areas with greater wolf abundance in GTNP. Coyote densities also declined 39% in Yellowstone National Park after the reintroduction of wolves. Wolves caused 56% of transient coyote deaths.

Another study, from southeast Canada, found that the selective killing of wolves resulted in wolf hybridization with the coyotes that consequently colonized the region (Rutledge et al. 2012). This concerned the Eastern Wolf (*Canis lycaon*) in Algonquin Provincial Park (APP), Ontario, Canada, where research culls killed the majority of wolves in 1964 and 1965 (about 36% of the park's wolf population) 'at a time when coyotes were colonizing the region' (Rutledge et al. 2012: 19). DNA studies found that this resulted in a decrease in an eastern wolf mitochondrial DNA (mtDNA) haplotype (C1) in the resident wolves and an increase in coyote mitochondrial and nuclear DNA. Later legislation that protected wolves outside the boundaries of the park in 2001 appears to have reduced the incidence of coyote DNA.

The study by Wilmers et al. (2003) took place in winter, when vultures were absent or rare in the Greater Yellowstone Ecosystem. Carcasses supplied by wolves and human hunters were attended by coyotes which were the dominant scavengers. The 'primary winter scavengers in Greater Yellowstone are, in order of dominance at carcasses, coyote, golden eagle, bald eagle, raven and magpie' (ibid. 997). The study found that the more dispersed carrion, mostly wolf kills were more often scavenged by coyotes.

The American Ornithologists' Union (1998) lists only the Turkey vulture as present in the Greater Yellowstone Ecosystem, but being a migratory bird it is probably absent most of the winter. The absence of Turkey and Black vultures from the study allows speculation on their position in such scenarios, but in an area where both are present, given the sense of smell of the Turkey Vulture and the social behavior of the Black vulture, it might be assumed they would arrive at carcasses (whether dispersed or localised) before most of the facultative scavengers.

5.3 Old World Vultures and Other Avian Scavengers

The particular avian relationships for the Old World vultures, apart those between each other, concern storks, corvids, *Larus* gulls, eagles, kites and other predatory raptors. In Northern Europe, facultative scavengers dominate as vultures are largely extinct. Apart from the *Larus* gulls, in northern Europe other scavengers are the White Tailed Eagle (*Haliaeetus albicilla* Linnaeus 1758), Red Kite (*Milvus milvus* Linnaeus 1758) and Raven (*Corvus corax* Linnaeus 1758) (Gu and Krawczynski 2012). In Asia and Africa, there are more avian competitors for vultures including the huge storks and far more eagle species than in Europe of North America.

Before the relationship between vultures and other scavengers can be understood, it is first necessary to examine the relationship between the vultures themselves, as this may shape the outlook towards similar species. Old World vultures may be classified, as described in Chapters 1 and 2, into into three groups, the large tearers, the large and medium-sized pullers and the small pickers (Kruuk 1967; Konig 1974). The tearing species, include the huge Cinereous and Lappet-faced vultures and the smaller Red-headed and White-headed vultures. The pullers are the Griffon vultures, with long necks. The peckers are the small Hooded and Egyptian vultures (Konig 1983).

In competition between vultures, size is a factor for dominance (Kwuk 1967; Lack 1971; Konig 1974; Houston 1975). There are various reports concerning aggressive interactions between these feeding groups or species. Studies by Kruuk (1967), Brown (1971) and Anderson and Horwitz (1979) ranked six species of African vultures according to size and dominance: (1) Lappet-faced vulture 7.5 kg; (2) White-headed vulture 5.9 kg; (3) Rüppell's vulture 6.4 kg; (4) African White-backed vulture 5.7 kg; (5) Hooded vulture 1.9 kg; and (6) Egyptian vulture 1.9 kg. In this list, the White-headed vulture (a tearer) is dominant over the Rüppell's Griffon (a puller), despite the latter's larger size.

All larger vultures dominate the Hooded and Egyptian vultures, while the slightly larger and more aggressive Hooded vulture dominates the Egyptian vulture. On most occasions, the larger Cinereous and Lappet-faced vultures dominate the large griffons, but there are reports of griffons dominating the larger birds (Konig 1973, 1974). There are cases where the larger species do not dominate the smaller species, when they do not assume a threatening posture. Konig (1983) gives the example of the Lappet-faced vulture, dominating smaller griffons only when a strong threatening posture is adopted, by displaying white feathers and red skin. Of 30 observed conflicts between the Lappet-faced vultures and griffons, in 24 cases the larger bird won by assuming a dominant posture, while in six cases when it was not in the posture, it was driven away. The assessment was that the motivating force of hunger was at least as important as a fixed dominance based on size.

Konig (1983) notes that there is little competition between two tearers; the larger Lappet-faced vulture and the smaller White-headed vulture in Africa. The former is aggressively dominant, while the latter is more solitary, observational and quicker in movement (due to its smaller size). It forages in the early, pre-thermal morning, earning access to kills. Konig further notes that there is little competition between the griffons and the Lappet-faced vultures, because the griffons eat softer internal organs and the Lappet-faced vultures eat the tougher hide, sinew and muscles. The Cinereous and Indian King vultures are described as similar, but they do not share ranges with species with similar feeding methods. A similar relationship exists between the Himalayan griffon, the smaller Griffon vulture and the yet smaller Indian White-rumped vulture, Indian vulture and Slender-billed vulture in Asia. The first two, larger birds both nest on cliffs, the Indian vulture mostly on cliffs, and the Slender-billed and White-rumped vultures mostly in trees (Konig 1983).

Studies have also focused on dominance and exclusion between the African White-backed vultures and the larger Rüppell's griffons in central Africa, and the African White-backed vulture and the larger Cape griffon vultures in southern Africa. The White-backed vulture weighs a little above half of the griffons (5.3 kg, compared with about 8 kg). The larger birds are near the weight for a bird 'at which prolonged flapping flight becomes impossible' (Konig 1983; see also Pennycuick 1969), while the lighter White-backed vulture easily uses flapping flight, an important ability for savanna country, due to the seasonally cloudy and rainy climate that may restrict the formation of thermals for soaring flight (Houston 1975). An important hypothesis is that the griffons have developed a larger body to cope with the scarcer food supply of the mountains, and to dominate other vultures.

There is also 'temporal segregation' between vulture species (Kendall 2014: 12). Carrion may be more available in the morning than in the afternoon, as there is less chance that diurnal scavengers have acquired it. Morning activity differs among vultures, as they rely on thermals to fly. These warm air zones develop slowly after sunrise heats the ground and heavier birds with higher wing-loading require stronger thermals than lighter birds. In a study of the largely solitary Lappet-faced vultures and their competitors, the commoner and social White-backed vultures, the former are heavier than the latter (mean, range: 6.78, 6.10–7.95 as against 5.46, 4.15–7.20 kg) and have wider wings, which are also longer (wing span: 2.80 vs. 2.18 m) (Spiegel et al. 2013). This means that the larger birds have lower wing loading (6.4 vs. 7.8 kg m^{-2}) (see also Pennycuick 1971; Mundy et al. 1992). Therefore the Lappet-faced vulture can utilize weaker thermals than the White-backed vulture and can search smaller more permanent foraging areas, while the latter requires wider foraging using fast direct flight (Pennycuick 1972). The Lappet-faced vulture was more efficient in searching, in terms of 'first-to-find, first-to-land, and per-individual-finding rate measures' (Spiegel et al. 2013: 102).

In a related study (Kendall 2014) tested the hypothesis that the larger Lappet-faced vulture would be more abundant at carcasses in the morning than White-backed vultures and Rüppell's vultures. The results showed the opposite, as the social White-backed and Rüppell's vultures were more abundant at carcasses in the morning, and Lappet-faced vultures were more abundant in the afternoon. Gyps vultures when full, fed less in the afternoon allowing the larger bird to feed. The Lappet-faced vultures had lower foraging success than the Gyps vultures. The results hinted that the factors for dominance of one species over the other must be critically assessed.

Size, wing-loading, foraging efficiency, numbers at a carcass and feeding specializations are important social factors. These are relevant to the relations between vultures and other species. For the large Old World vultures, the main competitors are Marabou storks (*Leptoptilos crumenifer,* Lesson 1831) and large eagles, kites and corvids. In the study above, Spiegel et al. (2013) noted that apart from the Lappet-faced and White-backed vultures, the other avian competitors were the Tawny Eagle (*Aquila rapax,* Temminck 1828), Bateleur Eagle *Terathopius ecaudatus* Daudin 1800 and Yellow-billed Kite (*Milvus aegyptius,* Gmelin 1788).

In studies of interspecific relations of assorted vultures and non-vulture scavengers, body size and social foraging may determine competitive ability at carcasses (Kendall 2013). This case study of the Masai Mara National Reserve included seven species of raptors: two eagles—Bateleur and Tawny eagles—and five species of vultures—Lappet-faced, White-headed, African White-backed, Rüppell's vulture and Hooded vultures. In the dominance relations, size appeared to be a factor, even between the vultures and the eagles. Weights of the listed species were, in order of smallest to largest: Hooded vulture 1.9 kg, Bateleur 2.2 kg, Tawny eagle 2.3 kg, White-headed vulture 4.3 kg, White-backed vulture 5.6 kg, Lappet-faced vulture 6.8 kg, Rüppell's vulture 7.6 kg. The Hooded vultures had low beak strength, the White-backed and Rüppell's vultures had medium beak strength, and the other species had high beak strength (the White-headed and Lappet-faced vultures and the eagles) (the source is adapted from Mundy et al. 1992, by Kendall 2014).

The Lappet-faced vulture affected the feeding of the smaller species, including the Rüppell's, White-backed and Hooded vultures, and also the Tawny eagle and the Bateleur eagle. The Bateleur eagle was also affected by the Rüppell's vulture and the Tawny eagle. It is noted that the Lappet-faced and African White-backed vultures were able to force leaving behavior in Bateleur eagles. The White-backed vulture negatively affected the Tawny eagle. Both the White-backed and Rüppell's Griffon vultures were social. White-backed vulture groups could dominate the larger, more solitary Lappet-faced vulture. The social behavior of the Rüppell's vulture did not have the same effect. Lappet-faced vultures were also affected by the presence of the much smaller Bateleur Eagles and Hooded vultures. Tawny eagle presence also increased the likelihood that White-backed vultures would feed.

Smaller species located more carcasses than larger species. Bateleur eagles discovered the largest number of carcasses and did not arrive at any carcass after another species. Superior search efficiency could be due to flight height in foraging or visual acuity, from variations in wing-loading among the different species (see also Pennycuick 1972). Other studies have found that Bateleurs tend to fly at lower heights than other raptors, while Rüppell's vultures tend to fly higher than average, possibly making these species less and more reliant on other birds' behavior respectively (Mundy et al. 1992; Watson 2000). The White-backed vulture had similar foraging capacity to the Tawny eagles and Hooded vultures, possibly due to the social foraging of the former species (they outnumbered Tawny eagles almost eight to one and Hooded vultures 35 to one).

Birds that were either good searchers (e.g., the Bateleur) or socially dominant (e.g., Lappet-faced, Rüppell's and White-backed vultures) generally were found in high quality landcover (where wildlife density was high, human settlements were low), while other species with low social dominance and lower search efficiency (e.g., Hooded vultures and Tawny eagles) were commoner in low quality land cover with low wildlife density. Some of the higher quality searchers, such as the Bateleur, White-backed, and Lappet-faced vultures were also found to be more common morning foragers, possibly increasing food access. Of these more dominant species, only the Rüppell's Griffon vulture was a late forager despite problems with food access when it arrived to well attended carcasses. The late arrival of this species has been attributed to its nesting on cliffs distant from the feeding grounds (Pennycuick 1983; Kendall 2013).

Eagles were also found to be aggressive towards Lappet-faced vultures in Saudi Arabia (Shobrak 1996). Six species of eagle were recorded in this study of the Lappet-faced vulture: Bonelli's eagle (*Aquila fasciata*, Vieillot 1822), Booted eagle (*Hieraaetus pennatus*, Gmelin 1788), Golden eagle, Imperial eagle (*Aquila heliaca*, Savigny 1809), Spotted eagle (*Clanga clanga*, Pallas 1811), and Steppe eagle (*Aquila nipalensis*, Hodgson 1833). The other common scavenger was the Brown-necked raven (*Corvus ruficollis* Lesson 1830), a slightly smaller bird than the Common raven (*Corvus corax* Linnaeus 1758), with a length of about 52–56 cm. In this study, there was pronounced conflict at carcasses. Lappet-faced vultures had 71.5% of conflicts with each other, and 29.5% with other species. For Brown-necked ravens intraspecific conflicts were 39.7% and interspecific conflicts were 60.3%. For the eagles collectively, intraspecifc conflicts were 63.3% and interspecific conflicts were 36.7%. Lappet-faced vultures won more interspecifc conflicts (70.9%, losing 29.1%) than the eagles (64.8%, losing 35.2%) or the ravens (25.8%, losing 74.2%). Only the eagles attacked these vultures (Shobrak 1996).

In a similar study looking at eagle-vulture relations, Mundy et al. (1986) found that the Black eagle (the African Verreaux's eagle) (*Aquila verreauxii* Lesson 1830) attacks the nests and young of the Cape Griffon vulture, which shares its principal, mountainous foraging and nesting areas. In some cases, the eagles which nest in pairs would fly into the vulture colony and attack the vultures. Five out of 13 attacks resulted in the removal of a vulture nestling by the attacking eagle. Six attacks were unsuccessful. There were also four attacks by eagles on both Cape vultures and Bearded vultures: these resulted in the death of the former species. The eagle attacks on vulture nestlings and eggs may be seen as predation (Mundy et al. 1986). However, these authors dispute the opinion of Pitman (1960) that the attack on the adult vulture that resulted in its death was predation, because the eagle did not alight on the ground and feed on the vulture. An eagle was once seen eating the nestling of the White-headed vulture, in the nest of the latter, despite the fact that the White-headed vulture (unlike the Cape Griffon Vulture) does not share breeding areas with the eagle.

The Marabou Stork also has important relations with African vultures. This is a huge bird with a height of 152 cm (60 in) and weighing up to 9 kg (20 lbs) (Likoff 1986; Stevenson and Fanshawe 2001). The wingspan has been disputed, with measurements up to 3.7 m (12 ft) and 4.06 m (13.3 ft) published, but there is no verified measurement over 3.19 m (10.5 ft) with most measurements averaging 225–287 cm (7–9 ft). The large bill measures from 26.4 to 35 cm. Females are smaller than males (Hancock et al. 1992).

Marabou storks may wait until the large vultures have opened a kill before they start feeding, but they may also be dominant over vultures, ranking with the Lappet-faced vulture as the most dominant bird at the carcass (Hancock et al. 1992; Cheney 2010).

In Africa, kites and crows are common facultative scavengers that interact with vultures. One study compared the Hooded vulture's foraging and competition with those of the Pied Crow (*Corvus albus* Statius Muller 1776) and the Black Kite (*Milvus migrans* Boddaert 1783), classified by some as the Yellow-billed Kite (*Milvus aegyptius* Gmelin 1788), subspecies *Milvus aegyptius parasitus*) (Campbell 2009). The findings of this study indicated strong competition between vultures, kites and crows in urban settings, centred around meat and waste production, and human presence.

Vultures competed with crows for meat from butchers' tables in abattoirs. In meat production areas, the vultures were more numerous than the crows or kites, but were less mobile, and tended to fly overhead or perch and stand more often while the crows entered windows and doors. Crows and vultures had similar flight distances of about 2 m (the distance between an approaching human and a bird before the bird takes flight). Vultures, crows and to a lesser extent kites competed in observational perching spots, space for hovering and close soaring and ground forays for snatching of human discards or refuse. Conflicts between birds erupted when people discarded meal remnants, garbage or unwanted trade items.

Both vultures and crows, but not kites, gathered in areas of human communal eating, but unlike the meat production areas, crows outnumbered vultures and were faster when snatching deliberately thrown food. Unlike vultures, crows were more adept at foraging in gutters, on sidewalks, public toilets, under vehicles, house windows and old clothing. In these locations, flight distances were longer than in the meat and waste production areas.

In rural areas, all three species were correlated with human presence, especially farm clearance work, bush fires and tractor ploughing. Vultures, crows and kites perched in branches, the ground or circled overhead. The flight distances were much longer than in urban areas, on average about 9 m, but where there was food crows and to a lesser extent vultures approached to within 1–2 of people. For the vultures and the rarer kites, feeding was largely on animals killed by farm clearers (mostly monitor lizards, rats and snakes), while crows took such food and also human discards, crop residues and insects on the ground and in flight (Campbell 2009).

5.4 Old World Vultures and Mammal Scavengers

This section looks at the relationship between Old World vultures and mammal scavengers, in particular hyenas, jackals, lions, tigers, leopards, wolves, feral dogs and bears. As noted by Kendall et al. (2014: e83470) for scavengers, unlike predators 'food availability is a factor of not just prey density but also prey mortality rate, which will vary both spatially and temporally.' Therefore, the vultures have a relation not only with mammal scavengers, but also with the predators that will kill the prey to provide the carcasses. Mammal scavengers such as hyaenas and jackals in Africa and Asia are also capable of killing herbivores, so their relation is also as predators.

The relations between vultures and mammal scavengers are essentially those of competitors, while the relations between vultures and mammal predators are based on the latter's status as both providers and occasional competitors. All these animals may also be competitors either over unfinished kills or when they take over carcasses upon which vultures are already feeding. The other relations are with scavenging hyaenas, jackals, hunting dogs and other carnivore/facultative scavengers, which compete with vultures over the remains of the predators kills and may fight for their own kills when vultures surround their unfinished meals. Mammal scavengers usually dominate vultures, although in some cases scores or hundreds of vultures, may intimidate smaller mammals such as jackals into leaving the carcasses. In some cases, the larger vultures may attack jackals at carcasses, as numerous photographs published online attest (see for example Perry 2007).

Vulture relations with mammal scavengers are most important in Africa (especially East Africa) because it still has high densities of large wildlife and few human-created carrion sources (Wilcove and Wikelski 2008; Dobson et al. 2010; Kendall 2012). Large mammal carnivores are now commonest in Africa, especially the game parks of eastern and southern Africa. In Asia, and especially Europe, large carnivores are rare, with wolves, bears and tigers extinct in most areas. In Europe and Asia, despite the persistence of a few large ungulates such as the elk, livestock have slowly become the main source of food for many vultures. There are also vulture restaurants where livestock carcasses are dumped for vulture consumption (Green et al. 2006; Gilbert et al. 2007; Donazar et al. 2009, 2010; Cortes-Avizanda et al. 2010; Margalida et al. 2011).

Vultures are commonly seen as having an advantage over mammal scavengers, because their flight allows earlier arrival at carcasses (Houston 1974b; Mundy et al. 1992). While this is true, Houston (1983) also records a study where 64 carcasses were placed in the Serengeti to record the sequence of arrival of scavengers. The result was that no mammals arrived at 86% of the carcasses, despite vultures taking an average of two hours to reach the kills. 'This implies that vultures do not gain an advantage over mammalian scavengers just because they get to carcasses first. Their advantage is that they find caracsses that the mammalian scavengers often never reach at all' (Houston 1983: 143).

Vultures also use their flying skills to follow potential food sources more easily than mammals can, as some mammals, e.g., lions and hyenas are territorial, and consequently cannot follow ungulate movements over long distances (Schaller 1972; Kruuk 1992). Griffons, by contrast do not have feeding ranges and forage very widely, and have adapted to the vast herds of past African environments (Houston 1974b, 1983). Kendall's (2012) study identifies vultures as following migratory herds of large ungulates in East Africa (see also Pennycuick 1972; Houston 1974a, 1974b; Pennycuick 1983). This indicated that social vultures (*Gyps* spp.) were dominant and their numbers were higher during the presence of migratory ungulates in the dry season. These ungulates may be killed by lions and to a lesser extent leopards and cheetahs. Lions are the largest predators, they harvest the most meat and leave the largest carcasses, such as buffalo, wildebeest, zebra and even young rhinoceros and hippotamus, which they must defend against scavengers like hyenas, rather than against the subordinate carnivores (Cooper 1991). Leopards and cheetahs usually hunt smaller prey, such as antelopes and gazelles, and occasionally the warthog. Leopards commonly hide their prey in tree branches to avoid larger predators or packs of hyenas or African hunting dogs. Cheetahs, which cannot climb trees due to the lack of retractile claws, must feed out in the open and are thus vulnerable to competition from the large carnivores and facultative scavengers.

Some studies have shown that vultures prefer carcasses that died of causes other than being killed by carnivores. Kendall et al. (2012) document scavenger competition in the Mara-Serengeti ecosystem in Kenya. An important finding was that 'regardless of the predator's identity, presence of a predator reduced the number of vultures, suggesting that vultures prefer carrion not killed by predators where available' (*ibid.* 523). This result was also found by Houston (1974b) who estimated that 88% of the food consumed by vultures was of animals not killed by predators, possibly as a prior predatory feast would reduce the meat available and increase the likelihood of conflict with mammalian predators and scavengers.

Predatory mammals in East Africa frequently engage in scavenging (Hunter et al. (2006). Examples include lions, spotted hyaenas (*Crocuta crocuta*), black-backed and golden jackals (*Canis mesomelas* and *Canis aureus*), and less commonly in leopards (*Panthera pardus*) or cheetahs (*Acinonyx jubatus*) (Kruuk 1972; Schaller 1972; Houston 1979; Caro 1994). Large carcasses may attract more scavengers, as they last longer than smaller carcasses (Blumenschine 1986). Large carcasses in open areas are also more likely to attract scavengers and hence scavenger competition, because they are more visible than those in forest (Blumenschine 1986; Domınguez-Rodrigo 2000). Although mammals have a sense of smell, Old World vultures, unlike the Cathartids have no sense of smell and rely entirely on vision.

Cheetah kills are different from lion kills, as the lions are much larger, they tend to hunt and eat in groups, and may defend the prey for long periods using their superiority over all scavengers (Hunter 2006). Cheetahs are smaller, usually solitary or in pairs, are not dominant over the large mammal scavengers such as the spotted hyena, they usually do not defend the carcass and they tend to leave their kills early (Houston 1979; Hunter et al. 2006). Leopards may hide their kills in trees. Kendall et al. (2012: 528) found that 'vulture abundance was particularly low at leopard kills.'

The study by Hunter et al. (2006) of 458 kills that were observed in entirety (from the cheetah killing to the end of the scavenging), showed that vultures appeared at 32% (147 kills), spotted hyaenas at 15.5% (71 kills) and lions at 3.3% (15 kills). For a subset of 282 kills, jackals were also present at 14.2%. For 431 kills scavenger arrival was observed and at 69 (12.9%) of these cheetahs were driven away from the kill by the scavengers. These dominant acts were by spotted hyenas (54 of the 69, i.e., 78%), lions (15%), other cheetahs (3%) and by non-scavengers, such as tourist interference, warthogs (*Phacochoerus africanus*), baboons (*Papio anubis*) (4%).

Scavengers were generally unaffected by the time of the day of the cheetah kills. They were more attracted to larger carcasses and were as likely to appear at carcasses on the open plains as in the forest savanna—boundary. One important finding was that medium-sized vultures were more likely to scavenge on kills on the plains than the forest savanna—boundary and hyenas and lions were more attracted to kills in shorter grass. All size classes of vulture—large (Lappet-faced and White-headed), middle (Rüppell's griffon and White-backed), small (Hooded and Egyptian vultures)—were positively correlated in arrival at carcasses, with middle-sized vultures slightly more likely to be first. All vultures could arrive before or after hyenas.

Hunter et al. (2006) found some contrasting results from other studies. For example, although larger vultures have been documented as late arrivals at carcasses, due to their need for strong thermals (Pennycuick 1972; Houston 1975), in the current study there was no evidence of this. Also, despite the fact that Kruuk (1967) described the medium-sized Gyps vultures as arriving after the Lappet-faced and White-headed vultures, Hunter et al. (2006) found the opposite result, with medium-sized vultures more likely to arrive before their larger competitors. However, the study agreed with Houston (1974) that the griffon vultures located carrion by sight, as the Gyps vultures were less likely to arrive at kills in the forest savanna boundary. There was also minimal support for the evidence that the smaller vultures arrived late at kills (Kruuk 1967).

The Spotted hyena is decribed in most studies as predominantly a scavenger (Pienaar 1969; Waser 1980). However, they actually hunt for most of their food (Kruuk 1972; Tilson and Henschel 1986; Cooper 1990;

Gasaway et al. 1991), despite a strong propensity for daytime scavenging (Kruuk 1972). Some studies have stated that hyenas may locate carrion by watching vultures descending to carcasses (Mills and Hofer 1998). Hunter et al. (2006) found that the spotted hyenas were likely to arrive earlier than vultures for kills which they took from cheetahs early (<5 minutes after the kill) possibly because they witnessed the kill. Where scavengers acquired cheetah kills after a longer period, hyenas were more likely to arrive after vultures, possibly because they were more likely to be attracted by the sight of descending vultures. This tallied with the work of Kruuk (1972: 146), where links between vultures and hyenas were described as 'a complicated relationship of mutual benefit and competition.' Hyenas may provide vultures with opportunities by driving the predator from carcasses and may also use vultures as visual cues for carrion (Hunter et al. 2006).

Black-backed and Golden Jackals are also common scavengers in the eastern African plains. Some studies indicate that jackals principally forage for small vertebrates and some plant materials, and are not primarily scavengers (Lamprecht 1978; Blumenschine 1986), while other studies indicate that jackals are in intense competition with vultures (Kruuk 1967). Houston (1979) used faecal evidence and found that only 3% of jackal diets comprised carrion (Houston 1979). Hunter et al. (2006) found them in some association with vultures, but unlike the lions and hyenas they did not follow the vultures to kills.

Lions are recorded as primarily predators, but also opportunistic scavengers (Schaller 1972; Houston 1974). In the study by Hunter et al. (2006) lions were invariably able to drive the cheetahs off the carcasses. Although this study did not record lions and hyenas together, other studies have shown that lions and hyenas compete for carcasses (Kruuk 1972; Schaller 1972; Cooper 1991). The relationship between the vultures and lions was similar to that of the vultures and hyenas, vultures tended to arrive before the mammals when the carcass was acquired late and after the carnivores when the carcass was acquired early.

In Europe, the main predators that scavenge on carrion are wolves, foxes and bears. Cinereous vultures have been described as dominant over foxes at carcasses (Brown and Amadon 1986). Few studies have examined the relation between wolves or bears and European vultures such as the Cinereous, Griffon and Egyptian vultures. The European Wolf subsists on livestock and garbage where human populations are high. It also eats moose (*Alces alces* Linnaeus 1758), red deer (*Cervus elaphus* Linnaeus 1758), roe deer (*Capreolus capreolus* Linnaeus 1758), wild boar (*Sus scrofa* Linnaeus 1758), mouflon (*Ovis orientalis* Linnaeus 1758), saiga (*Saiga tatarica* Linnaeus 1766), wisent or European bison (*Bison bonasus* Linnaeus 1758), ibex (Genus Capra Linnaeus 1758), chamois (*Rupicapra rupicapra* Linnaeus 1758), musk deer (Genus Moschus Linnaeus 1758) and fallow deer. They also prey on reindeer (*Rangifer tarandus* Linnaeus 1758) and fallow deer (*Dama dama* Linnaeus, 1758), but these species occur largely north of the range of any vulture species (del Hoyo et al. 1994; Peterson and Ciucci 2003).

Concerning the impact of vultures on facultative scavengers, Ogada et al. (2012) in a Kenyan case study investigated whether declining vulture numbers at carcasses would result in the longer existence of carcasses and if this would increase the numbers and contacts between the facultative scavengers. The evidence hinted that with a decline or absence of vultures, the mean carcass consumption rates almost tripled, there was a tripling of the average number of mammals at the carcasses and the mean time spent by these mammals at the carcasses also increased almost three-fold. There was also a near tripling of the average number of contacts between mammals at the carcasses. Other researchers report that vulture decline or extinction contributed to increased numbers of other scavengers (Selva and Fortuna 2007); these in the African savanna, would be hyenas (*Crocuta crocuta, Hyaena hyaena*) and jackals (*Canis mesomelas*) (Kruuk 1972; Kingdon 1997).

In India, researchers have found that vulture absence has possibly led to an increase in the population of rats and feral dogs (*Rattus rattus*) (Pain et al. 2003; Prakash et al. 2003; Selva and Fortuna 2007). These facultative mammalian scavengers perform a similar role as vultures in finding and consuming carcasses before decomposition (Sekercioglu et al. 2004). The change from vultures to mammals may affect disease transmission, as the decomposing carcasses serve as incubators for many pathogens, which infect contact mammals (Wobeser 2002; Butler et al. 2004; Jennelle et al. 2009). These carcasses may then serve as centers for the spread of diseases, as increased numbers of mammals may share carcasses, and contacts may involve aggression (Mills 1993; Ragg et al. 2000; Ogada et al. 2012). In Africa, hyenas and jackals may increase and these species play host to many pathogens that may infect many other wild and domesticated species (Alexander et al. 1994; Harrison et al. 2004). Diseases include rabies and canine distemper, the two common carnivore diseases of Africa. These may spread through contact between infected and un-infected animals often at carcasses (Mills 1993; Roelke-Parker et al. 1996; Butler et al. 2004).

5.5 Conclusions

The relationships between vultures and other birds and mammals are obviously complex. Vulture relations with other vultures are determined by size and numbers. Social vultures such as the Black vulture and the White-backed vulture can dominate other larger species such as the Andean condor and the Cinereous vulture by sheer numbers, but are usually unable to do so in one to one encounters. Less social, medium-sized vultures such as the Turkey vulture or Hooded vulture, unable to dominate by size of sheer numbers, must accept third place, except in some instances of one to one encounters with smaller vultures. In relation to other birds of prey, it is apparent that the eagles may dominate New World vultures more easily than the larger Old World Vultures. Golden Eagles easily dominated California Condors and Turkey Vultures, but Tawny, Bateleur and other eagles could not obtain a similar relationship with griffons and Lappet-faced vultures. Similarly, coyotes are dominant over New World vultures, but the relation between jackals and Old World vultures was more uncertain. These relations showed the importance of vulture biology in relations with facultative scavengers. Factors that enable or constrain vultures and other scavengers in their access to food, and therefore the likelihood of competition include landcover and climate variables, as when the different species have similar ecologies, contact is more or less likely. These habitat variables will be discussed in Chapter 6.

6

Climate, Landscapes and Vultures

6 INTRODUCTION

This chapter examines the relationship between vultures and environmental factors. These include temperature, rainfall, thermals, wind and seasonal climate change. Climatic and weather variables are crucial for the survival of vultures in all their life activities and are directly related to vulture physiology, form and behavior. The most important variables are air movements in orographic lift and thermals, that control soaring flight; temperature, which influences migration; and rainfall which may affect foraging. Sea and river proximity also affect vulture foraging possibilities, in terms of food sources and migration possibilities. Relief (mostly cliffs for nesting) and vegetation (mostly the distinction between open rocky terrain, grassland, savanna and mixed tree-grass environments, and closed forest vegetation for foraging and trees for nesting) are also important. Vegetation influences foraging, as open landcover is necessary for sighting carcasses. Unless the vulture has a sense of smell or can follow other species that do have a sense of smell, dense vegetation is unsuitable for foraging. The evidence presented so far points to the environment as crucial to vulture survival and lifestyles, to determine or influence foraging, nesting and survival outcomes for both New and Old World vultures.

6.1 Climate, Weather and Relief

Climate is a very important issue for vultures. This section will look at the daily weather regimes, namely thermal creation, times for roosting and average temperatures that affect movements; seasonal climate regimes that affect migration and breeding schedules; and climate zones that influence vulture distribution. Firstly we look at the components of climate most important to vulture ecology: climatic regions, temperature, rainfall, wind and seasonal change.

Climate is defined as the average weather conditions—precipitation, temperature, humidity, pressure, wind over usually at least 30 years. Weather refers to these variables over the short-term; anything from a few minutes to a time-frame approaching that climate. Weather is determined by altitude, latitude, marine proximity and orientation of mountain barriers to ocean currents and winds. A climatic region is determined by the particular combination of five environmental components; the atmosphere (air), hydrosphere (water), cryosphere (soil), land surface, and biosphere (plants).

The study of past climates, termed Paleoclimatology uses parameters such as lake bed sediments, ice cores, tree rings and coral. Three climate classification schemes in common use for modern climates are the Köppen system, the Thornthwaite system and the Bergeron and Spatial Synoptic Classification. Classification systems have been criticized as being too rigid as they imply distinct boundaries between different zones, rather than gradual transitional areas.

The Bergeron classification used air masses for classification, using three letters; the first letter for moisture properties (c) for continental air masses (dry) and (m) for maritime air masses (moist); the second letter for the thermal characteristic of the source region (A) for Arctic or Antarctic, (E) for equatorial, (M) for monsoon, (P) for polar and (T) for tropical; and the third letter for the stability of the atmosphere. Air masses warmer than the ground below are (w); those colder than the ground below are designated (k).

The Spatial Synoptic Classification system (SSC) is based on the Bergeron classification scheme. This system has six categories: Dry Polar (similar to continental polar), Dry Moderate (similar to maritime superior), Dry Tropical (similar to continental tropical), Moist Polar (similar to maritime polar), Moist Moderate (a hybrid between maritime polar and maritime tropical), and Moist Tropical (similar to maritime tropical, maritime monsoon, or maritime equatorial).

The Köppen classification uses average monthly values of temperature and precipitation. The common version has five categories: A, tropical; B, dry; C, mild mid-latitude; D, cold mid-latitude; and E, polar. These can be divided into secondary classifications such as rainforest, monsoon, tropical savanna, humid subtropical, humid continental, oceanic climate, Mediterranean climate, steppe, subarctic climate, tundra, polar ice cap, and desert.

Relief, also termed terrain, refers to elevation, slope, and orientation of features on the earth's surface. Relief is very important for vultures, because it affects climate. Climates become colder with altitude; thus species that prefer cool weather and climate nest or forage on high mountains or plateaus. There is also the rain shadow effect, as high mountains may block rain-bearing clouds from large areas such as the American southwest, creating a desert or dry savanna climate. Relief also channels wind according to the shape of the landforms, creating orographic windflow that may be useful for foraging soarers. Large, flat plains may also produce large thermals for these soarers during foraging. As was noted in the first three chapters, many vultures also nest in inaccessible cliff caves or crevices high in the mountains, hence their distribution is largely influenced by mountain presence. As terrain affects water flow and distribution, it also determines river and sea coast patterns, affecting those vultures that feed on animals frequenting such areas.

From the perspective of vultures, all the climate categories are valid except for subarctic climate, tundra and polar ice cap. Rainforests have high rainfall, with definitions setting minimum normal annual rainfall between 1,750 mm (69 in) and 2,000 mm (79 in). Mean monthly temperatures exceed 18°C (64° F) during all months of the year. A monsoon is a season with a rain bringing wind that may be dominant for several months, almost always in the tropics and often in rainforest regions.

Tropical savanna is a biome with varying proportions of grasses, shrubs and trees; wetter savannas have more trees, drier savannas have more grasses. Savannas are usually located in semi-arid to semi-humid climate regions of subtropical and tropical latitudes, with average rainfall between 750 millimetres (30 in) and 1,270 millimetres (50 in) a year and temperatures above 18°C (64°F). These are common in Africa, South and Southeast Asia and northern South America where they form the intermediate landcover between forest and desert. A steppe is a dry grassland with few or no trees. In the temperate zone of Eurasia and North America (in the latter steppe is usually called prairie) the annual temperature range is high; in the summer of up to 40°C (104°F) and during the winter down to −40°C (−40°F). In Africa, the Sahel savanna, a dry transition zone between more moist tree savanna and the Sahara desert, broadly corresponds to the steppe.

A desert usually exists in regions that receive minimal precipitation and therefore few plants exist. Due to the very low humidity, deserts generally have large seasonal and diurnal ranges in temperature; day temperatures may reach 45°C (113°F), and night temperatures may go down to 0°C (32°F). Many deserts exist due to the rain shadow behind mountains that block the moisture bearing winds from the ocean or more moist landcover.

6.2 Vultures, Thermals and Orographic Lift

The climate is usually related to the geomorphology, such as mountain ranges, valleys and water bodies. Thermals or rising bodies of warm air, strong winds and rain patterns are related to the morphology of the Earth's surface, with thermals commonly forming over flat plains and strong winds forming between mountains and in canyons. Leshem and Yom-Tov (1998) in assessing the relevance of Earth features to bird migration pointed to 'the combined effects of geomorphology and climate on migratory routes.' In this case, migration refers to the regular seasonal movement of birds seeking warmer climates during northern winters (in the northern hemisphere), from northern to more southerly regions.

As is evident from Chapters 1, 2 and 3, thermals are necessary for vulture soaring which is necessary for food foraging. Before we look into the issues of climate, we need to first examine the terms 'soaring' and 'wing loading'. Soaring may be defined as 'the use of air movements to sustain gliding flight' (Alexander 2003). Wing loading is defined as the ratio of the wing area to the weight of the bird 'The ability to fly slowly, and turn in small circles, is critical for exploiting thermals, and therefore the wing loading is a useful indicator of a gliding bird's capacity for this activity' (Pennycuick 2008: 266). This author also notes that the wing loading does not have any special significance for powered flights, i.e., when the bird is flapping. Slope soarers are birds that fly at high angles to the wind, such as albatrosses which fly in the sloping wind currents over the ocean waves. For these birds, high wing loading is optimal, because they need to glide faster than the wind and to glide fast they need high wing loading. For thermal soarers, such as vultures, low wing loading is needed because they can thus fly in smaller thermals. The Andean Condor is described as using both methods, thermal soaring in thermals and slope soaring over mountains (Alexander 2003).

Thermals form through convection, which is the transfer of heat through the movement of a fluid, either air or water (Fig. 6.1). Natural convection of air occurs when some parts of the Earth's surface absorb more heat than nearby areas. When this happens, the air molecules near the hotter surface begin to bounce on the surface, gaining more energy through conduction. The molecules turn into heated air that expands. When the air expands, it becomes less dense than the surrounding cooler air. The result is that the warmer air rises to form large, warm, rising bubbles of air that transfer heat upwards. Cooler air surrounding the warmer, rising air flows towards the source of the warm air, i.e., the warm ground that heated the air at the beginning. This cooler air also becomes warmer, and begins to flow upwards, following the warm air above it. This vertical exchange of air is called convection and the rising air is called a thermal. As the warm air rises it expands and cools, spreading outwards and sinking to the surface again, where is moves to the warm surface and rises, starting the cycle again. This circulation is called convective circulation or a thermal cell (Ahrens 2007).

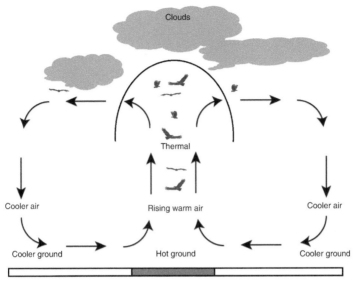

Fig. 6.1. Thermal Formation.

Bird weights and wing-loading (the ratio of wing area to body weight) are important for the use of thermals. For example, there are three weight categories of Griffon vultures. The heaviest is the Himalayan Griffon, which at 12 kg is similar to the Andean condor. It is a slope soarer, that rarely uses thermals. Houston (1983: 137) argues that 'wind speeds in the Himalayas are high, and the rising airflows over hill ridges provide the extremely good soaring conditions needed to support such a large soaring bird.' A heavy body and high wing-loading is an advantage in gliding flight as it increases flight speed in straight gliding flight, with only a small cost in the use of weak thermals.

The next weight category is the middle group, the Cape, Rüppell's Common, Indian and Slender-billed vultures (similar in weight, around 7–8 kg). 'All four species have comparable body proportions, presumably a compromise between the high wing loading that is efficient in gliding flight and the lower wing loading required for good performance when soaring in weak thermals' (Houston 1983: 138, see also Pennycuick 1972). The two lightest Griffons, the African White-backed and the Indian White-Rumped vultures (weighing about 5 kg each) live in flat savanna lands and rely more than the others on thermals, and need to use flapping flight which is very difficult for heavier birds. When the savanna climate is rainy and cloudy and the thermals are very weak or absent, flapping flight is necessary for foraging, and would be more difficult for a heavier bird (Houston 1983; see also Pennycuick 1972b; Houston 1975a).

Thermals may also determine the period that vultures remain in the roost. For example, in a study in Iowa, United States, McVey et al. (2008), found that Turkey vultures remained in communal roosts later from June to August than from in April, May, September and October. They attributed this to the fact that thermals are generated later in the morning during summer months than in spring or fall. As thermals are created by the contrast between ground and air, and in summer these contrasts are less pronounced, thermals form more slowly in summer and may require a higher angle of sun and longer exposure (Wallington 1977). The Turkey vultures (which need thermals for soaring) therefore remain at a roost when thermals are absent (Kirk and Mossman 1998). Vultures also varied their departure times, as opportunities arose (see also Mandel and Bildstein 2007). A study by Byman (2000) in the Pennsylvania winter found that Turkey vultures left their roosts earlier on warm sunny mornings than on cold mornings. The study by McVey et al. (2008) concluded that vultures left roosts later during the summer, but the lack of return time data prevents an assessment of whether the longer days in summer could be a factor for this, or it is only due to the later formation of thermals. Thiel (1976) reported a similar result for Turkey vultures. Hiraldo and Donazar (1990) also reported a similar departure behavior from the Cinereous vultures (*Aegypius monachus*).

A study by Xirouchakis (2007) of Griffon vultures in Crete found that Griffon departure times from roosts was determined by both thermals and wind. In one roosting colony, located near strong north-north-west winds, strong upward currents were produced; vultures used these winds for foraging for about 30–60 minutes after sunrise. In another breeding colony, where the winds were blocked by cliffs, the birds waited for 3 to 4 hours after sunrise for thermals before they could depart for foraging.

Temperature also has an effect on vulture physiology. An example is wing-spreading behavior in Turkey vultures. Ohmart and Clark (1985) found that wing-spreading during the morning pre-departure period was related to the intensity of the sun's rays, but not related to low overnight temperature. Wing-spreading was also more common when the vultures were wet than dry. Therefore, the birds would spread their wings to dry feathers and to 'ameliorate the thermal gradient between themselves and their environment' or both of these in combination.

Mahoney (1983) wrote that vultures are exposed to intense solar radiation and high day temperatures. Large birds also generate more heat from metabolism and store more heat than small birds, even when they are resting, and during exercise, e.g., flying body heat can increase ten-fold over the resting temperature (see also Kleiber 1961; Calder and King 1974). The black plumage may also retain heat (see Hamilton and Heppner 1967). This is probably a factor for the leg-wetting (urohidrosis) for evaporative cooling in Turkey and Black vultures. Turkey vultures also use panting when temperatures exceed their body temperature (Kahl 1963; Hatch 1970). Mahoney's (1983) study used four captive Black Vultures, and found that these birds had body temperatures and metabolic heat production similar to other similarly sized birds. The metabolic rates increased fivefold when the birds were running compared with resting rates, and flapping was predicted to be double this, with soaring and gliding double the resting rate. The study also found that resting Black Vultures in temperatures of 40 to 50°C were able to maintain body temperature at about 40 to 42°C for 45 minutes.

Few studies, however speculate how other vultures that do not use urohidrosis cool themselves, as Africa and Asia are at least as hot as the habitat of the Black vulture. In some studies, it has been suggested the bare heads of vultures exist as a thermoregulatory adaptation to avoid facial overheating. Many commentators suggest that bare heads also avoid the saturation of feathers with blood during feeding. This theory is however disputed (Mundy et al. 1992; Wilbur and Jackson 1983; Ward et al. 2008). Ward et al. (2008) used a mathematical model to study bare skin exposed by Griffon vultures in different postures, using heat flow through museum skins and estimates of exposed skin in different postures in hot and cold conditions as measureable examples. The skins studied had variable feather cover density, and could be used to estimate the coverage and hence the heat loss for the whole body. The results indicated that 'Postural change can cause the proportion of body surface composed of bare skin areas to change from 32% to 7%, and in cold conditions these changes are sufficient to account for a 52% saving in heat loss from the body'; the authors therefore concluded that 'the bare skin areas in Griffon vultures could be important for thermoregulation' (ibid. 168).

Rainfall affects foraging vultures. A study by Hiraldo and Donazar (1990: 130), found that 'the influence of rain on the flying of Cinereous vultures appeared to be very marked, decreasing activity and becoming practically nil for long periods of rain.' The effect of the rain was stronger during cold months, when unlike the warmer months no flights were recorded. For example, this study found that in one area of the western Sierra Morena in Spain, no vultures were seen on a strongly rainy day. On clear days in the same areas 10 or more vultures would be sighted. In another area, 130 vultures were recorded on a rainless day, while the next day with some rain only 68 were sighted. Also vultures tended to stop flying just before rain started and perched while the rain wet them. After the rain, the birds sunned themselves with spread wings before resuming flight. Some vultures also fly away from areas with rain, heading towards drier areas (Hiraldo and Donazar 1990).

Orographic lift is another weather phenomenon useful to soaring birds (Fig. 6.2). This occurs when an air mass, originating either over flat land or the ocean, moves up rising terrain, cooling as it moves up in altitude. This movement creates strong winds that blow up the slope and may also create turbulence over the uneven surface of the land. Soarers utilize winds that blow up the slopes, the turbulence created being sufficient to lift heavy birds such the condors and large vultures. Orographic lift is cited as a factor for condors and large vultures such as the Himalayan Griffon vulture favoring mountainous areas over adjacent flat lands from which strong thermals emerge (BirdLife International 2014; Rivers et al. 2014). The Bearded vulture also has a high wing loading and a long wedge shaped tail, possibly an adaptation for flight between mountain ridges using turbulent wind and orographic lift and updrafts (Ferguson-Lees and Christie 2001). Migratory birds also use such updrafts, for example over the parallel ridges of the Appalachian Mountains (Brandes and Ombalski 2004; Bildstein 2006; Mandel et al. 2011).

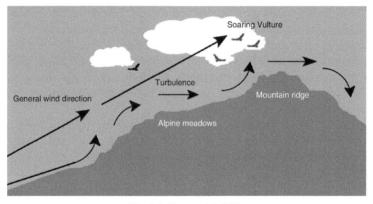

Fig. 6.2. Orographic Lift.

The California condor is also dependent on air flow in mountainous areas (Koford 1953; Wilbur 1978). The condor historically foraged over the flat Central Valley (Belding 1879; Stillman 1967). However, it slowly changed to mountain foraging as the San Joaquin Valley changed from grassland with herds of domestic livestock and native big game, to intensive cropland, and the urban areas of Los Angeles and San Francisco (Dooley et al. 1975). Also, in midsummer about 6 h per day may be suitable for soaring, but in winter, there may be only 4 soaring hours (Wilbur 1978).

Rivers et al. (2014) conducted the 'first quantitative assessment of habitat- and meteorological-based resource selection in the endangered California condor (*Gymnogyps californianus*) within its California range and across the annual cycle.' Condor use of habitat was influenced by meteorological conditions and the thermal characteristics (thermal height and velocity) were the most important. In addition, the study found that condors also used orthographic lift for soaring flight. This orographic lift is common in coastal areas (e.g., Big Sur region of western North America) and mountainous areas. Here, updrafts can support the condors' weight.

The California condor needs 50 to 60 feet for takeoff on flat land with no head wind (Koford 1953). It may be helpless if it accidently alights in dense vegetation, as large plants obstruct its movement in running and flapping. For this reason it usually does not access cougar kills or remains left by human hunters, as these are usually left in dense vegetation. The ideal landscape for feeding would be flat land with some wind, or small elevations from which to launch itself into the air.

Similar results were found for the Andean condor from the high slopes of the Andean mountains (McGahan 1973; Pennycuick and Scholey 1984; Lambertucci and Ruggiero 2013). The Himalayan Griffon, as mentioned above is similar, being fairly common in the high mountains of Nepal between 900 m and 4,000 m, sometimes up to 6,100 m (Inskipp and Inskipp 1991). It also occurs in the mountains of Bhutan, usually in the alpine and temperate zones from 1,400–4,000 m, occasionally down to 400 m and up to 4,800 m. The Cinereous vulture also frequents mountainous terrain in Israel (Shirihai 1996), Armenia (Adamian and Klem 1999) and Kazakhstan (Wassink and Oreel 2007).

Climate in combination with relief, including orographic lift has an important effect on vulture foraging. For example, in a study of Cinereous vultures, Hiraldo and Donázar (1990), compared two nesting nuclei, one in a mountainous area and the other in a lowland area, and found that in the former with higher slope lift, vultures initiated flights nearer sunrise, and concluded foraging nearer sunset. By contrast, in flatland, they depended only on thermals and foraged for a shorter period each day. Over flatlands, the thermals were only strong enough for foraging about seven hours after sunrise. Therefore, vultures which nested in mountainous area, but foraged on the plains would begin flying earlier in the mountainous area, fly around in this area, and glide down to the foraging area, where they would remain perched until the thermals were strong enough for soaring flight. Those vultures that nested in the flatland area also began foraging at this later time, about seven hours after sunrise.

This behavior was not limited to Cinereous vultures; Griffon vultures in the vicinity also exhibited this behavior. There were seasonal variations; for both nesters in mountainous areas and lowlands, the vultures left the mountain nests earlier in winter (February) and returned home latest in summer (June), when the flying hours were longest (Hiraldo and Donázar 1990). The authors note that 'The arrival of vultures at the colony occurred relatively closer to sunset in summer than in winter probably because of the high temperatures in the afternoon which permit thermal lift to continue until the last hours of the day' (ibid. 131). They however noted that slope lift did not seem to greatly modify vulture foraging times, as the birds in the study area preferred plains for foraging. It was hypothesized that in other more mountainous areas, slope lift might affect foraging.

6.3 Migratory Vultures

Weather conditions also have a strong affect on the migration of birds. Local weather conditions affect, among others, the start and duration of migration, energy cost, migratory routes, flight speed and flight strategies of raptors (Maransky 1997; Shamoun-Baranes et al. 2006; Shamoun-Baranes 2010; Vardanis 2011). Apart from temperature, other factors for migration are availability of food, day length and breeding. Studies have found that migrating birds use the sun and stars, the earth's magnetic field, and mental maps for direction. In some species, not all populations are migratory; this is termed *partial migration*.

Shamoun-Baranese et al. (2010: 280) point out that 'meteorology should be integrated into research on migration, especially when trying to understand natural variability observed in aspects like the timing of migration; migratory routes; orientation; use of stopover sites; or population trends such as effects on survival or breeding success as a result of changes in arrival time or physiological condition.' They further note the importance of the longer term meteorological dynamics, namely climate; 'meteorology is also of interest at longer time scales when trying to understand the evolution of particular migratory systems' (ibid.).

These authors list some recent, important literature that provides a background for further discussion of this topic for general lists of bird species. For general overviews, see Richardson (1978, 1990), Drake and Farrow (1988), Dingle (1996), Liechti (2006), Newton (2008). Other topics concern the impact of atmospheric conditions on migration onset (Shamoun-Baranes et al. 2006; Gill et al. 2009), migratory success (Erni et al. 2005; Reilly and Reilly 2009), flight speeds (Garland and Davis 2002; Shamoun-Baranes et al. 2003), flight altitudes (Bruderer et al. 1995; Schmaljohann et al. 2009), flight strategy (Pennycuick et al. 1979; Spaar and Bruderer 1997; Spaar et al. 1998; Sapir 2009), bird stopover decisions (Åkesson and Hedenström 2000; Dänhardt and Lindström 2001; Schaub et al. 2004), migration phenology (Hüppop and Hüppop 2003; Jonzen et al. 2006; Bauer et al. 2008, 2009), flight orientation and flight trajectories (Thorup et al. 2003), and the probability and/or intensity of migration (Erni et al. 2005; van Belle et al. 2007).

Medium and large birds, with wing loadings that are relatively high, tend to migrate by soaring, using horizontal or vertical (thermal) winds rather than active flight (Alerstam 1990). Bohrer et al. (2012: 96) investigated the migratory behavior and the meteorological conditions for thermal and orographic uplift 'to contrast flight strategies of two morphologically similar but behaviorally different species: the Golden eagle, *Aquila chrysaetos*, and Turkey vulture, *Cathartes aura*, during autumn migration across eastern North America.' They argued that 'it has not been shown previously that migration tracks are affected by species-specific specialization to a particular uplift mode' (Bohrer et al. 2012).

This study noted that both the Golden eagle and the Turkey vulture are soaring migrants that use both thermal and orographic uplift. Differences involved the higher wing loading (mass/wing area) of golden eagles (7.2 kg m²) than the turkey vultures (3.9 kg m²) (see also Pennycuick 2008). Golden eagles have 'more muscle mass and thus stronger flapping flight than turkey vultures.' The Turkey vulture's lower wing-loading is adapted to the use of weak updrafts and a scavenging (non-predatory) specialization, while the higher wing loading of golden eagles is for faster flight and a predatory specialization. The Golden eagle's preference for strong orographic lift favored flight paths above the windward side of the Appalachian mountains and ridges (see also Allen et al. 1996; Brodeur et al. 1996). By contrast, Turkey vultures used thermals, even when the sources of the thermal lift were scattered (Boehrer et al. 2011, see also Mandel and Bildstein 2007).

For the use of thermals for migration, problems emerge when they have to cross the ocean, as thermals form only over land. This results in large numbers of birds using 'bottlenecks'; narrow land links between continents (e.g., the isthmus of Panama) and narrow seas that partition landmasses (e.g., Gibraltar, Falsterbo, and the Bosphorus) (Fig. 6.3). Other migratory links include passes between high mountains; examples are the Central American migratory bottleneck and the Batumi bottleneck in the

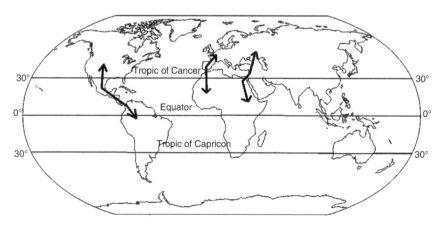

Fig. 6.3. Migration Bottlenecks for Raptors.

Caucasus, which allows avoidance of the vast Black Sea and high mountains (Maanen et al. 2001).

For the New World vultures, a crucial flyby is the Mesoamerican Land Corridor, connecting North America and South America, 'numerically the most important raptor-migration flyway in the world' (Bildstein and Sabario 2000: 197). These authors report that *Buteo platypterus* made up 76%, *B. swainsoni* 14%, and *Cathartes aura* 10% of one sample (see also Carriker 1910; Skutch 1945; Andrle 1968; Smith 1980, 1985a, 1985b; Thiollay 1980; Stiles and Skutch 1989, Tilly et al. 1990; Hernandez and Zook 1993; Hidalgo et al. 1995; Sutton and Sutton 1999; Zalles and Bildstein 2000). In this corridor, most birds 'soared in and glided between thermals; some engaged in slope soaring' (Bildstein and Sabario 2000: 200).

The results of this study suggested that migration is more likely on still, raining and overcast times in Costa Rica than in temperate regions. One study of migratory Swainson's hawks migrating between Argentina and North America show that they traveled 42% faster in the tropical part of the journey than in the temperate portions (Fuller et al. 1998); hence the dynamics of raptor flight may differ between tropical and temperate regions (Bildstein and Sabario 2000).

Griffon vultures have been documented using migratory bottlenecks, such as the Borsporus and Suez for migration (Bernis 1983). Bildstein et al. (2009: 382) studying the migrating flight behavior of Griffon vultures (*Gyps fulvus*) at another major migration bottleneck, the Strait of Gibraltar, found that the 14-km-wide sea channel 'significantly impeded the southern migration of the species into Africa, with many birds attempting repeated passage for weeks before crossing, and others not crossing at all and overwintering in Southern Spain.' Most attempts to cross the straits were between 11:00 and 14:00 hours, when there were either light or variable winds, or strong winds from the north or west. The study found that there were no crossing attempts when strong winds blew from the east or south. Furthermore, Griffons never tried to cross the straits alone; attempts were only made in large flocks.

Flight modes over water varied from those over land; above water, birds flapped more than ten times more often, and also did more flaps per flapping than when flying over land before the crossing attempts. The flapping was intermittent, combined with gliding or soaring flight. An important finding was that 'vultures flying over water that flapped at rates of 20 flaps or more per minute typically aborted attempted crossings and returned to Spain in intermittent flapping and gliding flight' (ibid. 382).Exhausted vultures fell into the sea and drowned and also collapsed on the beach after the crossing (see also Barrios Partida 2006). This indicated that Griffon vultures could not maintain flapping flight for even a short time, and that even narrow straits are thus serious barriers to migrating vultures.

Another study in another popular population bottleneck in Israel (Leshem and Yom-Tov 1998) found that migrating soarers use three main topographically determined routes in both directions in the autumn and the spring; an eastern route along the western edge of the central mountain range; an eastern route along the Jordan Valley and the Dead Sea towards the Sinai, merging with the western route in autumn; and a southern-Elat mountains route. Wind was also important, as in the morning breeze from the Mediterranean Sea shifted the migration path eastwards. This was also found to deflect the migration of the eastern route in the autumn. This route is favored when it moves over the eastern side of the central mountain range during the morning, as the sun creates thermals for soaring; in the afternoon the sea breezes deflect the path eastwards to the Jordan Valley. In the Elat mountains, the wind is affected by the southern Arava valley which moves the path in a north-south direction. Seasonally, it has been noticed that early autumn migrants move on an eastern route, but later migrants fly along the Mediterranean coast, possibly due to 'sub-optimal soaring conditions along the coastal plain during August' (*ibid.* 41). This is probably due to temperature differences between the land and sea, these differences declining and then disappearing by October. Species differences in routes were attributed to the avoidance of the Red Sea (Leshem and Yom-Tov 1998).

Another study, also focusing on this migration bottleneck in Israel (Spaar 1997) looked at the flight styles of migrating raptors. There was very little variation in the climbing rate of different species in thermal circling. The strength of thermal updrafts was the main determinant. Also, in interthermal gliding, heavier species had smaller gliding angles and glided faster, as gliding ability increased, than in lighter species. The Griffon vulture (*Gyps fulvus*) used soaring in a straight line whilst gliding. The Egyptian vulture was a soaring migrant.

Seasonal climate change may also contribute to mammal migration. This may influence movements of vultures that follow migratory mammals for food. For example, as noted above, Griffon vultures being pure scavengers follow migratory herds of ungulates in the African savannas, unlike the smaller, more adaptive Hooded and Egyptian vulture, which must find smaller prey items when the ungulates migrate. In the past colonial times, migratory herds were much larger than at present (Meinertzhagen 1957; Dorst 1970). In India as well, there were large migrations of the Blackbuck *Antilope cervicapra* during the 19th century (Blanford 1907; Seshadri 1969). These migrations may have shaped the migratory foraging behavior of the Griffons. Vultures are dependent on ungulate mortality, due to disease, accidents, old age and starvation and these are highest towards the end of the dry season. This is due to the lowered protein content of the foraged vegetation, which contributes to a decline in the strength and health of ungulates (see Sinclair 1977; Houston 1983).

The ungulate herds follow the movement of the Intertropical Convergence Zone. This zone is based on the movements of the sun and rain north across the Equator in April, May and June, staying north of the equator from July to September, and then moving south of the Equator again from October to December. This affects vultures, as ungulates are stronger during the rains, so vultures must follow the location of the period towards the end of the dry season, which varies in location. For example, the dry season in the Serengeti continues until August–October, when the rains slowly arrive as the Intertropical Convergence Zone moves south and the November rains begin. At this time the dry season is in Sudan. There, if vultures could fly distances, they could exploit weaknesses of the ungulates brought on by the regional occurrence of the dry season. This would mean a north–south migration, just in front of the rainbelt (Houston 1983). This study notes that in southern Africa, the seasonal rainbelt moves in a northwest–southeast line. The Griffons would have optimal feeding in the southeast (e.g., Zululand) in the dry season of June to September, and in southern Angola during the dry season of October to December.

6.4 Forest Vultures

Vegetation is a very important factor for vulture foraging and nesting. There are many classifications of global vegetation patterns. One of the most comprehensive is that of the World Wide Fund for Nature (WWF), which recognizes 14 biomes, where a biome is a region of similar climatic conditions with associated communities of animals and plants (Olsen et al. 2001). From the perspective of vultures, there are five particularly important biomes: Biome 1, Tropical and subtropical moist broadleaf forests; Biome 2, Tropical and subtropical dry broadleaf forests; Biome 7, Tropical and subtropical grasslands, savannas, and shrublands; Biome 8, Temperate grasslands, savannas, and shrublands; and Biome 10, Montane grasslands and shrublands (alpine or montane climate). Biome 13, the deserts, is also considered, as in some cases vultures may follow migratory ungulate herds or livestock herd across deserts and very dry savannas, especially in Africa.

The main issue for vultures in relation to these biomes is the ability to detect food. For the Old World vultures that lack a sense of smell and forage by sight, it is necessary to see the carcasses upon which they feed. For this reason almost all the vultures in Africa, Asia and Europe are absent from closed canopy forest. For some New World vultures, which do have a sense of smell (the Cathartes species), or weak or no sense of smell but can follow the Cathartes species to carcasses (the other New World vultures), the issues are both sight and smell. Hence, these species may be found in both forest and open cover. The first landcover discussed is forest, followed by the tropical savanna and desert, then the temperate grassland and montane vegetation.

Biomes 1 and 2, the tropical and subtropical moist broadleaf forests and the drier tropical and subtropical dry broadleaf forests are characterized by high rainfall and in the case of the drier zones, a dry season. Tropical rainforests have warm and wet climates, with the mean monthly temperatures exceeding 18°C (64°F) during all months of the year and the mean annual rainfall usually between 175 cm (69 in) and 200 cm (79 in). Some rainforests have very high rainfall, with annual rainfall between 250 cm (98 in) and 450 cm (180 in). There are generally four layers in the rainforest: the emergent, canopy, understorey and forest floor layers. The emergent layer is composed of emergents; a few very tall trees growing above the canopy, usually reaching heights of 45–55 m. The canopy layer is composed of the majority of the larger trees, usually between 30 (98 ft) and 45 m (148 ft) in height, forming a continuous cover of foliage. The understorey layer is composed of smaller, shrub-sized plants between the canopy and the forest floor. The forest floor has plants adapted to lack of sunlight, as only a small amount of light can pass through the dense canopy. Where a gap appears in the canopy, due to a blowdown in a storm, a forest fire or human action, the sunlight reaches the forest floor and usually contributes to a dense growth of shrubbery, grasses and saplings. Eventually, a new canopy may emerge (Fig. 6.4).

Vultures may not be able to see through the canopy to carcasses on the forest floor. Due to this difficulty, the ability to scent carcasses is important. Most vultures, unfortunately do not have a developed sense of smell. For this reason vultures are rare in forests, and all vultures except the Cathartid species hunt by sight rather than smell. As pointed out by Houston (1986: 318) 'forested areas of Africa or Asia do not support scavenging birds, while neotropical forest is the center of distribution for the cathartid vultures.' In

Fig. 6.4. Forest Structure Layers.

addition, 'Turkey vultures have a well-developed olfactory lobe and sense of smell which is used for finding food in forested areas...' therefore the Turkey Vulture and '.... the closely related Greater Yellow-headed vulture are the commonest vultures of neotropical forests' (see also Chapman 1929, 1938; Bang 1964; Stager 1964).

The close link between dense vegetation and olfaction or sense of smell in vultures, and disputes as whether vultures use sight or smell or both, have been recorded in the literature for more than a century. From the 19th century, researchers and debaters included Barrows (1887), Hoxie (1887), Sayles (1887) and Hopkins (1888). The debate continued into the early 20th century (see for example Taylor 1923; Leighton 1928; Lewis 1928; Earl 1929; Darlington 1930; Coles 1938; Vogt 1941; Rapp 1943). Studies hovered between belief in the birds' use of sight or use of olfaction, or both and used sometimes crude experimentation methods. Later studies were more sophisticated, using detailed biological examination and rigorous ecological assessments (such as those of Owre et al. 1961; Barros Valenzuela 1962; Stager 1964; Bang and Cobb 1968; Fischer 1969; Bang 1972; Wenzel and Sieck 1972; Houston 1982, 1987, 1990, 1994; Smith et al. 1986; Applegate 1990; Graves 1992; Gomez et al. 1994; Smith et al. 2002; Ristow 2003; Gilbert and Chansocheat 2006). The eventual concensus of these studies was that the Cathartes vultures were able to locate food by smell, while the other New World vultures were unable to do so, as they either had a weak sense of smell or none.

Vultures in forested areas use low flight just above the canopy, using the rising air on the windward side of tall emergents. Only when they detect food do they descend below the canopy, where they are agile, flying between the branches or even walking on the ground. To test the ability of Turkey vultures in detecting carrion in dense rain forest, Houston (1986) placed carcasses in different places in a dense forest in Panama, with varying visibility from the sky, some invisible under leaves, and at various stages of decay (fresh, one-day and four-day old carcasses) on the hypothesis that the birds used both vision and smell. Also recorded was the time taken for other scavengers to find the carcasses, as the success of the vultures would also depend on arriving at the carcass before other ground-based species. Dead chickens were used, partly because they are similar in size to normal vulture food, such as dead opossums, sloths and small monkeys.

The results of this study showed no correlation between carcass detection-time for vultures and the size of the gaps in the canopy. Also, there was no significant difference in vulture detection time for exposed carcasses and carcasses covered by leaves. For the age of carcasses, fresh carcasses were significantly more difficult than one-day old carcasses for vultures to detect, but there was no significant difference between a day-old and four-day old carcasses in speed of discovery (although there was a slight, but insignificant preference for the one-day old carcass over the four-day old carcasses). Turkey vultures consumed the bulk of the carcasses, followed by Black vultures, which arrived long after the Turkey vultures in larger numbers and were dominant in conflicts. A few mammals, such as opossums (*Didelphis marsupialisa,* Linnaeus 1758) and coatis (*Nasua nasua,* Linnaeus 1766) were also seen at the carcasses. 'The results show smell to be the major sense used by Turkey vultures to locate carrion and puts vision in a minor role, for completely hidden food was found as quickly as visible bait' (Houston 1986: 321). This author also wrote that 'Black Vultures do not have a sense of smell, and my observations in other study areas have shown that they are not found in undisturbed forest and cannot locate food in forest conditions unless led there by Turkey vultures' (ibid. 322).

In another study in dense forest in Venezuela, Turkey vultures or Yellow-headed vultures arrived first at carcasses, while groups of Black vultures were the most likely to arrive at large carcasses or carcasses in more open vegetation. King Vultures were recorded as equally likely to arrive at large or small carcasses. Turkey vultures were less aggressive than Black vultures and usually subordinate to Black vultures in aggressive interactions at carcasses (Stewart 1978; Wallace and Temple 1987; Houston 1988; Buckley 1996; Buckley 1998).

Another denizen of the forest is the Greater Yellow-headed vulture, often named the 'Forest vulture' to distinguish it from the Yellow-headed vulture or 'Savanna vulture.' Found in the dense forests of South America, it rarely roams into open country (Hilty 1977). This species flies low over the canopy, detecting small carcasses on the forest floor and along rivers (Robinson 1994). In the forested Amacayacu National Park, Colombia found this species was the most abundant vulture and usually the first to locate a carcass, in both in open clearings and under canopied forest (Gomez et al. 1994). These vultures located 63% of provided carcasses, while mammalian scavengers found only 5%. The Greater Yellow-headed vultures were however displaced from carcasses by Turkey vultures and King vultures.

6.5 Savanna Vultures

Biome 7, the tropical and subtropical grasslands, savannas, and shrublands, is a vulture habitat in Africa and also in the Americas and Asia. Savannas cover about half the African continent and one-fifth of the land surface of the world (Scholes and Walker 1993). Savanna landscapes vary in the relative proportions of trees, shrubs and grasses, and such variability is due to both human and natural elements (Defries et al. 2010; Miller and Hackett

2011). It is important to define exactly what the savanna vegetation looks like. Savanna vegetation is highly heterogeneous (Solbrig 1993; Mistry 2000; Campbell 2005; Parr and Andersen 2006). Harris (1980: 3) points out that 'the word savanna has had an ambiguous career since it entered the English language in the 16th century'. Scholes and Walker (1993: 3) argue that 'the central concept—a tropical mixed tree-grass community—is widely accepted, but the delimitation of the boundaries has always been a problem'. Therefore they envisage the savanna as representing a 'continuum of vegetation types which have trees and grasses as their main constituents', while 'acknowledging that at the extremes, the distinction between savannas and woodlands is unavoidably arbitrary' (ibid.).

Scholes and Walker (1993: 4) define the savanna as a 'tropical vegetation type in which ecological processes, such as primary production, hydrology and nutrient cycling, are strongly influenced by both woody plants and grasses, and only weakly influenced by plants of other growth forms', a definition based on that which emerged from the International Responses of Savanna to Stresses and Disturbance (RSSD) Program (cited in Frost et al. 1986, in Scholes and Walker 1993: 4). Solbrig and Young (1993: 3) add that savanna 'designates tropical grassland with scattered trees' and is the 'most common tropical landscape unit', a point supported by Werner et al. (1990: 343) who argue that savannas are 'delineated both structurally (specifying a wood/grass composition) and climatically (seasonality of water availability)', and also by Menaut et al. (1990: 471) who note that 'the structure and dynamics of savanna vegetation is generally expressed as the balance between trees and grasses, and more easily by tree density and distribution'.

Tropical thicket, another common vulture habitat is not always classified as savanna (Scholes and Walker 1993). Tropical thicket is described as vegetation dominated by multi-stemmed woody plants (of course some multi-stemmed woody plants owe their form to cutting for firewood), and in some cases with no grasses, supporting the contention that they are 'disqualified as savannas' (ibid.). Scholes and Walker (1993: 9) however, acknowledge that due to the dynamism of compositional forms, which often sees savannas becoming thickets and vice versa, 'in the absence of better understanding of the ecology of thickets, we treat them as a class of savannas rather than as miniature dry forests'. This position is supported by Cole (1986: 16), who includes 'thicket and shrub' in the vegetation types classified as savanna. Menaut (1983: 128) noted that thickets, which may be found in 'a great diversity of environments, close to the rainforest as well as near the desert edge', are common aspects of coastal savannas.

There is a broad range of climatic types within which savanna may develop, with average annual rainfall varying from 2000 to 250 millimeters, these being intermediate between moist forest and desert (Harris 1980; Stott

1994; Bucini and Lambin 2002). In Africa, the driest savanna type is the Sahel or Sahelian savanna. The Sahel is the ecoclimatic and biogeographic zone of transition, between the Sahara desert to the north and the Sudanian savanna to the south. Having a semi-arid climate, it stretches across the southernmost extent of Northern Africa between the Atlantic Ocean and the Red Sea.

The Sahel is mostly covered in grasses, shrubs and a few trees. Grass cover is dominated by annual grass species such as *Cenchrus biflorus, Schoenefeldia gracilis*, and *Aristida stipoides*. Acacia species are the dominant trees, with *Acacia tortilis* the most common, along with *Acacia senegal* and *Acacia laeta*. Other tree species include *Commiphora africana*, *Balanites aegyptiaca*, *Faidherbia albida*, and *Boscia senegalensis*. In the drier, northern part of the Sahel, areas of desert shrub, including *Panicum turgidum* and *Aristida sieberana*, alternate with areas of grassland and savanna.

The Sahel the habitat for large herds of grazing mammals, including the scimitar-horned oryx (*Oryx dammah*), dama gazelle (*Gazella dama*), Dorcas gazelle (*Gazella dorcas*), red-fronted gazelle (*Gazella rufifrons*), the giant prehistoric buffalo (*Pelorovis antiquus*) and Bubal Hartebeest (*Alcelaphus busephalus buselaphus*). Carnivores included the lion, African wild dog and cheetah. However, populations of these larger species have been decimated by over-hunting and competition with livestock (for example Dorcas gazelle and red-fronted gazelle), endangered (Dama gazelle, African wild dog, cheetah, lion) or extinct (the Scimitar-horned oryx is probably extinct in the wild, and both the Giant Buffalo and the Bubal hartebeest are now extinct). The reduction of the large ungulates and the carnivores that fed on them is a factor for the decline in the numbers of large vultures such as the White-backed and Rüppell's griffon in this region.

The Sudanian savanna is found to the south of the Sahel. This region has more trees, and a longer rainy season. Common tree species are *Combretaceae* and *Caesalpinioideae*, with some Acacia species. The dominant grass species are usually Andropogoneae, especially the genera *Andropogon* and *Hyparrhenia*. Vast areas of the Sudanian savanna have been farmed, and culturally useful trees, such as shea, baobab, locust-bean tree and others are left standing, while farms of sorghum, maize, millet or other crops are cultivated.

The Guinea savanna is a transition zone between the Sudan savanna and the dry deciduous forest to the south. This region has a longer rainy season than the Sudan savanna and more trees, hence it is sometimes called the tree savanna. Similar to the drier savannas, large herbivores and carnivores occur, and therefore vultures are fairly common in the more open areas. South of this region, the forest-savanna mosaic is transitory to the tropical moist broadleaf forests of Equatorial Africa where most vultures are absent, except for the small Hooded vulture (Campbell 2009).

Biome 8, Temperate grasslands, savannas, and shrublands is an important home of New World vultures in southern Canada and the plains of the United States and Western Eurasia. This biome may interface with the Mediterranean forests, woodlands, and scrub or sclerophyll forests of Biome 12. Grasslands are similar to the savannas, but tend to lack trees or shrubs. These are found interspersed with savanna in Africa. There are also grasslands in South America. Examples are the Llanos grasslands of northern South America. In such regions, foraging vultures include the Yellow-headed vulture of Central and South America, and the Black and Turkey vultures.

Grasslands are also found in colder temperate regions. These include the Prairie and Pacific Grasslands of North America, the Pampas of Argentina, Brazil and Uruguay, calcareous downland, and the steppes of Europe. These are classified with temperate savannas and shrublands as the temperate grasslands, savannas, and shrublands biome. These temperate grasslands are the habitat of large herbivores, such as American and European bison, antelopes and wild horses such as the Przewalski's Horse. Carnivores such as wolves, coyotes and leopards are also found in temperate grasslands. Other animals include: deer, prairie dogs, coyotes, snakes, foxes, owls and badgers. Hence, vultures are able to live in such ecosystems.

Denizens include the Turkey and Black Vultures in the North America, and the Griffon, Cinereous and Egyptian Vultures in Eurasia. For example, the Griffon vulture has been recorded in several types of open environments (del Hoyo et al. 1994). The Cinereous vulture is recorded in scrub and arid and semi-arid alpine steppe and grasslands up to 4,500 m (Thiollay 1994). Foraging takes place over bare mountains, steppe and open grasslands and even forest edges. The Egyptian vulture forages in lowland and montane regions over open, often arid, country, and in such open country, it nests in ledges or in caves on cliffs, crags and rocky outcrops, but occasionally also in large trees, buildings, electricity pylons and even on the ground (Sarà and Di Vittorio 2003; Gangoso and Palacios 2005; Naoroji 2006). Open terrain in Europe has expanded due to deforestation over the past centuries, as also has the decline in the population of larger wild ungulates that previously inhabited open temperate grasslands. Hence many such vultures rely on livestock free ranging on grassland pasture and also small mammals in open areas.

6.6 Desert Vultures

Biome 13, desert is found in areas drier than the driest savannas and grasslands, with very little rain and little vegetation. Xeric grassland, which has little clumps of grass may be a transition between desert and dry savanna. In many cases, due to overgrazing of livestock or intensive agriculture, dry savanna may become desert. Although desertification is generally defined as land degradation where land becomes increasingly dry, and may lose water, vegetation and wildlife there are many disputed definitions, causes and contributory factors. Geist (2005) notes a widely accepted definition is that of the *Princeton University Dictionary* which defines it as 'the process of fertile land transforming into desert typically as a result of deforestation, drought or improper/inappropriate agriculture'. According to Geist and Lambin (2004: 817) the most authoritative definition of desertification remains that of the Convention to Combat Desertification: 'land degradation in arid, semi-arid and dry sub-humid areas resulting from various factors, including climatic variations and human activities' (see UNEP 1994).

Desertification assumed prominence after the French colonization of West Africa (Mortimore 1989). During this period, the French Comité d'Etudes commissioned a study to investigate and explore the prehistoric expansion of the Sahara Desert. Desertification may be termed as the final stage of deforestation, after the transition passes through the stages of savanna woodland, tree savanna, grassland (also termed steppe in Europe and prairie in the United States) (Campbell 1998).

Geist and Lambin 2004 list some factors for desertification which are relevant to avian issues. These include: agricultural activities such as livestock production (nomadic/extensive grazing, intensive production), crop production (annuals, perennials); infrastructure extension such as watering/irrigation (hydrotechnical installations, dams, canals, boreholes, etc.); transport (roads); human settlements; public/private companies (oil, gas, mining, quarrying); wood extraction and related activities such as harvesting of fuelwood or pole wood (from woodlands/forests), digging for medicinal herbs and other collection of plant or animal products; and demographic factors such as migration (in- and out-migration), natural increment (fertility, mortality) population density and life-cycle features (Geist and Lambin 2004: 819).

Vultures have been observed in deserts, where there is likely to be a food source, such as cattle-herding, camel driving or seasonal migration of ungulates across patches of desert between savanna landcover patches. An example of the existence of vultures in the desert of north-central Africa is the instructive study by Wacher et al. (2013). In this study, the landscape was of Sahelian and Saharan habitats, including volcanic outcrops, plateaux and valleys of the Termit Massif, and wide, flat plains and fixed dunes of the ancient greater Lake Chad Basin and its margins. The southern part of the study area was the transition between the mainly grassland Sahelian and the light woodlands of the Sudan savannas. The northern part of the study area included Tin Toumma (Niger) and Eguey (Chad) in the southern savanna with fewer trees. The Sahelian grassland plains included trees such as Sclereocarya, *Balanites aegyptiaca, Acacia raddiana* and *Maerua crassifolia.*

Ungulates are few, the food of vultures being primarily livestock, mostly large herds of camels, cattle, donkeys, horses and smaller animals. The few wild ungulates were Dorcas gazelle (*Gazella dorcas* Linnaeus, 1758) and Dama gazelle (*Nanger dama* Pallas, 1766), Barbary sheep (*Ammotragus lervia* Pall., 1777) and Addax (*Addax nasomaculatus* de Blainville, 1816). Movements were for fresh grazing and seasonal rainfall.

In this study, six species of vultures were recorded; the Egyptian vulture, Hooded vulture, White-backed vulture, Rüppell's vulture, Lappet-faced vulture and the White-headed Vulture. The Egyptian vultures were recorded in Termit Massif and Koutous hills in Niger, and rocky outcrops of Dibella and Agadem. These locations were surrounded by treeless sand desert. These vultures could be 'residents, migrants, or opportunistic followers of nomads and their livestock temporarily based at the ancient wells and scattered clumps of Hyphaene palms that grow in the lee of the hills' (Wacher et al. 2013: 190). Hooded vultures were found closer to Sudan savanna, near human settlements and also abattoirs. White-backed vultures were also mostly associated with the Sudanian zone in the southern margin of the study area, but some moved north during the rainy season. Rüppell's vultures had colonies on prominent granite inselbergs in Sahel and Sudan savanna and used treetop nests, mostly *Balanites aegyptiaca* trees. They frequented livestock carcasses in the southern parts of the study area. The Lappet-faced vulture nested in either *Acacia raddiana*, *Maerua crassifolia*, or occasionally *Balanites aegyptiaca* in the sahel region and in the semi-desert/ sahel of the Termit Massif. The White-headed vulture was mostly found in the south in the Sudan savanna zone (Wacher et al. 2013).

Vultures are also found in the deserts of the New World. For example, the Turkey and Black vultures are common in the deserts, semi-deserts, xeric grassland, scrub, shrublands and mixed tree-grass landcover of California, Arizona, New Mexico, Nevada and Northern Mexico. A study by Dean et al. (2006) found that Turkey vultures 'are probably the most abundant avian scavengers in semi-arid shrublands' of central Mexico. Black vultures and Common ravens were also common. Dead ungulates were an important food source, and birds commonly foraged above roads. This study was echoed by the research of Hiraldo et al. (1991) in a desert border 2100 m above sea level in northern Mexico, where the diet of Turkey and Black vultures was principally small mammals and large domestic ungulates.

6.7 Mountain Vultures

Biome 10, montane grasslands and shrublands (alpine or montane climate), exists in the Asian Himalayan mountains and the South American Andes. The montane grasslands and shrublands above the tree line (the maximum height at which trees will grow, based on the temperature) are termed alpine tundra. The montane grasslands and shrublands below the tree line are termed subalpine and montane grasslands and shrublands. Stunted subalpine forests occur just below the tree line, as cold, windy conditions and poor soils are factors for small, twisted, slow-growing trees. Large areas of montane grasslands and shrublands are in the Neotropic Páramo of the Andes mountains and the montane steppes of the Tibetan plateau.

The Tibetan plateau has been described as the largest and highest highland in the world. The landcover is principally meadow, alpine shrub, and primitive forests. The main human livelihood is livestock raising, especially in the meadows. There are also several large wild ungulates; yak (*Bos grunniens* Linnaeus, 1766), Tibetan ass (*Equus kiang* Moorcroft, 1841) and Tibetan antelope (*Pantholops hodgsonii* Abel, 1826). In the Alpine shrub habitat human populations are dense, with farming in the riverine lowlands and livestock rearing in the nearby mountains. There is forest habitat on eastern and southeastern fringes (Lu et al. 2009).

Himalayan Griffon vultures, 'the only high-elevation Gyps species', have been documented in the plateau. The meadow supported the largest population of Griffon vultures, (71%), followed by the alpine shrub (23%) and forest (6%). Carcasses in the forest and shrubbery may be overlooked by foraging vultures and the vultures' habitat preferences are believed to be due to the differences in livestock availability in the three habitat types. Most of the livestock grazed in the meadows, due to the 'greater visibility of approaching terrestrial predators' (Lu et al. 2009: 171; see also Cramp and Simmons 1980; Gavashelishvili and McGrady 2006; Bose and Sarrazin 2007, for a discussion on the importance of prey size and open landcover for vultures). The vultures' diet was principally Domestic yaks (64%), horses, donkeys and also wild ungulates (1%) and human corpses (2%) (Lu et al. 2009). Considering the diet of the vultures was principally livestock, it is uncertain what the population density of the vultures would be in a natural environment. The human corpses were an aspect of the Tibetan culture, termed sky burials, where human bodies were cut up, bones broken into fragments and given to the vultures at certain sites. Cinereous and Bearded vultures were recorded with Himalayan Griffons at carcasses. The latter bird dominated the others in numbers and access to carcasses. Other scavengers were feral dogs and ravens. In some cases the feral dogs were able to exclude the vultures from the carcass.

Another species that occurs in montane environments is the Bearded Vulture. This species occurs in the mountainous regions from Europe (the Pyrenees, the Alps, the Caucasus region), Asia and Africa. The range in Asia includes the Zagros Mountains, the Alborzs, the Altai Mountains, the Himalayas, western and central China, Israel and the Arabian Peninsula. The African range includes the Atlas Mountains, the Ethiopian Highlands and the Drakensberg of South Africa (Ferguson-Lees and Christie 2001).

The Bearded Vulture favors alpine meadows, grassland and inselbergs with cliffs, crags, precipices, canyons and gorges, especially where human populations are low or absent, and predators such as Golden Eagles and wolves provide the carcasses from which the bones are derived.

A study by Hirzel et al. (2004) documented environmental preferences among the reintroduced Bearded vultures in the European Alps from which they became extinct about 100 years ago. The introduction programme began in 1986 and by 2003, 121 birds had been released from captivity. Factors for the species presence were the presence of ibex and the limestone substrates in the area. Ibex forage in Alpine ecosystems with meadows and grasses. The limestone landscapes, compared to silicate substrates, provided better ground for bone breaking and food storage and also better thermal conditions for soaring. The study concluded that population reintroductions would be successful if birds were released in areas with large limestone massifs. Another study by Bogliani et al. (2011) in the Italian Alps found that Bearded Vultures foraged at higher altitudes during warmer months, and at lower altitudes during snowfall. This followed the presence of commonest ungulates of the park, the alpine ibex (Capra ibex) and the alpine chamois (*Rupicapra rupicapra* Linnaeus, 1758), their main food. Habitats included vegetated cliffs and screes, forest-scrub mosaics, and agriculturaland to a lesser extent bare rocks and deciduous forests.

The Andean condor in South America has a similar lifestyle. This species prefers high montane canyons and peaks and forages in open grasslands, while communally on inaccessible cliffs and rocky crevices and outcrops. This species feeds off the carcasses of llamas (*Lama glama*, Linnaeus 1758), alpacas (*Vicugna pacos*, Linnaeus 1758), guanacos (*Lama guanicoe*, Müller 1776), rheas (*Rhea americana*, Linnaeus 1758) and armadillos (*Dasypodidae* spp., Gray 1821). More recently it has been recorded feeding on domestic livestock (cattle, horses, donkeys, mules, sheep, pigs, goats and dogs) and introduced species such as boars (*Sus scrofa*, Linnaeus 1758) and red deer (*Cervus elaphus*, Linnaeus 1758) (Newton 1990; del Hoyo et al. 1994; Swaringen et al. 1995). All of these are found in high and lowland meadows and pastures (Newton 1990; del Hoyo et al. 1994; Swaringen et al. 1995). When the California Condor was common, it also favored similar terrain, namely the temperate and subtropical prairies and grasslands of western North America, and also the Rocky mountain ranges of that area (Koford 1953).

6.8 Conclusions

Climate and landcover are plainly important issues for vulture distribution. These affect the flying ability and foraging, feeding and wellbeing of vultures. They are also factors for the distribution of competitors, providers and prey species. Apart from the olfactory abilities of the Cathartes vultures, there are few differences between the New World and Old World vultures. In both groups, the very large species favor mountains and highlands near lowland grassland, while the smaller species favor mixed terrain that may include forest or desert patches. Understanding the terrain requirements of vultures and associated species is important for their conservation and future. The next chapters will look at the relations between vultures and people and human modification of the landscapes covered in this chapter. People change the landscape, converting forest, savanna and mixed landcover into urban areas, farmland and transportation zones. How vultures adapt to these changes may be determined by their adaptive behavior, the similarity of the new landcover to the previous natural landcover, and also the likelihood of a continued, food supply.

PART 3

Vulture Ecology and Conservation

This part looks at the relations between New and Old World Vultures and people; through the lens of vulture ecology and human culture, politics, history and varied scientific outlooks. The first important issue concerns vulture presence in human landscapes in urban settings, and how this new interaction affects vultures and people. Other cultures created by agricultural activities are also important. These have altered both the landscapes and the vulture human relationship. One current example of these new or transformed engagements has the been the diclofenac disaster that has greatly reduced vulture populations. These issues have been hinted in the preceding chapters.

Before delving into the details of these topics, there are important topics to be tackled in this introduction. These concern; rates of urbanisation and the meaning of urbanisation for avian foragers and the impacts of agricultural intensification and technological advancements not only on scavengers, but also on the predators and herbivores that provide them with food. The political structures that underpin actions to save vultures are also important. The future of vultures may be assessed in relation to these dynamics.

Urbanisation refers to both the increased number of people living in urban areas and the development of the infrastructure of settlement and commerce, such as roads, parking lots, buildings, railways, parks, rubbish dumps and commercial centers. Urbanization is increasing throughout the range of the New and Old World vultures. Avian, mostly scavenger colonization of urban areas has emerged as a vital field of enquiry, especially as heterogeneous urban landscapes attract varied species, urban waste provides for avian foraging and links are perceived between avian ecology, sanitation and human quality of life (Campbell 2006, 2007; Gbogbo and Awotwe-Pratt 2008).

Central business districts, industrial zones, parks, gardens, urban riverine areas and (mostly in developing tropical countries) open sewers, rubbish dumps and open abattoirs, are important avian foraging areas (Marzluff et al. 2001; Schochat et al. 2006). This habitat status is derived from the land cover configuration, general human behavior (recreational food stands, garbage disposal sites, food markets and abattoirs), individual human perception and reactions to birds in that particular context, and how these may vary in different levels of population density, food and cover availability, land cover heterogeneity and socio-cultural understanding (Campbell 2006, 2007).

An important issue concerns the changes in the human-vulture relationship in different land uses within the urban areas (Philo and Wilbert 2000). In particular, people's attitudes and reactions to birds and human socio-cultural relations, combine with land cover factors as factors for species presence. Few studies have examined the spatial variations in avian–human relations (especially those with large scavengers), avian adaptation, local attitudes to wildlife and socio-cultural relations in both urban and rural areas (Brooks and Thompson 2001).

Avian species numbers and density are much higher in the tropics than in temperate countries (Marzluff et al. 2001). Tropical cities in developing countries are often characterised by open food markets, large garbage disposal areas and in some cases poor sanitation that may facilitate avian colonisation (Mundy et al. 1992). Large, tropical avian scavengers are relatively understudied (Borrow and Demey 2001; Marzluff et al. 2001; Schochat et al. 2006). Due to 'the expected increase in human populations and urbanization in the tropics and the rich biodiversity that characterizes this region', 'more studies are desperately needed to inform public policy so that the negative consequences of human development are mitigated' (Marzluff et al. 2001: 1; also Schochat et al. 2006).

In tropical cities, vultures and corvids are dominant foraging guilds (Borrow and Demey 2001). Factors for this include their aerial mobility, attraction to refuse, large size and tolerance by people (Gbogbo and Awotwe-Pratt 2008). Human perceptions of large avian scavengers may be practical, based on their cleaning role, aggressive behavior, defecation and/ or food stealing (Soewu 2008). These perceptions may also be spiritual, the birds being associated with witchcraft, demons, spirits or bad luck omens (Adeola 1992; Soewu 2008). Perceptions of avian scavengers are important, as they influence human responses to avian presence in increasingly dense urban settings and have implications for conservation (Marzluff et al. 2001).

As noted in Chapter 3, Black Vultures in the United States have been identified as affecting human quality of life in urban areas. Human quality of life is a broad concept, related on multiculturalism, historical changes and environmental dynamics (Campbell 1998; Schalock et al. 2002). Definitions vary among scholars and lists of QOL indicators have been attempted (Felce and Perry 1995, 1996; Hughes and Hwang 1996; Cummins, 1997a, 1997b; Schalock 1997, 2000). Frequently cited aspects are: (a) emotional well-being, (b) interpersonal relationships, (c) material well-being, (d) personal development, (e) physical well-being, (f) self-determination, (g) social inclusion, and (h) rights (Schalock et al. 2002). These vary according to other social variables such as age, culture and social outlook (Elorriaga et al. 2000). Implicit in these definitions is idea that animals may both enhance and degrade QOL (Campbell 2007; 2008a,b; Schaltegger and Beständig 2012). The relevant environmental variables in this case are shared physical spaces in urban areas (green spaces, car parks, river banks, beaches, airports, recreational facilities) and the biological actors (in this case scavengers).

In northern countries such as Canada, and northern Europe, Larus gulls and corvids are rare or the dominant urban avian scavengers, as vultures are absent. Some have considered these birds as affecting human quality of life, because of their large populations and invasive scavenging in densely populated urban areas (Campbell 2007, 2008, 2010; Ma et al. 2010). The principal foraging areas for these birds are green spaces, car parks, roads, waterfronts, riverbanks, beaches, eating places, transport spaces and suburban farms. These areas are used by people for recreation, feeding, transport, and social interaction, and by birds for feeding and sometimes nesting. Invasive behavior, including roof colonisation, nuisance scavenging, human life space intrusion, defecation, colonisation and obstruction of human communications and transport by birds are common problems (Campbell 2007; Camphuysen et al. 2011).

For Larus gulls, there are arguments for both increased bird presence (bird conservation and even more extreme preservation) and bird eradication. Waterbird conservation has pros and cons; visual biodiversity and wetland and coastal conservation are seen as beneficial for human emotional, physical and material well-being, and ecosystem health (Deluca et al. 2008; Jones and Kress 2012). Simultaneously, there is increasing recognition of the pest status of some waterbirds, especially gulls, which have serious impacts on human safety and well-being (Cooper 2002; Rock 2002; DeLuca et al. 2008; Douglas et al. 2010; Camphuysen et al. 2011).

Airline collisions are common; research on Bird-Aircraft Strike Hazard (BASH), is an extremely important aspect of avian and applied social studies (Mackinnon 1999; Gard et al. 2007; Campbell 2008, 2010; Camphuysen et al. 2011). The U.S. Federal Aviation Administration recorded 57,702 bird-aircraft collision reports within the United States between 1990 and 2004, with over $600 million damage (Gard et al. 2007). 'Large flocking birds and birds of large body size' are described as particularly susceptible to such collisions (United States Department of Agriculture 2003). Factors include bird migration patterns and bird attraction to the prey sources in the grassed areas near the airports (Gard et al. 2007).

Gulls have also been implicated in roof nesting, harassing maintenance personnel, and defecating on vehicles, blocking roof drain systems with feathers and excreta and causing chemical and structural damage to buildings (Belant 1997). They also contribute to human diseases through defecation and subsequent microbial colonization of the feces by bacteria (e.g., *Bacillus, Clostridium, Campylobacter, Escherichia coli, Listeria* and *Salomonella* bacteria), and general nuisance through noise and human harassment, food stealing, fouling of tables and park benches and frightening pedestrians (Belant 1997; Campbell 2010).

For vultures, there are similar problems. However, they are less researched, possibly because they are absent from most of the northern cities except those of the United States and to a lesser extent southern Canada (Borrow and Demey 2001; Marzluff et al. 2001; Schochat et al. 2006; Campbell

2009). In tropical cities, vultures and corvids are dominant foraging guilds (Borrow and Demey 2001; Gbogbo and Awotwe-Pratt 2008). The avian density in the tropics is much higher than in temperate countries (Marzluff et al. 2001). Tropical cities have more to offer bird scavengers; open food markets, large garbage disposal areas, poorer sanitation and faster rates of decomposition of organic discards (Mundy et al. 1992).

Due to 'the expected increase in human populations and urbanization in the tropics and the rich biodiversity that characterizes this region', 'more studies are desperately needed to inform public policy so that the negative consequences of human development are mitigated' (Marzluff et al. 2001: 1; also Schochat et al. 2006). People perceive vultures as cleaners, and also aggressive polluters or food thieves (Soewu 2008). There are also spiritual perceptions, connected to witchcraft, demons, spirits or bad luck omens (Adeola 1992; Soewu 2008). These perceptions have implications for conservation (Marzluff et al. 2001).

In rural areas, agriculture creates advantages and problems for vultures. It is a major factor for deforestation, savannisation and desertification, this enabling vulture foraging. As recorded in the earlier chapters, although some vultures are predominantly forest foragers (e.g., the Greater Yellow-headed Vulture) most vultures forage in open areas. Also, fires for land clearance may kill small wild animals for vulture consumption. Therefore, the savannaisation prevalent in many forested areas may not necessarily be negative for vultures (Mundy et al. 1992; Campbell 2009). Agriculture is negative for vultures if their predominant food source, large ungulates killed by predators, decline. In many cases, livestock rearing is good for vultures, especially if the vultures are permitted to eat the carcasses of dead domestic livestock.

Vulture relations with agricultural activities are particularly important, due to the recent catastrophic decline in vulture numbers in Asia and Africa. The drug Diclofenac, which kills vultures, has thrust vultures into the league of endangered animals and made vulture research a hot topic globally. The interactions between vultures and humans, as enabling or constraining vulture numbers or ecology is a socio-economic, political and ecological issue. This part of the book will address these issues in successive chapters. The final chapter will speculate on the future of vultures.

7

Vultures, Cultural Landscapes and Environmental Change

7 INTRODUCTION

This chapter focusses on vultures' adaptation to urban and agricultural features, and how the processes of deforestation, savannaization, urbanization and desertification affect vultures. The main issues concern the impact of urban areas on vultures in general, the parts of the urban areas that are favored by vultures or avoided by vultures and the different impacts of urbanization on different vulture species, Genera and families (i.e., the New and Old World Vultures). The smaller vultures, i.e., the Black, Turkey and Hooded vultures and to a lesser extent the Egyptian and possibly the White-rumped vulture have adapted variably to urban areas. The larger species generally avoid the urban areas. Vultures are also killed by pylons and wind turbines. Vultures also create problems when they collide with aeroplanes and when they invade human life spaces. Agriculture, the dominant modified landcover globally, has also had both positive and negative impacts on different vulture species.

7.1 Vultures and Urbanization

As noted by Campbell (2009: 341), 'a central, emerging research problem concerns the mutual co-evolution of the human–avian relationship, and how this is manifested in different land uses within the urban areas. Urban development impacts on vultures in several ways; positively in terms of food provision (road kills, human discards), buildings and green spaces for nesting, and negatively through habitat loss, noise, pollution, shooting, lower biodiversity and absence of food. Vultures in urban areas must adapt to intensive human presence and also to total landscape change, and in some cases new competition from corvids and *Larus* gulls. As was described in the first three chapters, different vultures react to urbanization in different ways. In this chapter we consider the species that are attracted to urban areas, followed by those that are minimally affected, and finally those that are not known to frequent urban areas, and may be locally extinct due to urbanization.

Among the New World vultures, the Black vulture has adapted well to urbanization. Black vultures are increasingly common in urban environments, increasing conflicts with people (Novaes and Cintra 2013; see also Buckley 1999; Avery 2004). Problems include property damage, roosting pollution due to defecation and aircraft collisions (Lowney 1999; Avery and Cummings 2004; Blackwell and Wright 2006).

Rubbish or garbage dumps are important factors for Black vulture presence. In some cases, a consequence of urban foraging by Black vultures is the ingestion of artificial products. For example, a study by Elias (1987) using defecated pellets found synthetic products, mostly plastic from bags in 39.1% of the total pellets examined. A study by Novaes and Cintra (2013) on Black vultures in Manaus (metropolitan population 1.8 million inhabitants), in the Central Amazon of Brazil, found that Black vultures scavenged in street markets, garbage dumps, open sewers and polluted streams. They also visited thermal power plants and roosted in vegetation remnants close to the feeding areas. 'Black Vultures adjusted to the nearest possible roost to the food source to reduce the cost of movement' (ibid. 1). Most roosting was in vegetation remnants, ranging from very small (0.31 ha) to very large (773 ha), and possibly because of the large number of such sites, nests in other areas were rare. Black vultures were able to roost in very small vegetation patches, where there was intensive human activity (such as street markets) when the roosts were near large garbage containers. By contrast, proximity to areas of thermal power for soaring was relatively irrelevant for roost site selection probably because the Black vultures approached the feeding site flapping rather than gliding on thermals (see also Buckley 1997).

Another study by Hill and Neto (1991: 173) found that, as skyscraper nesting is common in Brazil, 'The Black Vulture has benefited immensely from human post-Columbian colonization of the New World... feeding on the garbage of dumps, fish killed in polluted waters, livestock dying in pastures and animals killed along highways, the Black Vulture's numbers have increased so dramatically that it is now considered the most abundant bird of prey in the Western Hemisphere.' Cited examples of Black vultures in urban areas of Brazil include two nestling Black vultures on the 22nd floor window-ledge planter of a 23-story office building in downtown Sao Paulo, Brazil. The planter was 15 m long, 1.3 m wide and 12 cm deep. Other examples were of a nest with two eggs which later fledged on the 19th floor window-ledge planter (80 cm by 60 cm) of a condominium; and two other nests, one with two eggs in a roof recess of a tall Curitiba building.

Hill and Neto (1991; 175) point out that 'Skyscraper-nesting by the Black vulture appears to be common in Brazil. According to Dr. Werner Bokermann, Curator of Birds at the Sao Paulo Zoo and Dr. Lfizaro Puglia, Director of the Sorocaba Zoo (pers. comm.), 'nestling Black vultures are commonly brought into their facilities by the general public who report finding them on the window ledges of city buildings or underneath large roof-mounted water storage tanks.' They further note the possibility that the Black vulture population will increase further, with the increasing human population. The natural nesting sites are increasingly scarce. Hence, 'the increasing use of skyscrapers as nesting sites by Black vultures should be expected, due to the similarities that large, unkempt window planters and roof recesses have to natural cliff ledges and caves' (*ibid.* 176).

Some recommend the reduction of Black vulture in urban areas (Novaes and Cintra 2013). Methods include the reduction of their food supply, usually by more complete garbage collection, changing the sewage collection systems, the replacement of open garbage cans with closed garbage cans and the cessation of indiscriminate garbage disposal by local communities.

Similar to the Black vulture, Turkey vultures forage and roost in urban areas. The main distinction is the Turkey vulture's less communal nature. Although Turkey vultures are recorded as roosting in large groups, they are usually solitary or forage in smaller groups than the Black vulture. Kiff (2000) writes that urbanization may reduce Turkey vulture populations, for example in California (see also Garrett and Dunn 1981; Unitt 1984), although there are records of Turkey vultures nesting in buildings (Valentine 1873). However, some Turkey vultures in southern and south-central Saskatchewan were observed nesting in abandoned houses in 1982, in the Aspen Parkland and Boreal Transition ecoregions of Saskatchewan, a phenomenon that increased in the mid-1990s as vulture numbers increased and habitated zones expanded. Nesting in such areas produced similar numbers of fledglings as elsewhere in more natural surroundings (Houston et al. 2007). Igl and Peterson (2010) also recorded Turkey Vultures nesting in an abandoned car.

Airola (2011) examined Turkey vulture presence in a large 67-ha park named William Land Park with large, old ornamental trees in urban Sacramento. Trees in the park included Deodora cedar (*Cedrus deodara*), Atlas cedar (*Cedrus atlantica*), Italian stone pine (*Pinus pinea*); London plane (*Plantanus acerifolia*); coast redwood (*Sequoia sempervirens*) and red-gum eucalyptus (*Eucalyptus camaldulensis*). Turkey vultures used roosts in the park throughout the year, with average numbers peaking for October at about 200, and lowest in April-May at 10 to 15 birds. In the mornings, the vultures circled above the park to about 100 to 200 m and then flew directly southwards, mostly uncultivated grassland habitat. It was observed that vultures generally showed no response to the 'low-intensity human activities beneath and adjacent to roosting trees, including use of a walking and running trail and golf course, the feeding of ducks in the pond' (Airola 2011: 5). In this case the impact on people was minimal, despite the large number of roosting birds. The few problems included feces and regurgitated pellets during the migration season.

None of the other New World vultures are common in urban areas. To repeat the information from the earlier chapters, the King vulture lives mostly in tropical lowland forests (below 1500 m–5000 ft) from southern Mexico to northern Argentina, and also savannas and grasslands, and swamps or marshes near forests and nests on the ground (Wood 1862; Brown 1976; Houston 1994; Schlee 1995; Ferguson-Lees and Christie 2001). The bird is a ground nester. Mountainous, gallery and flooded forest are the main foraging areas (Ellis et al. 1983; Reid 1989). They also nest at the edge of undisturbed forest remnants. It remains when forest fragmentation occurs, if there are still large patched of forest (Stiles 1985). It does not adapt to human presence (Clinton-Eitniear 1986; Whitacre et al. 1991; Berlanga and Wood 1992).

The Yellow-headed vulture is also not known as an urban scavenger. It occurs in open terrain; freshwater or brackish and marshes, moist savannas, mangroves swamps, scrubland and scrub savanna, open and moderately forested riverine landcover and farmland (Ferguson-Lees and Christie 2001). In a case study in Brazil, it occurred only in savanna with lagoons (Zilio et al. 2013), while in Guyana it was reported in savanna grasslands, scrub or brush habitats, including white sand scrub, bush islands, and fresh water habitats, including lakes, impoundments, ponds, oxbows, marshes, and canals (Braun et al. 2007). Other habitats were forest edge and secondary pine-dominated savanna in Nicaragua (Martínez-Sánchez and Will 2010); uncultivated, mixed riverine and marsh vegetation in Brazil (Sick 1993); and near harvesting equipment in Mexico (Pyle and Howell 1993). Eggs are laid on in a tree hollow (Sick 1993) or on the ground in dense grass (Yanosky 1987).

The Greater Yellow-headed vulture is a forest species that avoids urban areas. It forages in the moist, dense, lowland forest, generally outside high-altitudes and non-forested areas (Hilty 1977; BirdLife International 2014). In Guyana, its habitat was lowland forest, including both seasonally flooded and non-seasonally flooded forest (Braun et al. 2007). Nest sites include cliff cave floors, in crevices or in tree hollows.

The condors are open terrain and upland specialists. The California condor is absent from urban areas rocky shrubland, coniferous forests, and oak savannas. When it was more numerous, an important former foraging habitat was the littoral zone of the Pacific Coast, which is gradually being recolonized by birds released from captivity (Nielsen 2006). Eggs are laid in cliff caves or crevices rocky outcrops or large trees are used as nest sites (United States Fish and Wildlife Service 1996). The Andean condor inhabits high, open terrain, such as grasslands, alpine meadows, rocky mountains up to 5,000 m (16,000 ft); and also lowlands in eastern Bolivia and southwestern Brazil, lowland desert areas in Chile and Peru and southern-beech forests in Patagonia (Sibley and Monroe 1990; Houston 1994; Parker et al. 1996; BirdLife International 2014). Eggs are laid on cliffs or rock ledges, or cavities scraped among boulders in relatively low areas or up to 5,000 m (16,000 ft) (Fjeldså and Krabbe 1990; Haemig 2007).

Old World Vultures are also divided into urban invaders and those that avoid urban and human occupied spaces. African vultures are known to be present in cities, especially near abattoirs (Brown et al. 1982) Ssemmanda (2005: 10) notes that 'within urban centers they are generalist scavengers utilizing all kinds of waste associated with humans including human excrement, carcasses, offal, bones and fresh meat associated with slaughterhouses', thus removing organic waste. The Hooded Vulture is a known denizen of urban areas. Wacher et al. (2013) recorded this species near or over villages and towns, in line with its well-known commensal habits in West Africa (Brown et al. 1982). In this study, the Hooded Vulture was the only species recorded in towns, despite others of the six species observed being more numerous in the countryside (Rüppell's Vulture, Lappet-faced Vulture). Ogada and Buij (2011) describe the Hooded Vulture as extremely adaptable to any environment; e.g., deserts, forests, savanna and urban areas. It is a 'human commensal' that favors rubbish dumps and slaughterhouses in urban areas, one factor being the lack of competition from other vultures (Ogada and Buij 2011: 101; see also Anderson 1999). However, the disadvantage of this close relationship with people is the frequent killing of vultures for food and traditional medicines, especially in West Africa (Anderson 1999; Sodeinde and Soewu 1999).

In Guinea, most Hooded vultures observed were in towns, especially in larger cities like the capital Conakry and rural areas; none in protected areas (Richards 1982; Walsh 1987; Halleux 1994; Rondeau et al. 2008). The Hooded vulture is also fairly common in towns in Côte d'Ivoire (Demey and Fishpool 1991) and was previously common but now rare or absent from towns in Mali and Niger (Thiollay 2006a), and also Senegal, Sierra Leone, Togo (Ogada and Buij 2011). Older reports record it as common in Chad and the Central African Republic (Carroll 1988; Scholte 1998). It is strongly associated with towns in Ethiopia and South Sudan (Dellelegn and Abdu 2010; Ogada and Buij 2011) and still occurs in Malawi (Dowsett-Lemaire and Dowsett 2006). It is rare even in cities in other countries, such as Kenya,

Tanzania, Zimbabwe, Lesotho, Mozambique and South Africa (Mundy 1997; Maphisa 2001; Parker 2004; Njilima et al. 2010; Ogada and Keesing 2010).

The Hooded vulture is also described as an urban vulture in Uganda (Ssemmanda and Plumptre 2011). In Uganda, the decline in vulture numbers is attributed to improved abattoir hygiene (Ssemmanda and Pomeroy 2010). Ssemmanda and Plumptre (2011: 17), note that 'across most of Uganda this is an urban dwelling species with the largest population in Kampala', this possibly partly due to 'new abattoirs in emerging towns that have attracted some of the birds.' It is also described as the only urban species among eight vulture species in Senegal and Gambia, the extreme western part of Africa (Barlow and Wacher 1997). Chemonges (1991) went further and wrote that 0.02% of the waste produced in Kampala, Uganda was consumed by Hooded vultures and marabou storks. Other scavengers were the Yellow-billed kite Gmelin, 1788, Pied crow (*Corvus albus* Statius Muller, 1776), Palm-nut vulture (Pomeroy 1975; Carswell 1986; Amuno 2001) and Egyptian vulture. The highest density of vultures were found near the Kampala Meat Packers (also termed the City Abattoir), where the largest slaughter house kills several hundred cattle and goats each morning. Items dumped at the back of the buildings are hooves, diseased parts, skin scrapings and surplus meat. The number of hooded vultures roosting in the vicinity increased from an estimated 124 in the early 1970s to 237 in 2001 (see also Pomeroy 1975; Amuno 2001).

In Burkina Faso, Hooded vultures and Black kites are more strongly associated with densely inhabited human settlements and refuse than with natural food sources and savannas (Thiollay 2006). Ogada and Buij (2011) describe past reports in which thousands of Hooded vultures were sighted near markets in Ouagadougou (the capital city of Burkina Faso) in the 1990s. These numbers have plummeted more recently. In Ghana, hundreds have been reported near the meat processing centres (Mundy 2000). Hooded vulture numbers increase with increased human presence during the academic session of the University of Ghana (Gbogbo and Awotwe-Pratt 2008). Vultures also nest in trees on the campus of the Kwame Nkrumah University of Science and Technology, in Kumasi, the second largest city in Ghana (Akyeampong et al. 2009).

A study by Campbell (2009) on the Hooded vulture in Ghana, found this species to be an urban specialist, that avoids rural areas. The hooded vulture is the only vulture species in the densely populated southern half of Ghana, with other, larger species confined to the northern half of the country (Mundy et al. 1992; Gbogbo and Awotwe-Pratt 2008). White-backed and White-headed vultures have been recorded in Mole in the northern savanna region of Ghana (Vermeulen 2011). Few studies have investigated the reason for the absence of the larger vultures (Rüppell's Griffon, White-backed vulture and White-headed vulture) from the more forested and populated southern half of the country, but possibly people and forests are a factor. In the south, the Hooded vulture's only scavenging rival is the Pied crow and to a lesser extent the Yellow-billed kite (Campbell 2009).

In this study Hooded vulture and Pied crow presence was positively correlated with human numbers, these species being most common in meat and waste production areas, and also places where there were street discards in non-food production and residential areas. Vultures were most common in meat markets, followed by vegetable markets, non-food markets, residential areas, rubbish dumps, central business districts, urban green spaces and finally, rural areas. Vultures and crows took meat pieces from butchers' tables, walked into windows and doors and flapped about people. In the meat production areas, vultures outnumbered crows, but were less mobile, circling or perching, rather than invading buildings as the crows commonly did. In the non-meat production areas, crows outnumbered vultures, and foraged in gutters, on sidewalks, public toilets, under vehicles, house windows and even piles of old clothing.

In the urban non-meat production areas, Hooded vultures and crows engaged in observational perching, hovering or low soaring, and dived to the ground to snatch human discards, refuse and dead rodents. Human discards included meal remnants, garbage and unwanted trade items. Vultures and especially crows also gathered in areas of human communal eating; crows appeared faster at snatching deliberately thrown food and outnumbered vultures three to one in such actions, creating sporadic nuisances, so people flapped cloths and papers at them, or even threw stones.

Regarding other species in Africa, information is minimal; possibly partly due to 'lack of active resident ornithologists' (Ssemmanda and Plumptre 2011). These authors note that much of the information is based on the Serengeti-Mara area (see also Kruuk 1967; Pennycuick 1972; Houston 1974, 1975, 1979). In the 1960s–1980s populations of vultures were larger, while in more recent times many species are now threatened. In Western and Eastern Africa, the Egyptian vulture is often migratory. There are also five resident species which, excepting the Hooded vulture are not recorded in cities. These are the Hooded vulture, African White-backed vulture, Rüppell's vulture, Lappet-faced vulture, and White-headed vulture.

Urbanization may be a factor for their extinction. For example, Margalida et al. (2007a) found that the Egyptian vulture was linked to availability of food and low human presence. Possibly other studies that showed that they were tolerant of humans were recorded in areas of higher vulture density (Ceballos and Donazar 1989). The Eurasian Griffon vulture was associated with the availability of food and landscape with low tree cover. The Bearded vulture was very selective, linked only to mountain environments with little vegetation cover for food, steep slopes for nesting and food preparation (ossuaries) and low human disturbance (an impossibility in and urban environment (Donázar et al. 1993; Margalida et al. 2007b).

In Asia, vultures are described as foraging in urban or other habitated areas, even if they roost away from habitation. Arun and Azeez (2004: 567) describe White-rumped vultures as using the waste dumps of urban areas as convenient 'fast food restaurants'. A study by Harris (2013) in Nepal describes the White-rumped vulture as foraging in human habitation and nesting in trees near habitation. The Indian vulture was 'found within cities, towns and villages associated with cultivated areas'. The Slender-billed vultures inhabited open forests up to 1,500 m in the vicinity of human habitation, scavenging at rubbish dumps and slaughter houses. The Red-headed vulture lived in open country near habitation and wooded hills.

7.2 Vultures and Aeroplane Collisions

A very important aspect of the relationship between vultures and urban development concerns the collisions between these birds and aeroplanes, usually near airports in large cities (Lowney 1999; Avery and Cummings 2004; Blackwell and Wright 2006). Unlike road kills, where mostly mammals collide with road vehicles and vultures may eat the carcasses, in aeroplane collisions, the vultures are the animals killed. Vulture collisions occur largely in southern countries, due to the tropical and subtropical distribution of most vulture species. In northern countries, *Larus* gulls and other seabirds compete with raptors. The literature on vultures appears more scanty.

Vulture collisions with aeroplanes are globally common, creating problems in both urbanized airport areas and the airspace over other landcover (Satheesan and Satheesan 2000). These authors record that between 1955 and 1999 more than 33 aircrafts (27 military and other civil) were destroyed, and 21 lives were lost because of collisions in seven countries in Asia, Africa, Europe and North America. In some cases, collisions have occurred at extreme altitudes. For example, on 29 November 1973 a Rüppell's Griffon collided with a commercial aircraft at 37,000 feet (11,280 m) over Abidjan, Cote d'Ivoire (Laybourne 1974). Curiously, the vulture in this incident was flying at an altitude where temperatures are about –50°C and oxygen is very thin.

Many accidents have been recorded in the United States, Spain, India, Pakistan, Kenya, Ethiopia and Tanzania and other countries. In the United States of America, aerpolane collisions with Black vultures cost more than US$25 million to the US Air Force (USAF 2009). Turkey vultures have also caused a lot of damage (Satheesan and Satheesan 2000). Also in Brazil, aircraft strikes are considered the most important vulture-related problem. The Aeronautical Accidents Investigation and Prevention Center (CENIPA) recorded more than 980 strikes involving vultures between 2000 and 2011. In Manaus, a total of 65 vulture-aircraft strikes were recorded between 2000 and 2012 (CENIPA 2012). In India, the losses were about $70 million annually from 1980 to 1994 due to vulture strikes (Satheesan 1994, 1996, 1998, 1999a). In India, all the crashes involved one species, the White-backed vulture; other species also had more minor encounters (Indian vulture, Slender-billed vulture and Egyptian vulture). The African White-backed vulture, the White-headed vulture and the Rüppell's griffon brought down aeroplanes in Kenya. In other nations, such as the USA and Spain losses totalled about $10 million to $17 million per accident (Satheesan and Satheesan 2000).

The principal reasons for vulture collisions are solitary and communal thermal soaring, gliding between thermals, large bird size and slow reflexes on the part of the birds. Generalizing, most collisions occur when the plane is cruising, and a few during descent, where the altitude of the accident was known, about half were below 200 m; fighter jets are more likely than transport aircraft to collide with vultures, due to the high speed of the fighter aircraft at low latitudes.

One study examined mid-air collisions between Black and Turkey vultures and military and civilian aircraft at the Savannah River Site (SRS) in South Carolina, USA (DeVault et al. 2010). It examined the hypothesis that vulture flight characteristics were predictable with respect to weather and time variables. The results showed that Black vultures flew at an average altitude of 169 ± 115 (SD) m, whereas Turkey vultures flew at an average altitude of 163 ± 92 m. These findings contrasted with other studies that recorded less frequent and lower altitude flights. The flight behavior of both species was only minimally influenced by weather and time variables. Food availability, inter- and intra-specific interactions, and physiological demands were more important. It was concluded that bird avoidance strategies should focus on the variable flight behaviors of Black and Turkey vultures across their ranges (i.e., factors contributing to differences in flight behavior among regions), rather than examinations of the effects of local conditions on flight behavior.

Several methods have been devised to reduce bird-aeroplane collisions. Satheesan and Satheesan (2000), citing India, suggest the removal of vulture food sources within 100 km of civil airports and 200 km of military airfields, bombing ranges or low-level high-speed exercise zones. In practice, shooting birds is more common globally. For example, from 1991 and 1997 two to five hired shooters at JFK International Airport USA killed 52,235 gulls (47,601 Laughing Gulls, 4,634 others) in 6369 person-hours of shooting (Dolbeer 1998). Such shooting has also been practised in India. Satheesan and Satheesan (2000) point out that such shooting, which may create a vacuum into which other birds may enter to fill, and also be shot, is a factor for the low populations of vultures and other avian scavengers near aerodromes in India. Another problem is the low sanitation near airports, that encourage scavengers and might lead to the replacement of birds depleted through shooting (Blokpoel 1976; Ali and Grubh 1984; Satheesan 1992a, 1992d, 1994, 1996).

7.3 Vultures and Wind Turbines

Wind turbines and power lines constitute another important problem for urban vultures, as these constructions are rapidly increasing in North America, Europe, Asia and Africa (Moccia and Arapogianni 2011; Martin et al. 2012). Large raptors, especially vultures and eagles are very vulnerable to collisions with wind turbines and power lines, despite their sharp vision (much sharper vision than human vision) (Land and Nilsson 2002). The vulnerability of these birds to collisions has become a factor for the decline of several species, an important international conservation issue and a topic at international conferences (Drewitt and Langston 2008; De Lucas et al. 2008; Carette et al. 2009; Jenkins et al. 2010; Martin et al. 2012).

The development of wind turbines was an important aspect of the policy to reduce greenhouse gas emissions by using renewable sources. The target of the European Commission was that 20% of the energy generated by European Union (EU) should have renewable sources by 2020. The Spanish government was committed to this by 2010 (AEE (Asociación Eólica Española), 2009; Telleria 2009a; IDAE (Instituto para la Diversificación y el Ahorro de la Energía) 2010). In Spain, renewable sources generated 14% of energy production (IDAE 2010). As wind was used as an energy source, the number of wind farms increased across Spain. About 16,842 turbines in 737 wind farms were operational, generating more than 16 GW (gigawatts) of electricity in 2009 (AEE 2009).

Wind turbines are an important factor for vulture mortality in modified landscapes (Telleria 2009a). De Lucas et al. (2012a). note that the wind turbine has developed into an important factor for avian mortality. In some cases, they kill more birds than power lines (Barrios and Rodríguez 2004). Examples occur in the Netherlands (Winkelman, 1990, 1992, 1995; Musters et al. 1996; Dirksen et al. 1998), USA (Orloff and Flannery 1993; Howell 1997; Morrison et al. 1998; Osborn et al. 2000; Erickson et al. 2002; Johnson et al. 2002; Thelander et al. 2003; Arnett 2005); Sweden (Peterson and Stalin 2003) and Spain (Barrios and Rodríguez 2004; Dorin et al. 2005).

Wind power plants are frequently sited near the breeding sites of vultures. Telleria (2009b) notes that there are no evaluations of the potential impact of the expanding wind power industry on the Spanish Griffon vultures which are described as more than 20,000 breeding pairs by Marti and Del Moral (2003). This is despite the documented incidents on wind turbines in Spain (Barrios and Rodriguez 2004). Spanish griffon vultures breed mainly in the northern half of the country, where there are thousands of turbines. Telleria (2009b) hypothesizes that the number of deaths may be high, as the wind power industry selects highlands and mountain rides or turbine location, precisely the places frequented by foraging vultures. Lekuona (2001) estimated a turbine generated death rate of eight griffons per year in one area near the Salajones wind plant, in Navarre, Southern Spain. Lekuona and Ursua (2007) also found the griffons represented 63.1 percent of bird fatalities near the wind plants of Navarre.

Another study by De Lucas et al. (2012b) studied Griffon vulture mortality at 13 wind farms in Tarifa, Cadiz, Spain, before (2006–2007) and after (2008–2009), when selective turbine stopping programs were implemented as a mitigation measure. The study found 221 dead Griffon vultures during the entire study. When turbines were stopped when vultures approached, the Griffon vulture mortality rate was reduced by 50% with a consequent reduction in total energy production of the wind farms by only 0.07% per year.

7.4 Vultures and Electric Power Lines

Electric power lines are also dangerous for vultures, either through electrocution or collisions. Bird electrocutions were first documented in the 1970s, when thousands of raptors were killed in North America (APLIC 2006). In the following years, research on this topic has been conducted in North America, Western Europe and South Africa (Lehman et al. 2007; Manville 2005). Electric power lines (both collisions and electrocutions) contribute to the deaths of millions of birds in the African-Eurasian region (Haas et al. 2005). It is estimated that up to 10,000 electrocutions and hundreds of thousands of collisions occur in most countries in the African-Eurasian region annually (Prinsen et al. 2011). Millions are estimated in Germany (Hoerschelman et al. 1988). In some countries in Northwest Europe, the electrocution problem is lower, as many voltage lines have been placed underground. This is the case in Germany, Belgium, the United Kingdom, Denmark and Austria (Tucker et al. 2008). Collisions are more common when the power lines bisect vulture foraging or nesting habitats (Hunting 2002). Features necessary for vulture presence may include alternative perch availability, vegetation form and carcass presence. Power lines and towers may be utilized as replacements for nesting or roosting trees, especially deserts, open plains and intermontane basins (APLIC 2006).

Lehman et al. (2007) conducted a literature review of raptor electrocution research, mitigation and monitoring, and also evaluated the results of 30 years of efforts to reduce deaths due to electrocution and collisions. Most studies focused on North America, Western Europe, and South Africa.

Many early studies from the 1970s focussed on the United States from the 1970s onwards, where there has been 'extensive research, product testing, design standards development, and mitigation' (Lehman et al. 2007; see also Boeker and Nickerson 1975; Nelson and Nelson 1976; Olendorff et al. 1981). There is evidence that these efforts have not stopped the killing of birds, at least in the United States (Franson et al. 1995; Melcher and Suazo 1999; Harness and Wilson 2001). In addition to bird deaths, there was 'negative media attention, increased scrutiny by regulatory agencies, and a landmark court conviction' (Lehman et al. 2007; see also Melcher and Suazo 1999; Williams 2000; Suazo 2000). The electrocution problems also cost the energy suppliers billions of dollars each year due to power interruptions, necessary equipment repairs, revenue losses and statutory compliance issues (Hunting 2002).

A 2001 review of 30 years of responses and projections in the United States that bird deaths would be eliminated found overly optimistic projections and no real results (Lehman 2001; see also Nelson and Nelson 1976; Wildlife Management Institute 1982; Phillips 1986; Gauthereaux 1993; Avian Power Line Interaction Committee [APLIC] 1996). More than 185 million wood distribution poles were operating in North America in 2005, with at least some threats to birds (Lehman 2001; American Iron and Steel Institute 2005). Actions to reduce deaths are mostly modifications of the poles after the event, rather than examination and elimination of the issues that contribute to the mortalities and ameliorating or eliminating them where possible (Lehman 2001).

Factors for the electrocution risk depend on ecological, physical, and landscape factors, i.e., vegetation structure and composition, which are also related to the species, time, location, or environmental conditions (Hunting 2002; Lehman et al. 2007).

Electrocution of the bird occurs when the bird bridges the gap between two energized components or an energized and an earthed (also called 'grounded') component of the pole structure. Electricity then flows through the bird and kills it. Low to medium voltage lines are usually the most dangerous, as the structures are more closely spaced. The conductors and ground/earth wires or earthed devices are usually too far apart for smaller birds to touch simultaneously (APLIC 1996; Janss and Ferrer 1998, 1999a, 2000; Janss et al. 1999). Therefore, most incidents involve large raptors and storks during the breeding season (Boeker and Nickerson 1975; Benson 1981; Olendorff et al. 1981; APLIC 1996; Kruger 1999; Harness and Wilson 2001; Prinsen et al. 2011). In South Africa vultures are the risk birds (Ledger and Annegarn 1981; Ledger 1984; Kruger 1999). Bird size is less relevant in Europe, as most poles are constructed of steel or steel-reinforced concrete, which can make all birds of any size vulnerable (Bayle 1999; Negro 1999).

Birds are more likely to perch on the power lines if there are no other perches, natural or artificial nearby (Olendorff et al. 1981; Janss and Ferrer 1999a). In forests with many natural perches, very few birds are killed by electrocution (Switzer 1977; O'Neil 1988; Harness and Wilson 2001). The same applies for ground-nesting species that may hunt in flight but perch on or near the ground (Pendleton 1978; Benson 1981). Other factors are relative pole height, pole-top configuration and clearances among electrical components (Lehman et al. 2007).

In the United States there are reported to be 50 regularly breeding species of birds of prey (thirty-one of diurnal raptors and 19 of owls) (Johnsgard 1988, 1990). Twenty six of these species have been recorded as victims of electrocution. Golden Eagles comprised between 50–93% of bird deaths in some reports (Smith and Murphy 1972; Boeker and Nickerson 1975; Ansell and Smith 1980; O'Neil 1988; Harness and Wilson 2001). Golden eagles in the United States are commonest in the shrub-steppe regions in the Western inter-montane region, where there are few natural perches (Harlow and Bloom 1989). The similarly sized Bald eagle inhabits forested areas with abundant perches (Stalmaster 1987). Vultures in the open African plains, such as the gregarious Cape and African white-backed vultures also crowd onto powerlines.

The design of the power lines and the associated hardware is also important for electrocution risk. Most electrocutions in the United States occur on low-voltage distribution lines (<69 kV) for mostly individual residential use, because prior to 1971, most of these lines were built with little or no insulation and narrow clearances between the energized components. Most poles are made of wood, which is nonconductive. Electrocution is more likely to occur when there is more possibility of links between energized components, e.g., with transformers or other auxiliary equipment (fused cutouts, capacitors, reclosers, jumper wires) (Olendorff et al. 1981; APLIC 1996; Harness and Wilson 2001).

The design in Europe is composed of steel or steel-reinforced concrete, which are conductive and grounded (Bayle 1999; Janss 2000). Therefore, unlike in the American model, a bird on a crossarm may be electrocuted on contact with a conductor (Janss and Ferrer 1999b). This results in much higher mortality levels, such as the 10,000 losses in the Slovak Republic (Adamec 2004). This is one reason for the transfer of power lines in Europe underground, which has been done in the Netherlands, and is being undertaken in Germany, The United Kingdom and Belgium (Bayle 1999). As the structures in Europe are usually steel or steel reinforced concrete, which are conductive, size is less important than in other continents (Bayle 1999; Janss 2000). Collision with cable lines also occurs, in which the bird may be killed by the impact or later through injuries. Collisions are common on multiple vertical layered, high voltage lines.

Another factor for electrocution is topography. Studies in the United States have shown that, as raptors use poles for surveying food possibilities, poles located on raised topography are perched more often and therefore have greater numbers of electrocutions (Benton and Dickinson 1966; Boeker and Nickerson 1975; Nelson and Nelson 1976; Benson 1981; APLIC 1996). The daily weather and seasonal changes also affect bird usage of poles and powerlines. For Golden Eagles in some western states, 80 percent of the deaths occurred during the winter (Benson 1981). Most other species recorded losses during the nesting period (Harness and Wilson 2001). Wet feathers increase the electrocution risk ten-fold and skin-to-skin contacts are also very conductive (Nelson 1979, 1980). Wind direction relative to the crossarms, may also be a factor for electrocutions. When the wind is diagonal or parallel to the crossarms, there may be more accidents than when the wind is perpendicular to the crossarm, because landings and takeoffs are more difficult in the former situation (Nelson and Nelson 1976; Benson 1981).

Bird behavior is also important. During fledging, the population increase may increase accidents (Harness and Wilson 2001). Physical contact between birds during nest defence or mating may link conductive components (Dickerman 2003). Nesting behavior and material lying across conductors have killed young birds (Hardy 1970; Gillard 1977; Switzer 1977; Vanderburgh 1993). Young, inexperienced birds are more likely to be electrocuted; for example at least 90% of Golden eagles killed in North America were immature or subadult (Boeker and Nickerson 1975; Benson 1981), with only slightly lower figures for young Spanish Imperial eagles (females also had more accidents due to their greater size) (Ferrer and Hiraldo 1992). Age has also been cited for smaller raptors (Fitzner 1978; Dawson and Mannan 1994).

In European countries, other than those mentioned above, accidents are common; the affected birds are mainly raptors and storks. Vultures are less affected, arguably because they were already rare in central Europe. In Eastern Europe, Bulgaria has been well studied for this topic. Forty- five thousand kilometres of power lines were a risk to birds (Stoychev and Karafeizov (2003). Twenty-two species were recorded dead, more than half being diurnal raptors, storks and crows (Demerdzhiev et al. 2009; Gerdzhikov and Demerdzhiev 2009; Demerdzhiev 2010). Studies are conducted on the effects the Cinereous vulture in Bulgaria by the Green Balkans Federation of Non-Governmental Organizations (Prinsen et al. 2011). In Hungary, 877 electrocuted birds of 46 species were found under one percent of the medium voltage electric poles in the country (6,500 poles) (Kovacs et al. 2008). The annual dead bird count has been estimated at over 30,000 (Demeter 2004). Affected species between 2003 and 2008, in order of decreasing number were the Golden eagle, Common kestrel

(*Falco tinnunculus*, Linnaeus 1758), Saker falcon (*Falco cherrug*, Gray, 1834) and European roller (*Coracias garrulus* Linnaeus, 1758) (Horvath et al. 2008). Other less affected species were the Imperial eagle (*Aquila heliaca* Savigny, 1809), the Eagle owl (*Bubo bubo* Duméril, 1805), White stork (*Ciconia ciconia* Linnaeus, 1758), Black stork (*Ciconia nigra* Linnaeus, 1758), Red-footed falcon (*Falco vespertinus* Linnaeus, 1766), (*Buteo buteo* Linnaeus, 1758) common buzzard, White-tailed eagle (*Haliaeetus albicilla* Linnaeus, 1758) and the Peregrine falcon (*Falco peregrinus* Tunstall, 1771) (Horvath et al. 2008; Horvath et al. 2011). In Central Europe, including Switzerland, White storks and Eagle owls are strongly affected (Lovaszi 1998; Marti 1998; Moritzi et al. 2001; Breuer 2007; Schaub et al. 2010; Schürenberg et al. 2010).

In Southern Europe, storks and raptors are similarly the species most affected by electrocution and to a lesser extent by collisions. In one study in France, the deaths were mostly by electrocution (96.5%) and the rest due to collisions (Sériot and Rocamora 1992 in Bayle 1999; see also Schürenberg et al. 2010). Raptors were seriously affected, e.g., Common buzzard, Common kestrel, Black kite, Bonelli's eagle (*Aquila fasciata* Vieillot, 1822) the Griffon vulture and the Short-toed eagle (Cheylan et al. 1996; Bayle 1999; Kabouche et al. 2006). In Italy, Common buzzard (*Buteo buteo* Linnaeus, 1758), Common kestrel, Griffon vulture, Osprey (*Pandion haliaetus* Linnaeus, 1758) Eurasian Sparrow hawk (*Accipiter nisus* Linnaeus, 1758), flamingoes, herons and storks were affected (Rubolini et al. 2005). Prinsen et al. (2011) note that the raptors and corvids were mostly affected by electrocution, and herons, flamingos and small passerines were more susceptible to collisions.

In Portugal, in a survey between 2003 and 2005, out of 945 dead birds mostly electrocuted in steppe areas, the most affected were the White stork (137 electrocuted) and common buzzard (146). Vultures included the Griffon vulture (12) and Eurasian black vulture (1) (Infante et al. 2005). In Spain, the Eurasian Black Vulture, the Griffon Vulture and the Egyptian vulture are affected (Martínez 2003; Palacios 2003). In the Doñana National Park, in Southern Spain, 233 electrocuted raptors included Griffon vulture (14 individuals) (Ferrer et al. 2001). Another study by Guzmán and Castaño (1998) in southern Spain for an eight year period (1988–1996) and 69 kilometres of lines and 1,629 poles, among 274 raptors and 14 species found dead were Eurasian Black Vultures (2) and Griffon Vultures (1).

In the Canary Islands, electrocution, in addition to poisoning, habitat destruction, reduction of food supplies and use of pesticides in the 1950s–1960s to eradicate locust plagues have been blamed for the decline of Egyptian vultures (Tucker and Heath 1994; Palacios 2000). In Fuerteventura, the second largest Canary Island, after Tenerife, Egyptian vultures roost on power lines, possibly due to the lack of trees in the desert environment (as this is not case in non desert environments) (Donazar et al. 1996; Sigismondi and Politano 1996). They are thus likely to be electrocuted (Janss 2000). In one area, six Egyptian vultures were found dead near 12 km of power lines—two by collision and four by electrocution (Lorenzo 1995). These incidents were important for vulture populations in the area (Ferrer 1993; Ferrer and Janss 1999).

In Asia, the electrocution and collision problems also exist. In the Kazakhstan steppes, larger raptors (including the Cinereous vulture), crows and gulls account for 93% of the dead (Haas and Nipkow 2006; Lasch et al. 2010). It is estimated that about 58,000 raptors are killed annually during spring migration along 9,478 kilometres of power lines (Karyakin 2008). In Kazakhstan, eagles (White-tailed Eagle, Steppe eagle (*Aquila nipalensis* Hodgson, 1833), Golden Eagle, Greater Spotted eagle (*Clanga clanga* Pallas, 1811) and the Short-toed Eagle) and falcons were particularly affected (Karyakin et al. 2006; Karyakin and Novikova 2006; Karyakin 2008; Lasch et al. 2010). In Mongolia, more than 60% of the electrocuted birds were raptors, especially near poles with closely spaced electrical equipment (Harness and Gombobaatar 2008; Harness et al. 2008). In the Russian Federation, above ground medium voltage power lines are estimated at 1,500,000 kilometres, with 0.5% equipped with isolated cables or modern facilities for bird protection. It is estimated that 10 million birds belonging to 100 species are killed annually through collisions and electrocutions (Matsyna and Matsyna 2011).

Concerning the Middle East, Prinsen et al. (2011) found no published information except for Israel. In Israel, mostly Griffon vultures were electrocuted (up to 5% of the population killed each year). Other affected species were Black kites, Ospreys storks and pelicans (Bahat 1997). Fewer deaths were recorded between 2007 and 2009, possibly due to pylon insulating near garbage dumps (Prinsen et al. 2011).

In Africa, as in Europe, sparse vegetation and absence of natural perching sites encourages use of power lines by any type of bird (Prinsen et al. 2011). In Egypt, a factor for bird deaths is the large number of low voltage power lines with short insulators and steel lattice towers near bird migration bottlenecks and food attractants such as rubbish dumps. In Sudan, in a 50-year old, 31 km section on the Red Sea coast 50 dead vultures were recorded in 1982 and two in 1983 (Niklaus 1984). Seventeen dead Egyptian vultures were also recorded during September 2010 (Angelov et al. 2011). The pole structures included a steel t-type and concrete staggered vertical type structures. Up to 5,000 vultures are estimated to have been killed in the past 80 years (Angelov et al. 2011). This is a possible factor for the declining population of Egyptian vultures in the Middle East, where many of the Sudanese birds originate.

In Ethiopia, the migratory raptors have been found dead near unsafe poles in areas with little vegetation for perching (Haas 2011). In Kenya, in the Magadi and Naivasha areas most pole designs were found to pose an electrocution risk to medium to large birds (Virani 2006; Smallie and Virani

2010). Species at risk included Egyptian vultures, White-headed vultures, Lappet-faced vultures, African White-backed vultures and Rüppell's vultures and also other species such as the Martial eagle (*Polemaetus bellicosus* Daudin, 1800), Augur buzzard *Buteo augur* (Rüppell, 1836) Grey crowned crane (*Balearica regulorum* Bennett, 1834), Lesser flamingo (*Phoenicopterus minor* Geoffroy Saint-Hilaire, 1798), White Stork, Secretary bird (*Sagittarius serpentarius* J.F. Miller, 1779).

In South Africa, there are several types of structures. There are 22-kV wooden T-structures, 88-kV steel kite transmission towers, terminal H-frame wood structures, and Delta suspension structures (Kruger 1999). The largest number of electrocutions of different sized birds occur on the T-structures and terminal H-Frames. Birds are killed from the Genera *Gyps, Polemaetus, Aquila, Buteo, Circaetus, Falco,* and *Bubo* spp. The steel kite transmission towers and the Delta suspension structures generally kill larger species such as Cape griffons, African White-backed vultures and Martial eagles (Lehman et al. 2007).

In this country, where vulture habitats are frequently open plains with few trees, the vulture species (e.g., Cape vultures and African White-backed vultures) roost or perch communally on power lines, resulting in multiple electrocutions (Bevanger 1994; APLIC 2006; Lehman et al. 2007; Prinsen et al. 2011). These two species have been strongly affected since the early 1970s (Markus 1972; Ledger and Annegarn 1981; Ledger and Hobbs 1999; Smallie et al. 2009). Kruger et al. (2004) report the electrocution of 46 Lappet faced vultures, 24 African White-backed vultures, 4 Cape griffons and 12 unidentified vultures in South Africa. For example, between August 1996 to May 2011, there were 1,504 electrocuted Cape griffons, White-backed vultures, Lappet faced vultures, and eagles, storks, corvids and other birds. For the Cape vulture, which is endemic to South Africa and a threatened species, it is estimated that about 80 are killed annually in the Eastern Cape (Boshoff et al. 2011).

Collisions are also a leading cause of non-fatal injuries to vultures. Where the birds survive, there are now organizations that attempt to rehabilitate them (Naidoo et al. 2011). Three organizations were involved in the rescue and/or rehabilitation (R&R) of vultures in the Magaliesberg mountain range of South Africa over 10 years; the De Wildt Cheetah and Wildlife Trust, the National Zoological Gardens of South Africa (NZG) and the Vulture Programme of the Rhino and Lion Wildlife Conservation NPO (R&L). The commonest birds in these centres were Cape griffon and African White-backed vultures. The available data identified the cause of injuries as collisions with pylons, resulting in soft tissue and skeletal injuries. The study concluded that 'urbanisation has had a major negative impact on vultures around the Magaliesberg mountain range' (ibid. 24).

Data management on bird deaths on or near power lines in South Africa is conducted by Eskom-EWT Strategic Partnership in a Central Incident Register. The result of these mortalities induced Eskom to stop building vertically configured medium voltage designs that were a particular risk to vultures. Vulture deaths were blamed on the birds' size, gregarious nature, limited alternative natural perching areas and the small size of pole-top perching area. Electrocution occurred when a vulture landed on the pole top, and slipped between the conductors. Insulating the conductors reduced but did not stop such fatal accidents. 'Furthermore, in some instances, the vultures started attacking the insulation by ripping it apart' (Kruger et al. 2004: 439). Vultures were found to be perching on one conductor while touching another. The study recommended the discontinuation of vertically configured medium voltage structures (Kruger et al. 2004).

Considering the problems created by wind turbines and power lines in terms of structures and collisions, the question is why cannot birds see these objects, especially considering most flying birds have good eyesight? Martin et al. (2012) set out to answer this question. The findings of their study are that 'the visual fields of vultures contain a small binocular region and large blind areas above, below and behind the head.' Studies of the visual ability of cranes, ibises, spoonbills and bustards have noted that even a small-amplitude forward pitch (for example when the bird scans the ground) results in inability to detect direction and any object in front of the bird; a reason for powerline and wind turbine collisions (Martin and Shaw 2010; Martin and Portugal 2011; Martin 2011a, 2011b). Martin et al. (2012) also found that the small binocular region and the large blind spot allows coverage of the ground below, protects the eyes from the sun above, and also allows extensive lateral vision during foraging. Vultures therefore have narrow, frontal binocular vision. By 'erecting structures such as wind turbines, which extend into open airspace, humans have provided a perceptual challenge that the vision of foraging vultures cannot overcome' and possible solutions should include attracting vultures away from such areas at the broad, landscape level, due to the wide foraging ranges of vultures (Martin et al. 2012: 5).

7.5 Vultures, Agricultural Development and Rural Environmental Change

Vultures have had a long interaction with people engaged in agriculture and other rural activities, either as perpetrators of disease, predators of small livestock or in a more positive light, as removers of carrion. Mateo-Tomás and Olea (2010: 520) refer to the 'huge dependence of vulture populations on farming activity' (see also Murn and Anderson 2008). Recently, human applied chemicals have also become important (Pain et al. 2003; Green et al. 2006), as have sanitary restrictions based on bovine spongiform encephalopathy in cattle, which affected vultures in Europe (Tella 2001; Camiña and Montelío 2006), antibiotics in livestock that affect vultures in Spain (Lemus et al. 2008) and attacks by Black vultures on cattle in the United States (Avery and Cummings 2004).

These issues applied to both the New World and Old World vultures. A major effect of agriculture and also urbanization, which may affect vultures is deforestation and either savannaization or in extreme cases, desertification as was described in Part 2. Savanna is an important foraging area for vultures. As noted in Chapter 3, in the Americas, the Yellow-headed Vulture has been called the savanna vulture, and with the exception of the other Cathartes vultures which have a sense of smell, all other vultures find food more easily in the open terrain, or like the King vulture they may follow Cathartes vultures to the carcasses in dense vegetation. In Africa and Asia, vultures forage principally in open terrain, mostly savanna. In West Africa, the small Hooded vulture is the main species found in the forest, while the larger vultures are principally denizens of the savanna. In Asia, open or mountainous terrain is the main vulture habitat. This section looks at the main interactions between vultures and rural people in terms of agriculture and food access change, as the other main environmental issues were discussed in Part 2.

Interactions between agriculturists and vultures concern not only the impacts of people on vultures, but also the impacts of vultures on human activities. The main issues concern the extent to which vultures may make use of the activities and products of farming and livetstock rearing for food access. Concerning crops, the main issue is environmental change, as most vultures (with the exception of the Palm-nut Vulture) are not fruit eaters. Open farmland may also attract herbivorous animals and ploughed soils may expose small animals that may attract vultures. Livestock rearing is generally more important, as vultures may feed on the caracsses of dead livestock such as cattle, and may even kill small or young livestock.

Kirk and Gosler (1994) studied migratory and resident Turkey and Black vultures in the Llanos of central Venezuela. The resident Turkey vultures were the smaller subspecies (*Cathartes aura ruficollis*) and the migrants were the Northern subspecies *Cathartes aura meridionalis* (that is in Venezuela for the northern Boreal winter, which approximately coincides with the dry season of northern South America, between October and April). The Turkey and Black vultures focussed mostly on different types of carcasses (Stewart 1978; Coleman and Fraser 1987). Turkey vultures favored carcasses of small mammals, while Black vultures favored mostly domestic livestock, which was 'a more predictable and abundant food source' (ibid. 940). The smaller resident Turkey vultures foraged over forest, where the larger migrant

Turkey vultures were rare, rather than savanna, where the migrants were common. The biomass of carcasses, especially of domestic livestock (which were restricted to the savanna) was higher in the savanna than in the forest. Migrants, being larger won almost all conflicts with residents over carcasses (470 to 1 of the conflicts observed). The unequal competition between the migrants and residents affected the health of the latter, as the study found that the condition of resident Turkey vultures was below average when they lived with migrants, and in above-average condition when the migrants were absent. The residents began to forage again in the savanna when the migrants moved northwards.

In northern North America, public attitudes towards the Black vultures has been partly shaped by the belief that the vultures affect farm animals. Kiff (2000: 180) noted that 'throughout the first half of the present century, many ranchers and farmers associated turkey and black vultures with the spread of diseases, especially anthrax, and therefore often sought to eliminate them, usually by trapping.' Black vultures were also believed to attack newborn calves. As a result, both Turkey and Black vultures were persecuted, using both private and government sponsored programs. These killed hundreds of thousands of birds in Florida and Texas during the 1940s and 1950s, before the later protection of both species by the Migratory Bird Treaty Act, signed by the United States, Canada and Mexico (Howell 1928; Parmalee 1954; Snyder and Rea 1998; Kiff 2000). In some areas, with large roosting centers, attempts have been made to disperse the birds on the premise of public nuisance 'generally by non-lethal means', but such measures are increasingly rare due to greater tolerance of wildlife in the USA over the years (Kiff 2000: 180).

The National Wildlife Research Center's (NWRC) field station in Gainesville, Florida, researches into vulture-related problems (Avery 2004). Roosting Black and Turkey vultures create nuisance, health, and safety problems (such as electrical power outages), but the tendency of Black vultures to prey on livestock is the most serious problem. Conflicts with Black and Turkey vultures are reported in at least 15 States. Where Black vultures killed livestock, ameliorative action included the dispersal of a large communal winter roost in Virginia. The elimination of the winter roosts did not substantially alter vulture activity in nearby livestock rearing areas, as the vultures had several alternate roost sites. Also, there were cattle and sheep carcasses in the vicinity that enabled vultures to feed and maintain their presence. Collaborative research between the NWRC, the Virginia Wildlife Services and the Florida Farm Bureau continued on the factors for vulture predation on new livestock.

The impact of Black vultures on livestock is worth examining in more detail. Avery and Cummings (2004: 59) described Black vultures as 'a problem species for many livestock producers': with evidence going back decades (see also Baynard 1909; Roads 1936; Sprunt 1946; Lovell 1947). An old publication by Baynard (1909) in Florida is cited, which noted that 'Hundreds of young pigs, lambs, etc., are annually devoured by them…. I have had them come into my yard and catch young chickens' (Avery and Cummings 2004: 59). The main compiler of information on these incidents, the United States Department of Agriculture's Wildlife Services Program documents an increase in domestic animal attacks since 1997. From 1997 to 2002, 84% of reports were from Virginia, Florida, Texas, South Carolina, and Tennessee, with 13 other states accounting for the rest. More than half of these attacks involved young cattle.

For example, in Virginia, there were 115 incidents of black vulture contact with 1037 livestock animals (1990–1996) (Lowney 1999). The main action was a group attack by about 20 to 60 vultures, which pecked the eyes, rectum, genitals and nose of young lambs and calves, and sometimes of cows giving birth. Avery and Cummings (2004) observed Black and Turkey vultures in a farm in central Florida foraging in pastures where there was active calving and feeding on afterbirth and fresh calf droppings. There were attacks on newborn calves., with 20 to 40 vultures feeding on the bodies. It was uncertain if the calves were stillborn or the vultures had killed them. The heifer in both events was so badly injured that it had to be euthanized by the rancher. There was a third event, in which three Black vultures pecked at the hooves of a calf while it was being born. The mother was however, able to drive the vultures away (Avery and Cummings 2004).

The killing of small- and medium-sized vertebrates is comparatively rarer for Old World vultures (Houston 1994). For example, before 1990, the reports of Giffon vultures predating livestock in Spain were described as 'anecdotal' but from around 2006 possibly more reliable reports were made (Margalida et al. 2014: 3; see also Camiña et al. 1995). These authors nevetheless acknowledge that the 'killing of livestock by griffon vultures is a relatively minor problem. Domestic species, mostly dogs, cause greater damage to livestock than vultures' (Margalida et al. 2014: 4).

Nevertheless, Margalida et al. (2014) document a similar problem to that of Black vultures attacking livestock in the United States, but in this case it concerns Griffon vultures in Spain. This is a problem that appears to have emerged in the mid 1990s. These authors acknowledge that there is little factual data on this problem, much of the information being media driven and politically, rather than scientifically relevant.

The changes in European agriculture over the 19th and 20th centuries, especially in livestock rearing contributed to the decline in the numbers of Griffon vultures (Donázar et al. 1996; Donázar et al. 2009a). Conservation policies were successful, as vulture populations have increased by about 200% over the last 20 years over Europe, especially in the Iberian Peninsula (Spain and Portugal, and also parts of southern France; see C. Arthur and V.

Zenoni, unpubl. data, on the increase in vulture-related complaints in the French Pyrenees 1993 and 2009 in Margalida et al. 2014, see also Donázar et al. 2009b; Margalida et al. 2010). Vultures were increasingly seen breeding and feeding near human habitations and infrastructure. The consequences were wildlife-human conflicts, angering farmers but neglected by scientists (Margalida et al. 2011a).

Also important were changes in farming practice. The traditional farming system in Europe used grazed livestock, protected from predators by shepherds and dogs (Kaczensky 1999). In many mountainous areas, there are predation possibilities by smaller avian and mammalian predators (such as ravens and crows, hawks and eagles, vultures or foxes). Smaller predators targeted placentas after lambing, and weaker, sick or young animals, which would formerly have been protected by guard dogs or shepherds. For the Griffon vulture, the increase in population and the food and farm policy issues increased the likelihood of contacts with potential prey. The definition of 'attack' also varies. Vultures may eat the kills of other predators, or calves or lambs that died at birth, making it seem as though the vulture made the kill. The large size and flock feeding behavior of the Griffon, and the small size of the livestock consumed, allows the rapid consumption of the carcass, making it impossible for the farmers to determine the cause of the death (Margalida et al. 2014). Basically, the question is, did the vulture kill this animal?

Margalida et al. (2014) studied 1,793 complaints on Griffon attacks on livestock collected by the relevant Spanish authorities (1996 to 2010). The location of this study is important, because about 95% of the Griffon vultures in the European Union are recorded in Spain. The information on the attacks concerned mainly sheep (49%) and cows (31%) and horses (11%) and the birthing times (between April–June) of these animals. Most the complaints of vulture attacks occurred during this period (60%), after closing the supplementary feeding stations (36% of complaints). A majority of the complaints (about 69%) were actually rejected as the course of death was uncertain. Nevertheless, compensation payments totalled EUR 278,590 from 2004 to 2010.

The case study of north-east Spain, recorded most of the livestock attacks. This area also had 7,433 pairs of Griffon vultures in 2008 (27.3% of the Griffon vulture population in the European Union) (Margalida et al. 2010). There are also many domestic animals (about 709,294 cows, 3,236,333 sheep, 109,118 goats and 24,772 horses; see MAGRAMA 2012). At this location, most of the livestock is free ranging, in open land in mountains and plains during June–September due to the seasonal movement of people with their livestock between summer and winter pastures. There are also supplementary feeding stations for scavengers, using carcasses (Donázar et al. 2009b). In the past, dead animals were abandoned for vultures, but the incidence of bovine spongiform encephalopathy (BSE) in 2001, prompted the EU to promulgate Regulation [CE] No. 1774/2002) that restricted this practice. Later legislation passed in 2009 and 2010 allowed occasional disposal of carcasses for avian scavengers. The action created a shortage of domestic carcasses and a decline in the carcasses supplied to the Spanish feeding stations (about 80%), as sanitary methods also improved (Donázar et al. 2009a; Cortés-Avizanda et al. 2010; Margalida et al. 2012; Margalida et al. 2014).Vulture behavior may have been altered by the reduction in the food supply, increasing tolerance of human proximity (Donázar et al. 2009b; Zuberogoitia et al. 2010; Margalida et al. 2010, 2012; Margalida and Colomer 2012). Human complaints about vulture behavior peaked from 2006 to 2010, coinciding with the period of food shortage.

Margalida et al. (2014) reported a clustering of the reports against vultures, suggesting that factors for attacks or complaints may be locally derived. Local husbandry practices may be important, and individual problem vultures may have been behind the attacks (Linnell et al. 1999). Negative media attention resulted, with some people calling for the poisoning of vultures, while others favored more feeding centers (Hernández and Margalida 2008, 2009; Margalida and Colomer 2012; Margalida et al. 2011a,b; Margalida 2012). The negative perception of the relationship between humans and vultures exemplified the development of conflicts between people and vultures, changing an ancient relationship (Margalida et al. 2014: 4). Suggestions for the amelioration of the conflict included the use of livestock protection dogs (Espuno et al. 2004; Shivik 2006).

A close relationship between vultures and livestock farming has also been recorded in Asia. In a study of Nepal, a nation with six resident vulture species (White-rumped, Slender-billed, Egyptian, Red-headed, Himalayan Griffon and Bearded vultures; the Cinereous vulture is a winter visitor and the Eurasian Griffon vulture a passing migrant), vulture populations have declined, due to several factors, one of which is the decline in large domesticated and semi-domesticated livestock (Shrestha and Prasad 2011). The Water Buffalo (*Bubalus bubalis* Linnaeus 1758) is a huge bovid common in southeast Asia, which may be wild, semi-domesticated or domesticated. Historically, vultures were scavengers of buffalo carcasses, which were large enough to satisfy their needs. Other animals on which vultures fed were domestic cattle and goats. In recent times livestock numbers have declined, as farmers have moved to other businesses. Only goat rearing has remained popular, due to the more rapid renumeration. Vulture numbers have declined, resulting in the establishment of a 'vulture restaurant' by the Kalika and Jhulke Community Forest User Groups (CFUGs) (Shrestha and Devkota 2011).

In Turkey, which has the second largest population of the Cinereous vulture (after Spain), livestock are the largest dietary source for this species. Of these, sheep were dominant (76.6% of all pellets examined), followed by wild boar (44.1%) and chicken (22.5%) (Yamaç and Günyel 2010). 'This shows that livestock plays an outstanding role in the diet of the Eurasian Black vulture, and underlines its dependence on extensive livestock farming and grazing' (ibid. 15). The relationship between vultures and livestock is complicated by the fact that it is illegal to dispose of livestock carcasses in natural areas, so dumping is done illegally in fields (Yamaç and Günyel 2010). In addition, there has been a recent decline in cattle farming (TÜIK 2010). There are virtually no alternatives to domestic livestock.

Moreno-Opo et al. (2010) also found extensive dependence on livestock among Cinereous vultures in Spain, and went further by examining what factors attracted this species to carcasses (see also Costillo et al. 2007b). The results showed that the number of vultures attending a carcass was related to the biomass and type of meat. The Cinereous vultures were attracted to carcasses with 'individual, medium-sized muscular pieces and small peripheral scraps of meat and tendon' (Moreno-Opo et al. 2010: 25); large morsels from large animals, and 'the carcasses of wild ungulates (natural and non-natural mortality through hunting practices)' (see also Donazar et al. 2009a,b). In the past, there was the traditional muladar, where carcasses of livestock were placed for the scavengers to remove. The food of the Cinereous vulture was in the past rabbits, sheep and wild ungulates, and also livestock (see Hiraldo 1976; Corbacho et al. 2007). Due to the environmental changes over the past three decades, the rabbit component of their diet has declined, and the livestock component has increased (see also Corbacho et al. 2007; Costillo et al. 2007b). The Cinereous vultures in the study were attracted by scattered, small- or medium-sized remains and large carcasses. At the former, the Cinereous vultures fed faster than at the latter. This favored communal feeders such as the Griffon vulture (Moreno-Opo et al. 2010). Therefore, both vulture species feed on livestock sized carcasses, but possibly the Griffon favors larger bovid carcasses.

In Africa, as was discussed in Chapters 1 and 2, and Part 2, vultures are primarily denizens of the savanna and are largely absent from the forests; the only common exception is the small Hooded vulture which may occur in forested towns such as Kumasi in Ghana (Campbell 2009). As described by Houston (1985: 856) 'The Old World vultures are confined to open habitats such as savannas, grasslands, and semideserts, and none of these species is found in any of the forested areas of Africa or Asia.' In the savannas of Africa, domestic cattle are a principal food source, especially in West Africa where there are fewer larger wild ungulates than in East Africa. This cattle herding culture in the savannas of West Africa enables the larger vultures to be present in large numbers (Scholte 1998; Wacher et al. 2013).

Cattle herding has been practiced in Africa for thousands of years. Most cattle are descended from *Bos taurus indicus* (Linnaeus, 1758), the humped Zebu from South Asia (Deshler 1963). Fossil, archaeological, historical and social sources have been used to uncover the origins of West African cattle (Doutressoulle 1947; Epstein 1971; Smith 1980; Muzzolini 1983; Epstein and Mason 1984; Shaw and Hoste 1987; Clutton-Brock 1989; Blench 1993). Traditional cattle herding, or pastoralism, was a subsistence lifestyle based on tending livestock herds, which produced meat, milk, blood, manure and traction. Traditionally, the nomadic pastoralists, mostly of the Fulani tribes, moved north into the Sahel and northern Sudan savanna during the rainy season, and southwards to the southern Sudan savanna during the dry season (Campbell 1998). An important consideration for cattle rearing in West Africa is the disease trypanosomiasis that kills animals and people, caused by protozoa of the species *Trypanosoma brucei* and carried by the tsetse fly (Murray et al. 1982). The protozoa lives in the blood of the animals and spreads with the bite of the fly. Tsetse flies include all the species of the genus Glossina, in the family, Glossinidae. Species that transmit the disease include *Glossina morsitans*, *G. swynnertoni*, *G. pallidipes*, *G. palpalis*, *G. actinides* and *G. fuscipes*. This disease is largely found in moist areas, such as the forests of central Africa, which is a major factor for the absence of cattle herding in the forests.

Some cattle breeds are more tolerant of trypanosomiasis than others. Trypanotolerant cattle breeds are the Hamitic Longhorns (N'Dama) and the Shorthorns. The N'Dama, indigenous to the Fouta-Djallon highlands of Guinea, is the main breed in western West Africa. Domesticated about 8,000 years ago, it acquired resistance towards trypanosomiasis (i.e., trypanotolerant) and other local diseases (Foy 1911; Chandler 1952; Desowitz 1959; Murray et al. 1982). Another breed, the Zebu, is found largely to the east of the N'Dama along the Sudano-Sahel belt. It has been crossbred with some other local breeds. Unlike the N'Dama, it is often affected and sometimes killed by tranpanosomiasis (Campbell 1998; Dwinger et al. 1992). Shorthorn cattle, also trypanotolerant are also present to the south of the range of the Zebu in the Guinea savanna belt. There are two variants. The larger savanna type is found in the Guinean or Sudano-Guinean savannas from Côte d'Ivoire to Cameroon; and the smaller Dwarf (Forest) Shorthorn is found in forested and coastal areas. Goats and sheep are also found in cattle rearing and farming areas. The historic breeds are the West African Dwarf goats (*Capra aegagus hircus* Linnaeus, 1758) and sheep (*Ovis aries* Linnaeus, 1758), which are mostly trypanotolerant (Adeoye 1984). Being fast breeders, these were important protein and supplementry income sources for the shifting cultivators and nomadic herders (Campbell 1998).

In the west-central African savanna of Chad and Niger, Wacher et al. (2013) found associations between Rüppell's Griffon vultures (the commonest), Lappet-faced vultures, White-headed vultures, White-backed vultures, Hooded vultures and Egyptian vultures and the density of livestock. In most parts of the study area, livestock outnumbered the wild ungulates; in a few areas wildlife reached similar density, mostly Dorcas Gazelle and a few Dama Gazelle, Barbary Sheep and Addax. The landcover in this area was of flat plains, including Sahelian grasslands and the light woodlands of the northern Sudan savanna.

In terms of individual vulture species presence, the Egyptian vultures in this study were present near rocky, desert outcrops and grasslands, in association with cattle herders, cattle near Hyphaene palms and livestock carcasses. The Hooded vultures were found in the southernmost areas, in association with more watered Sudan savanna, and close to human settlements and abattoirs. White-backed vultures were also found in the southern wetter, Sudan savanna, usually in association with livestock, and with other vultures mostly Rüppell's vultures and to a lesser extent Hooded and Lappet-faced vultures. They also moved north into the dryer Sahel during the rainy season. The Rüppell's Griffon vulture was found and associated with livestock throughout the Sahel and Sudan savannas under consideration, nesting in rocky inselbergs and trees.

This study points out that the close association between vultures and humans in the Sahel is due to 'high reliance on livestock carcasses.' Vultures were recorded at carcasses 21 times; the carcasses were those of cows (4), sheep or goats (6), camels (7), donkeys (3) and dorcas gazelle (1). The authors corrected the carcasses for body size differences by using the normal live weights of adults of each mammal species and concluded that 'livestock potentially contributed more than 99% by mass to vulture food sources' (Wacher et al. 2013: 197).

Hooded vultures in southern Ghana occur in close association with urban abattoirs and other meat processing areas, with little association with cattle herding in the coastal savanna. The cattle herds in the coastal savanna are small, with few carcasses for vulture consumption (Campbell 2009). Therefore, they congregate to eat the discards from abattoirs and meat markets in urban areas. The large wildlife of the coastal savanna of Ghana are largely extinct. The Hooded vulture is the only vulture species in this savanna, as the larger vultures such as the White-backed vulture and Rüppell's Griffon are confined to the northern savannas were ungulates are more common and cattle herds are more numerous.

Vultures, by eating cattle and other livestock carcasses, may also benefit people. The beneficial impact of vulture presence is most clearly seen when vultures are locally extinct, usually leading to increased but usually less effective scavenging by facultative scavengers, increased incidence of disease from the larger numbers of rotting carcasses, and possibly changes in human cultures that had adapted to vulture roles. For example, in the local extinction of vultures in parts of Asia, which will be examined in the next chapter in more detail, Pain et al. (2003) noted several unpleasant effects of vulture absence from areas where they were previously very common. Firstly, there was a great increase of small, predatory mammals, e.g., feral dogs and rats (*Rattus* spp.), which have high reproductive abilities. An example is cited of a carcass dump in western Rajasthan, India, where the dogs increased from about 60 in 1992 to 1200 in 2000. These created problems for nearby human settlements, increasing dog attacks on people, infectious diseases, livestock, and wildlife. Diseases, such as rabies and bubonic plague, are endemic and common in India and dogs and rats are the primary reservoirs. India has one of the highest rates of human deaths due to rabies in the world (World Health Organization 1998). Diseases that may spread from dogs and rats to wildlife and domestic livestock include canine distemper virus, canine parvovirus, and *Leptospira* spp. bacteria. As most facultative scavengers are predators, there may be higher predation pressure on wildlife when these scavengers replace vultures.

7.6 Vultures and Hunting

This section looks at the relationship between human hunting of wildlife and vultures. This concerns both the provision of carcasses and possible externalities. Human hunting of wild animals may have both positive and negative effects on scavengers; it may provide carcasses and kill off predatory competitors, but may also kill off the species that provide the carcasses in the first place (Mateo-Tomás and Olea 2010). A major problem concerning hunting is lead poisoning. For vultures, exposure to lead from spent ammunition in large game carcasses may lead to lead toxicity (García-Fernández et al. 2005; Cade 2007; Dobrowolska and Melosik 2008). This lead poisoning may have important effects at both individual and population levels (Gangoso et al. 2009).

Hunting of animals by people has a negative effect on biodiversity, which may impact vultures as well (Mateo-Tomás and Olea 2010). Impacts on predators may have serious impacts. Impacts may include the decimation of prey species (Lozano et al. 2007), the killing of predators and a consequent decline in carcasses they provide (Thirgood et al. 2000; Valkama et al. 2005) and animal behavior change (Casas et al. 2009). Nunez-Iturri et al. (2008) write that firearm hunting decimates primates that disperse the seeds of large-seeded trees; this reduction in primate affects the character of the forest in the case study in Peru. Effects on game species include overexploitation and possible extinction (Lindsey et al. 2007), threats to genetic diversity (Blanco-Aguiar et al. 2008) and possible modification of their behavior (Benhaiem et al. 2008). Mateo-Tomás and Olea (2010: 520) also point out that 'relatively few studies have examined positive effects of hunting on wildlife and/or ecosystem conservation' (see also Thirgood et al. 2000; Baker 1997; Lindsey et al. 2006, 2007; White et al. 2008).

In the United States, hunting remains an important attractant for obligate scavengers such as vultures and facultative scavengers such as coyotes (Smith 2013). In Alabama, where more than 447,000 hunters move into the woods annually for a deer kill, it is estimated the statewide average harvest is about 1.6 deer per hunter. Although most hunters dispose of the carcasses in a manner that limits odor, disease possibilities and environmental contamination, others do not do so. Unsuitable dumping pollutes water bodies, in fields and by roads (Smith et al. 2013).

Mateo-Tomás and Olea (2010: 520) further point out that although large animals are the principal food of Old World vultures and information on this topic is crucial for the current vulture population decline, 'To our knowledge, the influence that hunting could have in the ecology and conservation of this scavenger guild has not been assessed', with only passing references (see Sekercioglu et al. 2004; García-Fernández et al. 2005; Cade 2007; Murn and Anderson 2008). Mateo-Tomás and Olea (2010) therefore argue that considering the food sources favored by vultures, hunting of the animals whose carcasses they eat might be relevant to vulture populations. Their case study was of the Griffon vulture and its relations with hunting in a part of Spain with no intensive farming or vulture-feeding stations. This allowed an examination of hunting as the main resource activity (Camiña and Montelío 2006; Junta de Castilla y León 2006).

The study showed that Griffon vultures were highly dependent on shot ungulates, mainly red deer (*Cervus elaphus* Linnaeus, 1758) and wild boar (*Sus scrofa* Linnaeus, 1758). The vultures in the study area numbered about 350 breeding pairs and 1,100 others, and the results showed that the products of hunting could feed about 1,807 during the 5–6 hunting season. More vultures were recorded feeding of red deer and wild boar (69%) than on livestock (31%). Hunting could therefore play a role in maintaining the vulture population; as it would be the main food source during winter, outside the breeding season (Elosegui 1989; Olea and Mateo-Tomás 2009).

Vultures may also be hunted for food or religious use. For example, the decline of the Hooded vulture in East, West and Southern Africa, is mostly due to its over-exploitation for food and traditional medicines predominantly in West Africa (Anderson 1999; Sodeinde and Soewu 1999; Ogada and Buij 2011)). It is now rare in the southwest of Burkina Faso (where it is commonly used for traditional medicines and sorcery is common and it is routine to see their body parts on sale at markets), Mali and Niger (Thiollay 2006a). In Nigeria, the Hooded vulture was the most frequently traded bird for traditional medicine (Sodeinde and Soewu 1999). The head is used to protect against witches and the whole body may be used for good fortune (Sodeinde and Soewu 1999). Large vultures are more desired but more difficult to obtain (Nikolaus 2001). They are also eaten (Rondeau and Thiollay 2004).

7.7 Conclusions

This chapter has examined the relationship between vultures and human cultural patterns such as farming, hunting and urbanization. For urbanization, it is evident that there is a positive effect for some smaller species, and a negative effect for larger species. From a conservation perspective this has serious implications for the more threatened and larger species, such as the condors and larger Old World vultures (e.g., Lappet-faced, White-headed and the Gyps species). The smaller vultures, such as the Black and Turkey vultures in the New World, and the Hooded and Egyptian vultures in the Old World which are attracted to urban unsanitary places and abattoirs relay on the lack on sanitary procedures; improved sanitary management may spell their doom, especially if they have lost the acumen for foraging in natural habitat. For collisions, the problems are pronounced, for there is currently no clear solution beyond minimal improvements from changed designs and strategies for distracting vultures from the relevant localities. For farming and hunting, the issues are similar for both New World and Old World vultures, and cluster around three main issues: increased dependence on livestock carcasses, improved sanitation and ingestion of impurities. The attraction of vultures to livestock carcasses, though ancient, is a manifestation of a larger problem; the decline of natural species, partly due to over hunting and also due to habitat destruction. The next chapter will look at the diseases that have affected vultures.

Vultures, Chemicals and Diseases

8 INTRODUCTION

This chapter examines the chemicals and diseases that have affected vultures. Poison, whether lead, pesticide or veterinary, has emerged as possibly the most important contemporary determinant of vulture futures, pushing some species to near extinction. Major problems are indirect poisoning, from pesticides in carcasses, lead poisoning from shot animals, and veterinary chemical products such as diclofenac. These issues are important enough to merit their own chapter, especially with the recent devastating effect of the diclofenac drug on vultures in Asia. The occurrence of these effects are examined in Asia, where the effect was greatest, and in Africa, Europe and the Americas. Old World vultures have been affected much more than their New World counterparts. Therefore, the chapter also reviews the facts presented in the earlier chapters concerning the physical and behavioral characteristics and differences of Old and New World vultures and their ecosystems. It also examines public attitudes to the vulture diseases and extinction possibilities.

8.1 Poisons and other Diseases that Affect Vultures

Vultures are highly vulnerable to a range of threats, largely due to their specialized lifestyles. Avian specialization has been described as a possible factor for increased extinction risk (Sekercioglu et al. 2004). 'Avian scavengers have the highest percentage of extinction-prone species among avian functional groups' (Ogada et al. 2012: 453; see also Sekercioglu et al. 2004). Vulture declines are blamed on poisoning (Pain et al. 2003; Hernandez and Margalida 2008), food scarcity, hunting, and land-use changes (Rondeau and Thiollay 2004). Vultures are biologically specialized to deal with certain pathogens, as their stomach acids can detoxify bacterial toxins such as anthrax, reducing the risk of disease spread (Houston and Cooper 1975).

One important problem is ingestion of artificial products, such as plastic products (termed junk) discarded in the environment, which may be mistaken by vultures for bone fragments. When swallowed, such items may cause 'choking, poisoning, intestinal obstruction, malnutrition and death' (BirdLife International 2008c). Research shows that vultures and condors actively scavenge for bone fragments for both calcium needs and food pellet regurgitation, a habit that can influence them to swallow undesirable items that may harm them. Nestlings are at greater risk because they cannot regurgitate the indigestible objects, and may thus be subjected to choking, poisoning, intestinal obstruction and malnutrition (Ferro 2000). For example, for the reintroduced California condor, 'junk-induced nestling mortality is seriously threatening the re-establishment' of this species. One study found up to 650 junk items in nests, including glass, metal and plastics (Mee et al. 2007).

This is similar to the lead poisoning that brought the species to near extinction before 1987 (Cade 2007). Records indicate that six of the eight nestlings that either died in the nest or where taken from the wild since 2002 had swallowed pieces of plastic piping, cloth and rubber, and also glass, metal bottle-tops and even ammunition cartridges. Impacts on the nestlings included zinc poisoning, retarded feather development resulting from malnutrition (this caused by distended and blocked digestive passages), and in some cases, death (Mee et al. 2007). An important factor is the increased level of such discards in the environment since the reintroduction of the California condor, usually near urban areas (Mee et al. 2007).

For other vulture species, examples are the nests of the White-rumped vulture in Pakistan which frequently contain bits of china, metal and glass, possibly swallowed by nestlings. Nestling mortality from swallowing metal objects has also been cited as a factor behind the low breeding success of Griffon vultures in Israel and Armenia (Ferro 2000; Houston et al. 2007; BirdLife International 2008).

Lead poisoning from bullets in carcasses is dangerous to vultures (García-Fernández et al. 2005; Cade 2007; Dobrowolska and Melosik 2008; Gangoso et al. 2009). Mateo-Tomás and Olea (2010) note that hunting may also have negative effects; with lead poisoning being the most important. Therefore, it is recommended that the ingestion of lead is monitored for deleterious effects (Patte and Hennes 1983; Pain and Amiard-Triquet 1993; Miller et al. 2000).

Turkey vultures (*Cathartes aura*) show marked individual variation in lead poisoning. Symptoms of lead poisoning include weakness and lack of coordination. High levels may result in lead toxicosis (Carpenter et al. 2003).

One study of hunting and its impacts on Turkey vultures in California, noted that although lead ammunition has been banned from waterfowl hunting in North America for nearly 20 years, lead ammunition is still widely used for big and small game animal hunting (Kelly and Johnson 2011). The lead exposure in turkey vultures was found to be significantly higher during the deer hunting season compared to the off-season. The lead exposure of the vultures was also positively correlated with increased intensity of wild pig hunting.

Lead-related mortality was a factor for the decline of the California condor population during the 1980s. This also affected the precarious, reintroduced wild population, and some condors required medical attention (Snyder and Snyder 2000; Kelly and Johnson 2011). Several studies have found high levels of lead in the blood of condors (Pattee et al. 1990; Hall et al. 2007; Hunt et al. 2007; Sorensen et al. 2007), although the high blood concentrations have also been found in condors outside the hunting season (Pattee et al. 1990). The impact on condors and the fact that the condor is an endangered, highly observed species resulted in major actions that culminated in the ban on the use of lead ammunition 2008 for most hunting activities in the condor's California range. There was also input from groups desirous of similar bans to protect other species and those who question the links between lead exposure in wildlife and hunting activities. Higher blood levels of lead were also found in other species, such as golden eagles within the condor's range (Pattee et al. 1990) and in ravens in the Greater Yellowstone Area, the latter especially during the big game hunting season (Craighead et al. 2009).

The poisoning of Turkey vultures through lead has also been reported in Canada (Clark and Scheuhammer 2003; Martin et al. 2008). In the former study, 184 individual raptors from 16 species were found dead across Canada. One Turkey vulture had 'highly elevated bone-lead concentration' (Clark and Scheuhammer 2003: 23). Other affected species included Red-tailed hawks, Great horned owls, and Golden eagles (131 of the 184 specimens) and 3 to 4% of the deaths were attributed to lead poisoning. The evidence pointed to dietary reliance on game birds and mammals killed by lead projectiles from ammunition used in upland hunting. In the study by Martin et al. (2008), 225 individual dead birds from 19 species of terrestrial raptors in southern Ontario were dissected for analysis of bone, kidney and liver tissues. Turkey vultures were found to have the 'highest mean concentrations of lead in bone and kidney compared to other raptor species'; although only one bird was found to have lead levels described as acute. There were also levels above the subclinal level in four others (Martin et al. 2008: 96). The continued use of lead shot for upland hunting is a primary source of lead and a continued risk to the raptors and scavengers.

Old World vultures are also affected. In a Spanish study, vultures were found to be strongly affected by spent lead ammunition in carcasses (Rodriguez-Ramos et al. 2009: 235). 'This exposure may have increased after the ban on abandoning carcasses of domestic ruminants in the field due to the bovine spongiform encephalitis (BSE) crisis, both because the vultures consume hunting bag residues more frequently and because malnutrition may lead to mobilisation of lead stores' (Rodriguez-Ramos et al. 2009: 235; see also Iñigo and Atienza 2007). There is little information on the potential correlation of blood lead levels and clinical signs and potential subclinical effects of lead in vultures despite the evidence of clinical intoxication in Griffon vultures (*Gyps fulvus*) and Cinereous vultures, and other species (Mateo et al. 1997; Mateo et al. 2003; Hernandez and Margalida 2008). Rodriguez-Ramos et al. (2009) conclude that the high blood lead levels were derived only from the ingestion of spent lead ammunition (see also Garcia-Fernandez et al. 2005). The clinical symptoms of the lead toxicosis included disorientation, ataxia and impaired landing, posterior paresis, and severe hypochromic anemia. The higher lead levels were in birds admitted to the rehabilitation centers from mid-August to mid-February; correlated with the game hunting seasons.

In a study by Donazar et al. (2008: 89) on Egyptian vultures in the Canary Islands, blood samples showed high frequencies of lead poisoning; sub-clinical and clinical intoxication levels; 'these were probably caused by the ingestion of lead shot' (ibid.). The study found that 16% of birds had >0.2 ppm of Pb, these levels were high enough to cause declines in productivity (Ochiai et al. 1992; Burger 1995), physiological injuries and possibly death. Although the lead may be derived from scavenging on bird carcasses containing lead, in some cases some of the lead shot are rejected within pellets, but this is difficult to detect without a radiological study of the pellets (see Medina 1999, for another study of Egyptian vultures and Mateo 1998). It is suggested that lead pellets should be replaced by bullets of steel or molybdenum/tungsten alloys for hunting.

Another example was of vultures interacting with illegal elephant hunters in Namibia (Salisbury 2013). Illegal poaching of African elephants and White and Black rhinoceroses has increased, and carcasses are poisoned to kill vultures, because the large flocks of attendant vultures reveal carcass locations. In a recent (2013) incident, 600 vultures are reported to have died when they fed on a poisoned elephant carcass near Namibia's Bwabwata National Park. Salisbury (2013) quotes Leo Niskanen, Technical Coordinator, IUCN Conservation Areas and Species Diversity Programme; 'By poisoning carcasses, poachers hope to eradicate vultures from an area where they operate and thereby escape detection. The fact that incidents such as these can be linked to the rampant poaching of elephants in Africa is a serious concern. Similar incidents have been recorded in Tanzania,

Mozambique, Zimbabwe, Botswana and Zambia in recent years' (ibid.). In the cited incident, the vultures were incinerated before detailed study, so the details were not conclusive. However, for two of the birds, the tags indicated that the birds originated from Kimberley in South Africa, about 1000 km distant from the place of their death.

Pesticide poisoning is another danger for vultures. For example, Hernández and Margalida (2008), found incidents of pesticide poisoning of the Cinereous vulture (*Aegypius monachus*) in Spain during the period 1990–2006. Two hundred and forty one incidents related to 464 vultures were investigated. The possible classifications of the pesticide use were: approved use, misuse, or deliberate abuse. The method of application, spatial and temporal variation and reasons for the pesticide abuse were also investigated. Up to 98% of the incidents were intentional poisoning. Approved use was responsible for only a minor fraction (1.3%). Pesticide mortality affected vultures (83%). Three chemical compounds carbofuran, aldicarb, and strychnine, accounted for up to 88% of the cases, but eight others were also used. The illegal control of predators was the main factor for application. The Cinereous vultures were not seriously threatened, but the availability of highly toxic pesticides could increase illegal usage with more serious effects. The authors suggest that the elimination of the few frequently used compounds would benefit vultures and other wildlife without seriously affecting agriculture.

Agricultural pesticide poisoning is also serious in South East Asia. A study by Clements et al. (2012) found that vulture populations in Southeast Asia declined before the calamitous diclofenac episode in South Asia (Pakistan, India and Nepal). Common SE Asian vultures were the White-rumped, Slender-billed and Red-headed vultures, which ranged from southern China southwards into Myanmar, Thailand, and Peninsular Malaysia (Pain et al. 2003). By the 1980s, however, these species were either depleted or extirpated from southern China, Thailand and Malaysia (Round and Chantrasmi 1985; Round 1988; Zheng Guangmei and Wang Qishan 1998; Wells 1999). After the end of military conflicts in Cambodia, Lao People's Democratic Republic (PDR) and Vietnam, the vultures were restricted to areas bordering Cambodia in Lao PDR and Vietnam (Barzen 1995; Desai and Lic Vuthy 1996; Le Xuan Canh et al. 1997; Brickle et al. 1998; Timmins and Men 1998; Thewlis et al. 1998; Goes 1999; Long et al. 2000; Timmins and Ou 2001; Eames et al. 2004). It is speculated that the decline in vulture populations was due to the decline in numbers of wild ungulates, changes in husbandry of domestic stock and direct persecution.

The three species (White-rumped, Slender-billed and Red-headed vultures) were found only in Myanmar (Hla et al. 2011) and northern and eastern Cambodia by the 2000s (Hla et al. 2011), despite evidence that they had not been affected by diclofenac. Clements et al. (2012) argue that as these Cambodian birds represent one of the few populations of the three species outside South Asia where diclofenac was not used, their conservation is crucial. The populations of each of the three species are estimated at 50–200+ individuals.

Mortality was inferred to be the result of accidental poisoning, from strychnine or organophosphates from hunting and fishing (73%) normally used in agriculture and hunting or capture for traditional Khmer medicine (15%) (Timmins et al. 2003). Most poisoning incidents were intended to kill scavenging species (e.g., storks) and frugivores (fruit-eating animals), and also feral dogs, but vultures became the unintended victims (Cheke 1972). Killings using guns also occurred, as in the 1970s–1990s, during and just after the Kampuchean and Vietnamese wars, firearms were common (Wille 2006). However, the number of guns declined from the late 1990s (EU ASAC 2009). Some vultures also were affected by the HPAI H5N1 virus and at least one bird died of this ailment, but a survey found that some vultures carry antibodies to the avian encephalomyelitis virus (a picornavirus causing neural deficits in young poultry) (Gerlach 1994; Clements et al. 2012).

Deliberate and accidental poisoning contribute to the decline of Hooded vultures in Africa (Ogada and Buij 2011). Steep declines have been recorded in Cameroon, Uganda and Kenya and Southern Africa, largely through poisoning (Zimmerman et al. 1996; Mundy 1997; Thiollay 2001; Ogada et al. 2010; Ssemmanda and Pomeroy 2010; Virani et al. 2011). Poachers poison vultures in Botswana, in the belief that the vultures will reveal their illegal kills (Hancock 2009a, 2010). In Namibia, vultures are principally killed by poisons intended for problem mammals such as lions, hyenas and jackals (Bridgeford 2004). This also true of other parts of southern Africa (Anderson 1999; Verdoorn et al. 2004; Roche 2006). Pain et al. (2003) also report deliberate poisoning in southern Africa. Poisoning campaigns were intended to eliminate scavenging birds and eagles from the farming areas in Namibia in the 1980s. Where there was no poisoning in the nearby national parks of Kalahari Gemsbok and Etosha, the species remained at the previous levels (Mundy et al. 1992).

In another study in South Africa, van Wyk et al. (2001: 243) examined 'whole blood, clotted blood, heart, kidney, liver and muscle samples' from White-backed, Cape and Lappet-faced vultures and found 'the presence of quantifiable residues of 14 persistent chlorinated hydrocarbon pollutants.' They also compared the pesticide levels between nestlings from natural breeding colonies, adults from a wildlife area and also captive birds. The researchers found statistically significant differences between these populations with lower amounts for the captive birds for certain chemicals in pesticides: gamma-BHC (lindane), alpha(cis)-chlordane and alpha-endosulfan. The researchers acknowledged that the 'respective biocides measured in vulture samples were generally low in comparison to results documented for a number of avian species' and 'although no threat is posed by any of the organochloride pesticides' there should be continuous monitoring of breeding colonies and White-backed vulture nestlings may be used as bioindicators (ibid. 243).

Among the New World vultures, disease rarely contributes to serious declines in population. Kiff (2000) reports that populations of Turkey and Black vultures are even spreading northwards from the United States into Canada. Snyder and Snyder (1991) note that Turkey vultures are resistant to some natural pathogens, and even some rodenticides such as the compound 1080 (Sodium fluoroacetate). However, other poisons, such as cyanide and strychnine, which were formerly used to poison coyotes and any other predators of cattle and sheep, are serious threats to the vultures. Wilbur (1978), Kiff et al. (1983) and Coleman and Fraser (1989) report that DDE (dichlorodiphenyldichloroethylene) induced eggshell thinning. Wilbur's report noted 11%, 12% and 18% thinning of eggshells in post DDT (1947) California, Florida and Texas, respectively. Kiff et al. (1983) found thinning of 9.7%, 15.9% and 10.1% in the same states. Coleman and Fraser (1989) also found thinning among Turkey vulture eggs in southeastern and mid-Atlantic states during the period 1947–1964. Kiff (2000) notes that there is no evidence that DDT (dichlorodiphenyltrichloroethaneuse) hampered later expansion of vultures towards the northern United States, despite some evident population reductions in the southern states.

8.2 The Diclofenac Epidemic

Another major problem, probably the most serious for vultures in history and prehistory concerns the diclofenac issue (Woodford et al. 2008). Van Dooren (2011) notes that today, six of the 23 recognized vulture species are threatened or critically endangered on the ICUN Red List of Threatened Species, with the largest declines being of five species in India. This event has been termed the 'global vulture crises' (Olea and Mateo Thomas 2009). Some vultures in India have declined by as much as 90–99% of their previous numbers, approaching local, regional and possibly even global extinction. Prior to this catastrophic event, Griffon vultures were very common throughout South and South-East Asia. Seven species of vultures are resident in South Asia, and three of these seven—the White-rumped Vulture, Slender-billed vulture and Red-headed vulture—are also resident of South-East Asia (Houston 1985; Clements et al. 2012).The White-rumped vulture was considered one of the most abundant large birds of prey in the world. Vulture populations had declined marginally in much of Asia in the first half of the 20th century, but were still common on the Indian subcontinent, largely supported by a large supply of livestock carcasses.

Records of the sudden deaths of vultures were first made in India, in studies of Gyps vultures in Keoladeo National Park, Rajasthan, India (Prakash 1999; Gilbert et al. 2004).

This study was supported by later research across India (Prakash and Rahmani 1999; Prakash et al. 2003) and Nepal and Pakistan (Virani et al. 2001; Gilbert et al. 2002). In the late 1990s, it was evident to observers that the Indian populations of White-rumped vulture, Indian vulture and Slender-billed vulture had totally crashed, with similar declines also observed in Nepal and Pakistan. The decline of the vultures was also noted in the media (BBC 2004; PIB 2005; Gill 2009).

At first three species were affected, the White-rumped vulture, Long-billed vulture and Slender-billed vulture. The decline, which took place between 1993 and 2000, led to the classification of these vultures in 2001 as Critically Endangered (BirdLife International 2001). Over 15 years (1992 to 2007) the population of the White-rumped vulture was one thousandth of the previous population, that of the Red-headed vulture was less one tenth and that of the Slender-billed vulture was reduced to about 1,000 birds (Cuthbert et al. 2006; Prakash et al. 2007).

However, at that time Gilbert et al. (2004) argued that 'attempts to quantify the Asian vulture decline have been complicated by the lack of pre-decline data on population size in these formerly abundant species'. A noted exception was the study by Galushin 1971, and that of Prakash 1999. Nevertheless, the evidence until at least until 2008, was that the populations were still in decline (Green et al. 2004; Gilbert et al. 2006). The White-rumped vulture, formerly possibly the commonest vulture in Asia was particularly affected, as it declined by 99.9% since 1992 (Prakash et al. 2007). Non Gyps vultures were also affected, as the Egyptian vulture and Red-headed vulture were classified as Endangered and Critically Endangered respectively (Cuthbert et al. 2006; BirdLife International 2008).

The deaths of millions of vultures on the Indian subcontinent had a marked impact on the natural ecosystems and human quality of life. Facultative scavengers, mostly feral dogs (*Canis lupus familiaris*) greatly increased, but they were unable to eliminate carcasses with the same efficiency as the vultures. Large numbers of decaying carcasses littered the landscape contributing to diseases. Large numbers of dogs were attracted to the carcasses, leading to the incidence of rabies that affected thousands of people (Swan et al. 2006; Gill 2009). Gross (2006) points out that if rats increase, bubonic plague and other rodent-transmitted diseases may also increase. For the sky burials, in which Zoroastrian Parsis and Tibetans left their dead on platforms for vultures to eat, in the past hundreds of vultures could clean a corpse in less than a hour. With local extinction of vultures, alternatives had to be found.

Before the diclofenac identification was made, Pain et al. (2003) investigated all the possible causes of the population decline. This study noted that during the 1970s and 1980s, Gyps vultures were rare or absent in much of Southeast Asia, but *G. bengalensis* and also *G. indicus*, were very common in India, especially in and near urban centers. Their abundance was principally because of the religious prohibition of killing cattle in northern and central states of India. This resulted in the dumping of large numbers of cattle carcasses in rural landscapes (Grubh et al. 1990). The vulture populations grew so dense, they were an aircraft risk (Grubh et al. 1990). In Southeast Asia (Cambodia, Laos, Vietnam, Malaysia, Thailand, and even Yunnan Province, China), the vultures were much rarer (see also Timmins and Ou Ratanak 2001). Several reasons were advanced for the low populations and/or declines in southeast Asia.

Hunting was blamed for the decimation of wild ungulates that provided vultures with food (Srikosamatara and Suteethorn 1995; Duckworth et al. 1999; Hilton-Taylor 2000). In addition, changes in husbandry of domestic stock contributed to the reduction of food sources (Cambodian Wetland Team 2001). 'It seems likely that food supplies are no longer predictable enough to allow regular breeding' (Pain et al. 2003: 661).

Habitat loss was seen as less important, except for a minor effect on the wild ungulates (see also Thewlis et al. 1998). Vultures may also have been killed by people, when the vultures were gorged with food on the ground (Thewlis et al. 1998). Pain et al. (2003: 661) wrote that 'The role of agrochemicals remains unclear; there is no persuasive indication that they can explain region-wide losses in Southeast Asia, although they may have caused local declines' (see for example, the dated example of Cheke 1972). The role of infectious diseases was also ruled out, or as 'circumstantial evidence.' The decline of raptors, which was extreme in Laos, also affected other scavenging birds, e.g., such as the Greater Adjutant (*Leptoptilos dubius*, Gmelin 1789), Black kite (*Milvus migrans* Boddaert, 1783), Brahminy kite (*Haliastur indus* Boddaert, 1783), and Large-billed crow (*Corvus macrorhynchos* Wagler, 1827) (Lekagul and Round 1991; Thewlis et al. 1998; Wells 1999; Duckworth et al. 1999, 2002; Round 2000).

Laos, with some of most extreme declines, also had large areas of suitable habitat for vultures, with low human population densities, and has little evidence of environment—contaminating chemicals. The decline in open-country wild ungulates during the late 20th century from hunting and changes in livestock husbandry were the only known factors for vulture decline (Duckworth et al. 1999). Cambodia, with Gyps populations had large areas of open landscape, sparse settlements, wild ungulate populations and extensive free-ranging of domestic cattle (Timmins and Ou Ratanak 2001). Therefore, despite the lack of solid evidence for the reasons for the vulture declines in Southeast Asia, one initial explanation was that 'food shortage appears to be the most credible general explanation, although other factors including persecution and contaminants may have played a part locally' (Pain et al. 2003: 662).

India was differentiated from Southeast Asia, because of the high provision of cattle carcasses. Nevertheless, there were drastic declines of *G. bengalensis* and also *G. indicus* in India, for example in the study area of Keoladeo National Park where declines were about 95% between the mid-1980s and late 1990s (Pain et al. 2003; see also Prakash 1999; Prakash et al. 2003). The reported reductions were the result of very high mortality of both adults and juveniles. Sick birds were described as having a slumped posture after which death occurred. At this time, the cause was unknown, as Pain et al. (2003: 664) report the 'The population decline and high mortality was unexplained' even though there was no alteration of the food supply. The Bombay Natural History Society conducted surveys in 2000, on the model of earlier surveys in 1991–1993 covering states in north, west, and east India. The lowest declines of the Gyps vultures were of *G. bengalensis* and *G. indicus*; 96% and 92%, respectively. Note that at the time *G. indicus* was a combination of two currently recognized species, the Indian vulture (*G. indicus*) and the Slender-billed vulture (*G. tenuirostris*) (Rasmussen and Parry 2000). Other scavengers did not decline, and feral dogs increased in population over India (Cunningham et al. 2001). There was also no evidence that starvation was a factor for vulture deaths in both India and Pakistan (Gilbert et al. 2002; Prakash et al. 2003).

Pesticides were also investigated as possible factors for the collapsing populations. Although large amounts of pesticides are utilized in India, no chemical or pesticide was identified that could have impacted vultures so severely. For this to be the case, it would have to be applied across a vast area or used in a new way that increased its contact with vultures within the previous 10–20 years. Numerous toxicological analyses were conducted of vulture carcasses from Pakistan, testing for organochlorines, organophosphates, carbamates, and heavy metals; however none were found at toxic levels (Oaks et al. 2001). There were also no regional patterns and declines in other species. These suggested that contaminant poisoning could not be the sole factor (Pain et al. 2003).

The hypotheses were that the declines were the result of 'a simultaneous subcontinent-wide exposure to a toxic contaminant or a rapid spread of disease through the Gyps vulture population'; with the 'international, cross border nature of the problem hinting at the latter possibility' (ibid. 665). The extinction of colonies of vultures was also seen as possibly due to either vulture mortality or desertion by survivors when the population declined below a certain level. Nevertheless, at this point, prior to 2004, some studies of dead birds began to notice renal and visceral gout (crystallization of uric acid in the tissues) in most of them and enteritis in many of the Indian birds (Pain et al. 2002) and *G. bengalensis* from Pakistan (Oaks et al. 2001; Gilbert et al. 2002).

Visceral gout in most birds indicated similarity in the death factor. As the gout was acute, appearing a few hours before the bird died, it was hypothesized to be the result of another disease rather than the main disease (Pain et al. 2003). This evidence at the time was seen as indicating an infectious disease. 'Although we cannot be certain that an infectious disease is responsible until a causal agent has been identified, this is currently the most tenable hypothesis, so it is important to consider the implications of this explanation' (Pain et al. 2003: 666).

This infectious disease could have spread to all the eight Gyps species due to the social nature of the Gyps species in breeding, roosting and feeding, the very wide foraging ranges, the lack of geographical isolation of any Gyps species, and migratory behavior between Europe, Africa and Asia, sometimes through areas where the disease is prevalent (Houston 1974, 1983; del Hoyo et al. 1994; Griesinger 1996; Ferguson Lees and Christie 2000; Prakash et al. 2003). The spread could be from the three affected Gyps species (*G. bengalensis, G. indicus, G. tenuirostris*), through *Gyps fulvus* and *Gyps himalayensis*. Pain et al. (2003: 666) gave a detailed list of potential routes which we describe below.

The first routes of disease transmission would be (A) west through southern Iran into the Zagros mountains (*G. fulvus*); (B) northwest from Afghanistan and northern Iran to the Caucasus (*G. fulvus*); (C) north through the Pamir Knot and to the Tien-Shan of the former Soviet Union (*G. fulvus* and *G. himalayensis*); and (D) northeast from the Himalayas onto the Tibetan Plateau (*G. himalayensis*). A fifth route (E) would be across the Strait of Hormuz from southern Iran to the United Arab Emirates (*G. fulvus*).

The second stage of disease spread, through *G. fulvus*, would extend (F) into the southern Alps and Pyrenees in Europe and (G) through Middle Eastern mountains and thence to Ethiopia and sub-Saharan Africa. It would also enter Africa from Jordan and Israel into Egypt and into sub-Saharan Africa, or pass through across Saudi Arabia and Yemen and to Djibouti. Any gaps between Gyps species distributions would be overcome by species mixing, the result of the extreme broad foraging and migratory ranges of the sub-species, up to 1000 km (Houston 1974, 1976; Ferguson Lees and Christie 2000). However, at the time of the study by Pain et al. (2003: 666), concerning *Gyps fulvus* it was 'not yet clear, however, whether they are susceptible to the mortality factor affecting other Gyps vultures.' The disease hypothesis already seemed a little weak, as it did not seem to contribute to mortality of Gyps vultures in other parts of the world (Benson 2000).

Soon after this study by Pain et al. (2003), new ground was broken and the 'infectious disease' was identified. It was not an infectious disease. Gross (2006) records that in 2004 scientists from the United States–based Peregrine

Fund found the cause. The role of diclofenac was discerned, despite other possibilities such as reductions in food availability and pesticide poisoning which had also killed many vultures in the past. In the mid-1990s, livestock farmers in India and later in Pakistan and Nepal, began treating their cattle and water buffaloes with the non-steroidal anti-inflammatory drug (NSAID) diclofenac.

Diclofenac, a non-steroidal anti-inflammatory drug, apparently was so toxic to vultures that the deaths could have been due to a miniscule proportion (between 1:130 and 1:760) of livestock carcasses, which exist at carcass dumps, the traditional method of livestock disposal in South Asia (Green et al. 2004; Oaks et al. 2004; Gross 2006). Diclofenac was used widely on the Indian subcontinent from the early 1990s (Green et al. 2004; Oaks et al. 2004). This coincided with the period of vulture deaths. Evidence emerged that vultures and some other scavenging birds were fatally affected after eating the flesh from an animal that died after recent treatment with a veterinary dose of diclofenac (Green et al. 2006; Cuthbert et al. 2007; Green et al. 2007). This would be the case even if a very small proportion of the livestock carcasses had been treated with the drug (Green et al. 2004).

A study by Naidoo (2007) found visceral gout in dead vultures in India, Nepal and Pakistan (see also Cunningham et al. 2001, 2003; Virani et al. 2001; Oaks 2001; Gilbert et al. 2002). Research found that the visceral gout was caused by diclofenac (Oaks et al. 2004). However, visceral gout may occur after several types of infectious and non-infectious diseases that contribute to renal dysfunction. The main medical effect on the birds was renal failure and visceral gout. Vultures do not have an enzyme to break down diclofenac. Hence, they died within 36–58 hours of ingesting toxic doses, which were actually normal veterinary doses in cattle. This disease occurs when kidney failure—severe renal dysfunction—causes a buildup of urates in the internal organs, with resultant anorexia and emaciation. The reduction in renal filtration increases blood uric acid levels (Lumeij 1994). Uric acid precipitates form a chalky-white coating over the visceral organs, especially the liver and heart and this may be deposited in the body tissues.

In 2006, the Drug Controller General of India issued a notification to all the State drug controllers to withdraw the licences for the manufacture of diclofenac within three months (May 11 to August 11, 2006). After the ban on the use of diclofenac, the populations of the Slender-billed vulture in Pakistan increased by up to 52% in 2008 (Chaudhry et al. 2012). In the study, the largest known breeding colony of Slender-billed vultures in Pakistan was monitored before the ban from 2003 to 2006 and also from 2007 to 2012 after the ban. The number of vultures were recorded to have declined by 61% before the ban and increased by 55% after the ban. Similar increases have been reported for Slender-billed vultures in India and White-rumped vultures in Nepal. The report notes that the numbers may grow to the level before the diclofenac epidemic on condition that the adult mortality remains at a low level, more birds are attracted to nesting and there is sufficient food.

In this study, the evidence showed that after the diclofenac ban, the breeding population regained the numbers of 2003, but the nest occupancy and productivity of the birds did not recover as fast. The authors speculate that there was a change in the age structure of the population between 2003 and 2012; the ratio of breeding adults to subadults/juveniles changed from 1.8 to 1.2 in 2012. Therefore, the lower nest occupancy results from the higher proportion of subadults and juveniles. Possibly the adults were more affected because they accessed more of the poisoned meat than the younger birds and hence died in larger numbers (Chaudhry et al. 2012).

These authors also acknowledge that local livestock producers possibly changed carcass disposal methods after the vulture population decline, e.g., using burial or burning as disposal methods. If these methods continued during the recovery of the vulture population, the lowered food supply would result in the lowered habitat carrying capacity.

African vultures have also declined, mostly due to factors other than diclofenac. As pointed out by Swan (2004: 205) 'To date, only diclofenac has been identified as a risk for vultures India and Pakistan..., but diclofenac, as well as other NSAIDs, pose a danger to Gyps vultures across their geographic range. Swan (2004: 205) further notes 'it is not yet known if diclofenac is similarly toxic to other Gyps species (i.e., *G. africanus, G. ruppellii, G. coprotheres, G. fulvus* and *G. himalayensis*). The African White-backed vulture *G. africanus* is considered the nearest relative of the Oriental White-backed vulture (*G. bengalensis*), thus potentially the most likely to be similarly affected.'

Botha et al. (2012) describe African vultures 'as under severe pressure from a range of factors and populations of some species have been in drastic decline over the last 30 years.' They acknowledge the factors for decline in African vultures are more complicated than those for Asian vultures. In West Africa some species have declined by as much as 85% and there is an average decline of about 42% (see also Rondeau and Thiollay 2004). In East Africa as well, the Lappet-faced vulture, Egyptian vulture, African White-backed vulture and Hooded vulture have seriously declined (Ogada and Buij 2011; Virani et al. 2011). However, vulture populations in southern Africa are comparatively stable, at least over the last 20 years (Monadjem et al. 2004). Problems indentified on the horizon include the increased demand for vulture body parts for the juju trade, and veterinary medicines that are lethal to African vultures (Naidoo et al. 2009). However, a major problem with the African context is the lack of monitoring (Anderson 2004). There are few qualified observers, limited funding and logistical challenges (Botha et al. 2012).

An important finding was that Turkey vultures are not affected by diclofenac (Rattner et al. 2008). A toxicological study involving oral applications of diclofenac to selected Turkey vultures found no toxicity, visceral gout, renal necrosis, or elevate plasma uric acid, even at concentrations more than 100 times the estimated average dose that kills Gyps vultures. When the Turkey vultures received 8 or 25 mg/kg of diclofenac, the authors estimated the plasma half-life of diclofenac to be 6 hours, as examination showed it cleared from their systems with no residues in the liver or kidneys. The study recommended further studies of this important finding.

8.3 Solutions for Vulture Diseases

Alternatives to diclofenac have been suggested (Gross 2006). The main objective has been to find an anti-inflammatory drug that could treat livestock without killing vultures. For example, toxicologist Gerry Swan and a team from South Africa, Namibia, India, and the United Kingdom argued for meloxicam treatment as an alternative (Swan et al. 2006). Meloxicam was the only NSAID that gave no evidence of kidney damage in vultures. This compound was tested on the African White-backed vulture, which has similar reactions to diclofenac as the Indian vultures. The results showed statistically significant differences in meloxicam and diclofenac. All vultures with the former lived and all those with the latter from a previous study died. The diclofenac-treated vultures had a marked and dose-dependent elevation of uric acid levels compared to controls; meloxicam-treated vultures showed no such differences. The authors acknowledged the small size of the sample and could not rule out the possibility of risk from the safer drug. A study of a larger number of wild and captive White-backed vultures and also Asian vultures affirmed this result. The conclusions were that 'meloxicam is of low toxicity to Gyps vultures and that its use in place of diclofenac would reduce vulture mortality substantially in the Indian subcontinent. Meloxicam is already available for veterinary use in India' (Swan et al. 2006: 0395).

Another solution is the development of 'Vulture Restaurants'; places where carcasses are provided for vulture feeding (Clements et al. 2012). These are located in several countries; for example in Europe (Sarrazin et al. 1994; Carrete et al. 2006); Israel (Meretsky and Mannan 1999); and in some African countries (Mundy et al. 1992; Piper et al. 1999). The first such restaurant was developed in Giants Castle Game Reserve in the high Drakensberg of Natal, South Africa in 1966 for Bearded and Cape vultures (Friedman and Mundy 1983). Similar structures have also been developed for condors in North America (Wilbur et al. 1974). The impacts on vultures may be positive or negative, varying regionally depending on contextual factors (Piper 2006).

The provision of safe food benefits vultures (Gilbert et al. 2007; Oro et al. 2008). Other positive impacts include improved survival rates for adults and young (Oro et al. 2008), and possibly improved nesting success (see González et al. 2006 for a study of the Spanish Imperial Eagle) although this may vary (e.g., a study by Margalida 2010 for the Bearded Vulture). Piper (2004a: 218) lists positive actions that benefit vultures, which are commented on by other researchers. These include supplementary food provision (Verdoorn 1997b; van Rooyen et al. 1997); provision of bone fragments as a source of calcium supplementation, which reduces osteoporosis (Mundy and Ledger 1976; Richardson et al. 1986); the provision of poison-free food; the attraction of vultures back to previous habitats, as in the case of Cape Griffon, African White-backed, Bearded and Lappet-faced Vultures (Verdoorn 1997b); stabilizing of vulture populations, for example increased nestling survival for Cape griffons (Piper et al. 1999); and hypothetically diverting vultures from lamb-predation (Verdoorn 1997b).

Negative results may occur when some species are favored by superior adaptation or dominance over others (see Margalida 2010; Moreno-Opo et al. 2010). For example the highly social Gyps species may dominate other more solitary vultures (Cortés-Avizanda et al. 2010; see Chapter 1 of this book for a discussion of the sociality of the Gyps vultures). The predictable artificial feeding may also encourage expectation from vultures, possibly changing their behavior (Robb et al. 2008). A possible result would be the alteration of vulture behavior to suit the new pattern (Deygout et al. 2010; Zuberogoitia et al. 2010), affect fecundity (Carrete et al. 2006). A possible solution would be varying the locations and dates of the restaurants, to avoid adaptation (Deygout et al. 2009; Clements et al. 2012). Piper (2004a: 218; see also Piper 2004b) lists negative effects of restaurants. These are: injuries due to fences; the attraction of unwanted species to vulture restaurants, including feral and domestic dogs, Black-backed jackals, Brown hyenas, *mongoose* spp., porcupines, chacma baboons, warthogs, bushpigs and *fly* spp., rare cases of cattle illness due to eating bones (i.e., osteophagy); meat theft by local people; power-line collisions and electrocutions when close; bird drownings in reservoirs especially in dry areas; and poisons in the meat, e.g., barbiturates or non-steroidal anti-inflammatory drugs (NSAID's) (Oaks et al. 2004).

Another solution is more effective monitoring. Danielsen et al. 2008 point out that for successful monitoring, accurate data must be compiled at the level of precision required to detect population changes in a particular context at a particular magnitude. The technical, logistical and financial expertise must be available. In some cases such resources are unavailable. For example, in India monitoring of vultures is usually done using road surveys (Prakash et al. 2007), but Clements et al. (2012) noted that in Cambodia the population density of vultures was too low for this method to be effective and the roads were few in areas of vulture habitat. Capture-mark resighting methods were also ineffective, as few vultures could be captured. Also, counting of nest sites does take into account the visibility of the nests. Counting of vultures that visit the vulture restaurants is also problematic, because the number of vultures not attending the feeding points is not known.

Clements et al. (2012) note that increased numbers of vultures recorded in a census might imply three possibilities: increased habitation to regular artificial feeding, declining food sources in other places, or an increasing population. The most important point is the possibility that vultures are habituated to the supplemental feeding sites. This means that more would be counted at these sites, even if the total population remained unchanged. Other factors may also be important. For example, comparing the vultures species that might attend a vulture restaurant in Cambodia, Clements et al. (2012) point out that the Red-headed vulture has a more varied diet than the White-rumped vulture and Slender-billed vulture, preying in some instances on live prey such as reptiles (see also Naoroji 2007). On this point they are therefore less likely to be dependent on vulture restaurants. They may also be more territorial and therefore the location of the feeding centres may be relevant.

8.4 Impacts of Disease on Vultures

This section looks at the impacts of the chemicals and diseases on individual species of vultures. Also examined are other factors for population declines, to discern the impact of chemical compounds and disease among other factors. The first group examined are the Eurasian vultures: Red-headed, Slender-billed, Indian, Griffon, Himalayan Griffon, Egyptian, Cinereous and Bearded vultures. The second group examined are the African vultures: Hooded, White-backed, White-headed, Ruppells's Griffon, Palm-nut, Cape and Lappet-faced vulture. After these, the last group are the New World vultures: Andean and California condors, King, Turkey, Greater Yellow-headed, Yellow-headed and Black vultures.

In general, the decline of the Eurasian vultures has been blamed on several factors: increased agricultural activities, the decline of wild ungulates, increasingly efficient sanitation, persecution and disease. The Red Headed vulture has been described as less affected by the drug diclofenac than other Asian vultures. This species may have avoided the worst effects, because of the dominance of the other larger species at the carcasses (Cuthbert et al. 2006) and the non-use of diclofenac in parts of its range, such as Myanmar (Eames 2007a) and Cambodia (Mahood 2012). In

Cambodia, studies have shown that it is occasionally poisoned due to the use of such chemicals for catching fish and waterbirds at waterholes (ibid.).

The White-rumped, Slender-billed and Indian vultures were severely affected by diclofenac, especially in Nepal, Pakistan and India, dying from renal failure resulting in visceral gout (Oaks et al. 2004a; Shultz et al. 2004; Swan et al. 2005; Gilbert et al. 2006). Other causes were ketoprofen, which was also lethal (Naidoo et al. 2009), changes in the processing of dead livestock, non-target poisoning (Wildlife Trust of India 2009), avian malaria and pesticides (Poharkar et al. 2009).

The Red-headed vulture was already in decline in South-East Asia before the advent of diclofenac, the decline due to the serious decline of large wild ungulate populations, lower numbers of livestock, improved sanitation of livestock carcasses and use of poisons (Mahood 2012; Clements et al. 2013). For the Common Griffon, possibly because of the wider range, more factors are mentioned for its decline. These include poisoning and persecution (del Hoyo et al. 1994; Snow and Perrins 1998; Ferguson-Lees and Christie 2001), changes in livestock management practices (del Hoyo et al. 1994; Ferguson-Lees and Christie 2001) and wind energy development (Strix 2012).

For the Bearded vulture, the main causes of current declines are non-target poisoning, direct persecution, habitat degradation, disturbance of breeding birds, inadequate food availability, changes in livestock-rearing practices and collisions with power-lines and wind farms (Ferguson-Lees and Christie 2001; Barov and Derhé 2011). Simmons and Jenkins (2007) note that population trends in southern Africa may be correlated with climate trends. Other threats include habitat degradation and breeding disturbance (Ferguson-Lees and Christie 2001).

For the Egyptian vulture, European problems include disturbances, lead poisoning (from shot animals), direct poisoning, electrocution (by powerlines), collisions with wind turbines, reduced food availability and habitat change (Donázar et al. 2002; Kurtev et al. 2008; Zuberogoitia et al. 2008; Carrete et al. 2009; Sarà and Di Vittorio 2003; Sarà et al. 2009; Dzhamirzoev and Bukreev 2009; BirdLife International 2014), and EU regulations from 2002, managing carcass disposal reduced food sources, especially by the closure of traditional 'muladares' in Spain and Portugal (Donázar 2004; Lemus et al. 2008; Donázar et al. 2009, 2010; Cortés-Avizanda et al. 2010; Cortés-Avizanda 2011).

New regulations permit the operation of feeding stations for scavengers (BirdLife International 2014). In Spain and the Balkans, the main threat is illegal poisoning for carnivores (Hernandez and Margalida 2009). Fatal levels of contamination also affect birds in Spain (Lemus et al. 2008) and Bulgaria (Angelov 2009). Antibiotic residues in the carcasses of intensively-farmed livestock also may increase diseases among nestlings (Lemus et al. 2008) (an example being avian pox in Bulgaria) (Kurtev et al. 2008). Diclofenac is another important factor, but possibly less for the Egyptian vulture than for the larger vultures due to competitive exclusion of smaller vultures from carcasses (Cuthbert et al. 2006). African threats for the Egyptian vulture include: the decline of wild ungulates and livestock overgrazing (Mundy et al. 1992); poisoning targeted at terrestrial predators (Carrete et al. 2007; Carrete et al. 2009; Cortés-Avizanda et al. 2009c); power line collisions, especially in the Canary Islands, other parts of Spain (Donazar et al. 2002; Donazar et al. 2007, 2010) and in Port Sudan (Nikolaus 1984, 2006); and killings for traditional medicine, e.g., in Morocco; and also competition for nest sites with Griffon vultures (Kurtev et al. 2008).

For the Cinereous vulture, problems include: accidental ingestion of poisoned baits for predator extermination; deliberate shooting and poisoning; and decreasing food sources. Shooting, poisoning and trapping occur in Mongolia and China (BirdLife International 2014). Diclofenac may increase due to wintering in northern India. Antibiotics such as quinolones damage the liver, kidneys, lymphoid organs and bacteria flora, increasing infections among vultures in central Spain (Lemus et al. 2008). Other problems are European Union legislation on carcass disposal, reductions of livestock in Georgia and Armenia, due to the cancellation of subsidies for sheep-herding in the post-Soviet era, wild ungulate decline (e.g., the Saiga antelope (*Saiga tartarica* Linnaeus, 1766), declining from over one million to 30,000–40,000 in ten years, due to excessive hunting and severe winters); food source declines in South Korea (Lee et al. 2006); and habitat loss and climatic instability, which kills nestlings (Batbayar et al. 2006).

For the African vultures, the Hooded vulture faces threats from non-target poisoning, capture for traditional medicine and bushmeat, and direct persecution (Ogada and Buij 2011). Hooded vulture meat may be sold as chicken in some places. Poachers may also poison vultures to hide the locations of their kills. Carbofuran poisons in livestock baits intended for mammalian predators may also kill vultures in East Africa (Otieno et al. 2010). Other threats are land conversion through development and improvements to abattoir hygiene and rubbish disposal in some areas (Ogada and Buij 2011). The species may also be threatened by avian influenza (H5N1), which may also be acquired from discarded poultry carcasses (Ducatez et al. 2007).

For the White-backed vulture, threats are habitat conversion to agro-pastoral systems, declining populations of wild ungulates and less carrion, hunting for trade, persecution and poisoning (especially the highly toxic pesticide carbofuran) (Western et al. 2009; Ogada and Keesing 2010; Otieno et al. 2010). Diclofenac was also found on sale at a veterinary practice in Tanzania (BirdLife International 2014). Killings for medicinal and psychological benefits are also common in Nigeria and southern Africa

(McKean and Botha 2007; Birdlife International 2014). These may lead to local extinction in some parts of southern Africa (McKean and Botha 2007). International trade in captured birds is also a problem, as is electrocution on powerlines and human disturbance in nesting trees (Bamford et al. 2009; BirdLife International 2014).

For the White-headed vulture, threats include reductions in wildlife, especially ungulates, and habitat conversion (Mundy et al. 1992; BirdLife International 2014). Some birds are poisoned by bait for jackals and larger carnivores such as lions and hyenas, and secondary poisoning from carbofuran (Otieno et al. 2010). The White-headed vulture is possibly less susceptible to these problems than other vultures as its diet is comparatively broader. It is also captured for the international trade in raptors and for traditional medicines in Southern Africa. Human disturbances may contribute to nest desertion. It is sensitive to land-use and hence concentrates in conservation areas (Hancock 2008b; BirdLife International 2014).

For the Rüppell's Griffon vulture, the threats are similar as it suffers from habitat conversion to agro-pastoral landcover, wild ungulate decline, persecution and poisoning and hunting for trade (Western et al. 2009; Ogada and Keesing 2010; Otieno et al. 2010). In West Africa it is exploited for trade in traditional medicines, possibly leading a decline and near extinction in Nigeria (Rondeau and Thiollay 2004; Nikolaus 2006). Examples are cited of the Dogon people of central Mali climbing the Hombori cliffs to take eggs and chicks of this species (Rondeau and Thiollay 2004). In Mali, breeding birds may be affected by the large numbers of tourists visiting climbing routes in the Hombori and Dyounde Massifis (Rondeau and Thiollay 2004).

For the Cape vulture, assumed to be declining throughout much of its range (Boshoff and Anderson 2007), threats include a decrease in the amount of carrion (particularly during chick-rearing), inadvertent poisoning, electrocution on pylons or collision with cables, loss of foraging habitat and unsustainable harvesting for traditional uses (Mundy et al. 1992; Barnes 2000; Benson 2000; Borello and Borello 2002; Diekmann and Strachan 2006; Boshoff and Anderson 2007; Hancock 2008; Boshoff et al. 2009; Jenkins 2010). Other lesser threats include disturbance at colonies, bush encroachment and drowning (Anderson 1999; Borello and Borello 2002; Bamford et al. 2007). In southern Africa, like other vulture species, Cape vultures are killed for perceived medicinal and psychological reasons (McKean and Botha 2007).

If the White-backed vulture declines or faces extinction as a result of these factors, the increase in hunting on Cape vultures could be devastating (Beilis and Esterhuizen 2005; McKean and Botha 2007). The reduction of a colony in eastern Botswana was blamed on human disturbance, including tourism (Borello and Borello 2002). Other problems are urbanisation around Hartbeespoort dam and the Magaliesberg mountains, South Africa, which has reduced natural food sources and increased reliance on vulture 'restaurants'

(Piper et al. 1999; BirdLife International 2014). Bush encroachment into former grassland has also increased difficulties for foraging birds (Schutz 2007). Some birds in southern Africa drowned, possibly due to bathing or drinking attempts (Anderson et al. 1999; Boshoff et al. 2009). There are also fatalities due to powerline collisions and electrocutions, especially in Magaliesberg, South Africa (BirdLife International 2014). Climate also affects their habitats (Simmons and Jenkins 2007).

For the Lappet-faced vulture, declining populations are largely due to accidental poisoning, mostly from strychnine, which is used for predator control, and more recently carbofuran (Brown 1986; Komen 2009; Otieno et al. 2010; BirdLife International 2014). Some persecution is due to the belief that it kills livestock (Brown 1986; Simmons 1995). Other issues include reduced food availability, electrocution on power lines, agricultural pesticides, habitat loss, human nest predation and nest disturbance (Steyn 1982; Mundy et al. 1992; Shimelis et al. 2005; McCulloch 2006a,b). This species is especially sensitive to nest disturbance, with such a problem emerging due to forest settlement in Ethiopia (Steyn 1982; Bridgeford 2009; BirdLife International 2014).

Population reductions in West Africa have been blamed on ungulate extinctions through habitat modification and over-hunting, higher nest disturbance, intensified cattle farming without abandonment of dying animals and increases in accidental poisoning (Thiollay 2006a,b; Rondeau and Thiollay 2004). Available carcasses have been reduced, due to more effective livestock vaccinations and also the sale of carcasses to abattoirs rather than the disposal of such carcasses in rural areas. Hunting for medicine or food has also been suggested (Rondeau and Thiollay 2004). In Mozambique, the armed conflict in the 1970s and 1980s decimated wild ungulates and more recently the over-exploitation of game by poachers continues (Parker 2005b). Similar to other species, this species may be poisoned by animal poachers who believe that hovering vultures will expose their activities (Hancock 2009b).

For the New World vultures, the pattern is different. There is no drastic decline due to pesticides or other poisons. The larger species are rarer than the smaller species. The rarest was the California condor. For this species, threats that decimated the population during the 20th century were persecution, including shootings and lead poisoning from accidental ingestion of lead bullets and shot from carcasses (Parish 2012). Campaigns for condors appear to have reduced direct persecution (BirdLife International 2014). As already mentioned, lead poisoning is still a problem, after the reintroduction of captive birds into the wild, and several birds have died from poisoning (Cade 2007; Parish et al. 2007; Walters et al. 2010). The vulnerability of the condor also stems from its vast foraging range, in which any shot animals may be a risk factor and its slow breeding and long life, which allow a buildup of toxicity (Hunt et al. 2007).

Recent research has indicated that over 90% of released condors in Arizona are suffering from lead poisoning (Toops 2009). In January 2010 three birds died from lead poisoning in northern Arizona (Flagstaff 2010). In 2006, 9 of 13 birds released at the Pinnacles National Monument, California were recaptured for testing for lead poisoning, as they had eaten lead shot squirrels. Also in California, a study of samples from 2004–2009, found about one-third of condors had toxicological levels of lead from ammunition (Finkelstein et al. 2012). There are efforts to ban lead ammunition from areas within the species's range in California, and possibly there will one day be a ban on the use of lead across the United States.

Pesticide poisoning from DDT also affects the reintroduced condor population along the Central Pacific Californian coast, despite the lack of additional DDT inputs. The incidence of this poison may decline through time as more DDT is not being added to the environment. The result is reduced eggshell thickness causing problems with reproduction (Burnett et al. 2013). Possibly, the DDT is a derives from the carcasses of predatory marine mammals exposed to the DDT years previously (Burnett et al. 2013).

As also already mentioned, another threat to the California Condor is anthropogenic material, namely plastics and other manufactured products in the environment. For example, two nestlings died after swallowing plastic cartridge cases, glass fragments, wire, and other objects (Mee et al. 2007). Some hand-reared condors in the early 1990s died from collision with power-lines, necessitating a training programme using fake power poles to condition the birds to avoid the poles (Kiff 2005). Birds are also vaccinated against the possibility of the west Nile virus. The survival rate of the released birds is nevertheless unsustainable without treatment of lead-contaminated birds (Walters et al. 2010).

For the Andean Condor, the main threats are human persecution, partly based on the idea that condors attack wildlife (Houston 1994); and illegal poisoning of carcasses for the control of mountain lions and foxes (Imberti 2003). Changes in livestock rearing practices affect condors, which may be dependent on their carcasses (Lambertucci et al. 2009). There is also interspecific competition for carcasses with Black vultures (Carrete et al. 2010). For the King vulture, the main threat appears to be habitat destruction. For the Turkey vulture, the population is rather expanding, despite problems with lead poisoning as noted above. Populations of Turkey vultures are estimated to have increased by at least 100 percent over the last 40 years. The Black vulture has also expanded its range and population (BirdLife International 2014). For the Greater Yellow-headed vulture, the main issues are ongoing habitat destruction and over-hunting of prey species (del Hoyo et al. 1994; Ferguson-Lees and Christie 2001). For the Yellow-headed vulture, the population is described as stable (see chapter 3).

8.5 Solutions for Vulture Decline

Numerous actions have been initiated to halt the extinction of vultures due to diclofenac (mostly in India, Pakistan, Nepal and parts of South East Asia) and also to ameliorate other problems such as pesticide and lead poisoning (mostly in Europe). To solve the diclofenac problem, the governments of India, Nepal and Pakistan passed legislation in 2006 that banned the manufacture and importation of diclofenac for veterinary uses. In 2008, India went further and banned the sale, distribution or use of veterinary diclofenac. The Drug Controller General of India in 2008 informed scores of drugs firms to cease selling the veterinary form of diclofenac. Human diclofenac containers were ordered 'not for veterinary use' (BirdLife International 2008). The government of Bangladesh in October 2010, banned the production of diclofenac for cattle. The distribution and sale of the drug were ordered banned during 2011 (BirdLife International 2014). These laws have reduced diclofenac use for cattle, but diclofenac contamination remains common. Also, the forms of the drug for human use are still available for use in cattle (Cuthbert et al. 2011a,b). Studies of alternatives are intensive, and one alternative, meloxicam, was tested and found to have no ill effect on vultures (see above) (Swan et al. 2006; Swarup et al. 2007; Cuthbert et al. 2011c). In Nepal, diclofenac has been in replaced with meloxicam near breeding colonies; in addition there is diversionary feeding with diclofenac-free carcasses (Chaudhary et al. 2010).

In addition supplementary feeding of vultures with safe meat, combined with effective monitoring has increased. Seven vulture restaurants were set up in Cambodia by the Cambodia Vulture Conservation Project, which involves a partnership between the Royal Cambodian Government and national and international NGOs (Masphal and Vorsak 2007; Rainey 2008). The vulture restaurants are in some cases used as ecotourism attractions, especially in Cambodia for financial gain and also to promote public knowledge of the issues (Eames 2007a; Masphal and Vorsak 2007). Vulture restaurants were studied in Myanmar in late 2006 and early 2007, and more research was conducted into factors for vulture mortalities, nesting locations of nesting colonies and diclofenac use (Eames 2007b).

Diversionary feeding has met with moderate success. Vultures are highly mobile, and more movement detection studies are necessary for restaurant locations to be effective (Pain et al. 2008). Some vultures are satellite tagged for information on their movements (Ellis 2004). There is evidence of social support for vulture conservation, as local people are aware of the ecological benefits of vulture feeding, such as removal of rotting meat, reduction of diseases and reduced numbers of facultative scavengers such as feral dogs (Gautam et al. 2003).

Captive breeding is another strategy for vulture rescue. For example, the Report of the International South Asian Vulture Recovery Plan Workshop in 2004 recommended the development of at least three captive breeding centres, each capable of holding 25 pairs of vultures (Bombay Natural History Society 2004; Lindsay 2008). In Uthai Thani, Thailand, a five-year captive breeding and reintroduction was initiated by the Zoological Park Association and Kasetsart University. In Pinjore, Haryana, India, two chicks were hatched in 2007 and three more in 2009 at a breeding centre established by the Royal Society for the Protection of Birds (RSPB) and the Bombay Natural History Society (Bowden 2009). By April 2008, there were 88 captive birds in three Indian breeding centres, 11 in a centre started by WWF-Pakistan in Pakistan and 14 captive birds in Nepal (Pain et al. 2008). These populations of captive birds increased over the next year, to 120 in India, 43 in Nepal and 14 in Pakistan (Bowden 2009). By the end of November 2011, there were a total of 221 vultures in the breeding centres affiliated to SAVE (Saving Asia's Vultures from Extinction) (SAVE 2012). Some of the vultures in these centres were found poisoned and treated to recovery. An example of a rescue centre is the Centre for Wildlife Rehabilitation and Conservation, Assam, conducted by the Wildlife Trust of India and the International Fund for Animal Welfare (IFAW).

In Europe, the ameliorative actions largely concern the Cinereous Vulture and the Egyptian vulture. The EU Birds Directive has assisted in the recovery of Cinereous Vulture in Spain. Here the population increased from an estimated 290 pairs in 1984 to at least of 1,845 pairs in 2007 (De la Puente et al. 2007). Poisoned baits, which affect the vultures were also controlled by work involving Spanish government agencies and conservationists in the 'Antidote Programme'. Anti-poisoning plans were developed by the Spanish and the Andalusian Governments that may yet be implemented. Cinereous vultures were reintroduced in Grands Causses, Southern France, which increased to 16 breeding pairs in 2006 (Eliotout et al. 2007).

Supplementary feeding and captive breeding programmes have been initiated in Spain and France (Tewes et al. 1998). Other supplementary feeding centres are in Bulgaria and South Korea (Lee et al. 2006). A colony of vultures in the Dadia forest reserve in northern Greece, is supported by the World Wildlfe Fund. The Balkan Vulture Action Plan also attempts technology transfer on vulture conservation from western Europe to eastern Europe.

For the Egyptian vulture, conservation actions include monitoring programmes (in both Europe and Africa), supplementary feeding in Europe (Cortés-Avizanda et al. 2010) and campaigns against illegal use of poisons, including awareness-raising (BirdLife International 2014). This species may also benefit from the diclofenac ban by the Indian government. Research has also been conducted in Tanzania on the extent of Diclofenac use for veterinary purposes (BirdLife International 2007; Woodford et al. 2008; Iñigo et al. 2008). In Europe, there are national species action plans in France, Bulgaria and Italy.

The Egyptian vulture is also included in the Balkan Vulture Action Plan (BVAP) for eastern Europe. Captive breeding is also being developed in Italy. Satellite-tag surveys are conducted in Spain, France, Italy, Bulgaria and Macedonia for information on juvenile dispersal, migratory movements and wintering areas (García-Ripollés et al. 2010). Nests are also guarded against poachers in Italy and Bulgaria. As the Egyptian vulture winters in northern Africa, there have been studies on the factors affecting migrants in Mauritania, Senegal, Ethiopia, Sudan and Turkey. These studies are being conducted with local organizations in the relevant countries.

In Africa, despite an extensive literature on the decline of most vulture species (Campbell 2009; Ogada and Buij 2011) it appears most of the actual conservation actions are for the Cape vulture in South Africa. Non-governmental organizations have been able to elicit support for vulture conservation among farming communities in South Africa (Barnes 2000). This species enjoys legal protection throughout its range, and some breeding colonies are in protected areas (Barnes 2000). Nestlings were color-ringed in the 1970s and 1980s for information on their movements and survival (Botha 2006). Due to pylon related deaths and injuries, the national electricity suppliers in South Africa replaced some pylons with safer designs (Barnes 2000), and breeding numbers have increased in one area (Wolter et al. 2007).

Supplementary feeding has been successfully developed and may be a factor for minimal population recovery in some local areas (Barnes 2000). Two restaurants, one in Nooitgedacht and another in Magaliesberg appear to have assisted in local species recovery (Borello 2007; Wolter et al. 2007). In at least one area, supplementary feeding increased the survival rate of young birds in the Western Cape Province of South Africa (Piper et al. 1999).

Captive breeding is also practiced successfully. There were 37 birds in captivity in Namibia in 2011. Sixteen captive South African vultures were released in Namibia in October 2005. Two of these birds were fitted with satellite transmitters, and five more were fitted with these transmitters in 2006, although there is evidence that some released birds died (Diekmann and Strachan 2006; Komen 2006; Bamford et al. 2007).

Public education on vulture problems has also yielded dividends. In Namibia, farmers have learned of the benefits of vulture presence and the disadvantages of poisoning carcasses from publication education programmes, an education centre and a programme for schools (Diekmann and Strachan 2006). There was a conservation workshop in March 2006, with 19 people in attendance, the result being planning for the management of the main vulture colonies in southern Africa (Komen 2006). Topics receiving attention by recent actions in southern Africa include: the dangers of using diclofenac in the treatment of cattle; awareness of vulture drownings in reservoirs and actual modifications of these reservoirs to reduce such deaths; possible re-establishment of monitoring at the Cape vulture's only colony in Zimbabwe; the impacts of hunting for medicinal and cultural reasons in southern Africa; and general conservation matters (Boshoff and Anderson 2006, 2007; Komen 2006; McKean and Botha 2007; Wolter et al. 2007).

In other parts of Africa, there have been directives and studies, but fewer actual actions possibly due to lack of funds. For the Rueppell's Vulture, there may be a benefit from a survey in 2007 to determine diclofenac use for veterinary purposes in Tanzania (BirdLife International 2007). For the White-backed Vulture, a press release was circulated in July 2007 to raise awareness of the impacts of hunting for medicinal and cultural reasons in southern Africa (McKean and Botha 2007). For the White-headed Vulture, possibly a more threatened species, birds were marked with patagial tags in Fouta Djallon Vulture Sanctuary, Guinea, in 2007 to monitor movements and for a toxicological assessment of the vulture population of the park (Rondeau 2008). For the Lappet-faced vulture, there are studies on its status, especially in Saudi Arabia (Newton and Shobrak 1993). Also, a five-year international action plan for the species was published in 2005, aiming to stabilize or increase its population, increase ecological knowledge and minimize human impacts (Shimelis et al. 2005). For Botswana, there was a study in 2007 (Hancock in litt. 2006), and 221 chicks were marked with patagial tags between 2006 and 2009 (Bridgeford 2009). Also in 2008 a conference of the World Organisation for Animal Health in Senegal led to a resolution asking 'Members to consider their national situation with the aim to seek measures to find solutions to the problems caused by the administration of diclofenac in livestock' (Bowden 2008; Woodford et al. 2008).

8.6 Conclusions

This chapter has looked at chemical impacts on vultures from both the Old and New Worlds, and the ameliorative actions that have been employed through laws, policies and research. The relevant tools for the rescue of declining vultures exist. However, the evidence shows that more information is needed on the status and impacts of the actions. There is currently little information on the success or otherwise of the numerous actions described. What is clear is that, despite problems with the condors, the New World vultures are better off than the Old World vultures. Whether this is permanent remains to be seen. The factors for Old World Vulture decline go beyond the diclofenac problem, and include other issues such as changing environments and landuse. Even if diclofenac is eliminated, searching problems remain. These include the more effective disposal of cattle carcasses, cleaner cities, more sanitary abattoir disposal methods, ungulate population decline and competition from the increased numbers of facultative scavengers. Based on these issues, the question asked of the condor may be asked here again; are vultures from a bygone era, now to be fed like captive animals? Will the vulture restaurants be permanent, ensuring that vultures become like domestic or zoo animals rather than free agents? These questions have yet to answered, but the primary investigation must be of people rather than only of vultures. How determined are we to restore vultures to their former glory? This depends on both past and current attitudes to vultures, which will be examined in the next chapter.

9

Vultures in Social History and Conservation

9 INTRODUCTION

This chapter looks at the attitudes of people in various cultures to vultures in both historic and modern times, in both the New and Old Worlds, and also at the legislation for their protection and also other legislation that has had an effect on vultures. These include laws and policies on more effective sanitation, hunting, urban development and conservation of associated animals. New and Old World vultures are not just physically similar, it appears that they were also appraised similarly in historic times, the main issues being their size, admired soaring and sometimes possible evil association. Current attitudes are also similar across continents; the vulture elicits repugnance, sympathy, admiration and fear. Basic questions emerge; how have the attitudes changed, how deserved are they, will legislation be sustainable considering these attitudes and will the new laws and policies make the attitudes worse or better?

9.1 Historic Public Attitudes to New World Vultures

New World vultures had a strong cultural relationship with people in the prehistoric (Late Pleistocene/Holocene, 15000 BP) and more recent, pre-Colombian past. The ethnographic evidence hints the condors role was largely in religious or quasi-religious, ceremonial activities. In prehistoric, far western North America, evidence includes unmodified condor remains and also clear evidence of 'intentional burials', bone and feather artifacts and vulture representations in rock art (Simons 1983).

Significant numbers of artifacts were discovered from thirteen sites in California and Oregon, dated 10,000 years to early historic times (Simons

1983). There were unmodified remains at Five Mile Rapids site, the Lone Ranch Creek Shell Mound, the Hotchkiss Mound, the Avila Bridge site, and site 261 on San Miguel Island (cited in Miller 1942; Berreman 1944; Miller 1957; Cressman 1960; Miller 1963; Landberg 1965; Guthrie 1980). The largest number of these unmodified remains (about 63 birds, rather than one or two) were found at the Five Mile Rapids site, where condors comprised about 10% of the bird skeletons.

There are also 'questionable archaeological occurrences' of the California condor from this period. These include Smith Creek Cave in White Pine County, Nevada; Gypsum Cave in Clark County, Nevada; Boulder Springs Rock Shelter in Mohave County, Arizona; Stanton Cave, Tsean Kaetan Cave and Luka Cave in Coconino County, Arizona; Rocky Arroyo Cave in Otero County, New Mexico; Howell's Ridge Cave, Grant County, New Mexico; Conkling Cave in Dona Ana County, New Mexico; and Mule Ears Cave in Brewster County Texas. These findings are documented in a diverse literature (Miller 1931; Wetmore 1931, 1932; Howard and Miller 1933; Wetmore and Friedman 1933; Howard 1952; De Saussure 1956; Howard 1962a; Parmalee 1969; Hevly et al. 1978).

There are also questionable condor remains from sites which show evidence of human presence from the Pleistocene. These are found in Potter Creek Cave, Samwell Cave in Shasta County, California; Santa Rosa Island, Santa Barbara County, California; Rancho la Brea, Los Angeles County, California; and Friesenhahn Cave in Bexar County, Texas (Miller 1910, 1911; Howard and Miller 1939; Wormington 1957; Stock 1958; Evans 1961; Howard 1962b; Kreiger 1964; Brodkorb 1964; Heizer 1964; Bryan 1965; Orr 1968; Berger et al. 1971; Payen and Taylor 1976). The evidence supports natural, non-cultural deposits of condor remains, as there appears to be no association between these remains and the human artifacts, and none of the condor remains seem to be subjected to human use (e.g., manufacture of artifacts) (Simons 1983).

The evidence for intentional bird burials is the planned nature of the grave, resembling a human burial (Wallace and Lathrap 1959, 1975). These include Emeryville shell mound and the West Berkeley Shell mound near the eastern shore of San Francisco Bay (Wilbur and Jackson 1983), with similar findings in the West Berkeley shell mound.

Bone and feather artefacts are quite common. Bone artefacts include tube whistles, some created from the bones of the wing, found in the Berryessa–Adobe site, the McCauley mound (Schenk and Dawson 1929) and the Hotchkiss mound. Other items include humeri 'wands' as at the Old Bridge site (Johnson 1967; Wilbur and Jackson 1983) and a mandible, closely associated with human burials and possible ceremonial artefacts, possibly from a condor skin dance suit worn by the owner of the skeleton. Feather artefacts, including decorative bands from the feathers of condors and other species have been found in Bower's Cave, San Martin mountains, northern Los Angeles county (Elsasser and Heizer 1963).

Condors are also represented in prehistoric rock art. An example is Condor Cave in the San Rafael mountains, Santa Barbara county, which has been described as a winter solstice observatory (Hudson and Underhay 1978; see also Smith and Easton 1964; Grant 1965; Heizer and Clewlow 1973; Wilbur 1976; Hudson and Underhay 1978; Smith 1978). In this cave is a pictograph of a condor in flight. Other sites with such rock art are considered linked to the former habitation of the Cuyama and Ventureno Chumash, such as Painted Rock site (Smith and Easton 1964; Grant 1965; Hudson and Underhay 1978; Lee and Horne 1978).

Condors and other raptors appear to have been important for cultural activities during this period in Western North America, especially for Shamans who might have mutifaceted religious, medical, and social roles. An important factor was their association with supernatural powers (Bean 1975; Simons 1983). There is evidence that condors were seen as sources of power in some societies, for example the Hupa, Sinkyone, Yuki, Hill Patwin, Konkow, and central Sierra Miwok (Curtis 1824b; Kroeber 1925, 1932; Loeb 1933; Driver 1939; Foster 1944; Freeland and Broadbent 1960). Condor feathers might be used for initiation ceremonies for new shamans, which sometimes involved the thrusting of feathers into the throat of recruits, as in the Yuki and Wailaki (Kroeber 1925; Loeb 1932; Gifford 1939). In some cases, it was believed that diseases were removed from the human body by the feathers in the throat (e.g., among the Wiyot) and in other feathers were attached to a stick to pass over a person's head for medical help (Curtis 1824a; Kroeber 1925; Loeb 1932).

Simons (1983) describes three religious-ceremonial systems as dominant in prehistoric California. There was the Kuksu system, based on secret societies, rites of passage, and costumes, and the condor dance, which is well described for the Pomo, River Patwin, Valley Nise and Konkow (Curtis 1924b; Kroeber 1925, 1929, 1932). For the Pomo, a dancer who inherited the privilege, was dressed in the full skin and feathers of the condor, with the wings extended by sticks on the sides of the dancer. To acquire a condor skin, the bird was trapped by inserting a sharpened bone, attach to a string, into rabbit meat so that the points of the bone protruded on both sides of the meat. When the condor swallowed the bait and was unable to disgorge the morsel, it was killed with a blow to the head. Apart from the skin which was used for the dance, bones were used for whistles and ear ornaments and fat from the body cavity was used as medicine.

Another system was the Toloache system, of south and south-central California. In this system there were several subgroups: the Northern Complex (Southern Valley Yokuts, Kitanemuk, Tataviam, Gabrielino); the Antap-Yivar Cult (Chumash, Gabrielino, Tataviam); the Southern Complex

(Gabrielino, Luseno-Juaneno); and the Chinigchinich religion (Gabrielino, Luiseno-Juaneno, Cupeno and Tipai-Ipai). In some cases, the condor was used as a sacrifice and in some cases the skin and feathers were worn. Narcotic plants were also used (e.g., *Datura* sp.), which facilitated altered consciousness and beliefs in supernatural powers (Bean and Vane in Heizer 1978: 663). Some cultures within this system, such as the Miwok, Monache and Yokuts) saw the condor as a totem (Gifford 1916; Kroeber 1925; Driver 1937; Gayton 1948b). In some cases, such as the moloku (condor) dance performed by the Central Sierra Miwok (Kroeber 1925; Grifford 1926, 1955; Smith and Easton 1964; Smith 1978), (similar to the Pomo dance celebrating the killing of a condor) a dancer would wear a condor skin and mimic the bird, while being supported by a singer and drummer (Powers 1877; Gifford 1916, 1926, 1932; Kroeber 1925; Gayton 1930, 45, 48a, 48b; Driver 1937; Aginsky 1943). Condors were also used in ritual sacrifices, with symbolic dances to the ceremonial killing of the birds (Simons 1983).

Ceremonies were also performed for living and dead birds (among tribes such as the Miwok, Monache and Yokuts) with the living bird being raised in captivity, traded between villages and usually set free rather than killed. For dead birds brought to the settlement by hunters, the skin, bones, feathers, down and fat would be removed and the rest ceremonially buried after a mourning demonstration (Gifford 1916, 1926, 1932; Gayton 1948a). Many neighboring tribes, such as the Niseman and Tubatulabal also captured and raised condors, the former killing them eventually but the latter freeing them (Kroeber 1925; Beals 1933; Voegelin 1938). Other tribes on the south-central Californian coast also sacrificed condors and other raptors ceremoniously, as evidenced in records by the members of the Portola Expedition (1769) (Harris 1941; Smith and Easton 1964; McMillan 1968; Smith 1978) although some have questioned the identity of the bird (Simons 1983).

Other tribes in southern California, such as the Luiseno-Juaneno, Cupeno, Serrano, Cahuilla and Tipai-Ipai also sacrificed raptors (a chronology of literature on this topic is the following: Du-Bois 1905, 1908; Kroeber 1908a, 1908c, 1925; Sparkman 1908; Waterman 1910; Hooper 1920; Benedict 1924; Curtis 1926; Strong 1929; Drucker 1937; Harrington 1942; Johnston 1962; Bean 1972; Bean and Smith in Heizer 1978, 546: 573, 589; Bean and Shipek in Heizer 1978, 556; Bean in Heizer 1978, 583; Luomala in Heizer 1978, 604).

A detailed description of these sacrificial ceremonies, noted as part of the Chinigchinich religion, was written by Friar Boscana (1933: 57–58) about the Luiseno-Juaneno, living near Mission San Juan Capistrano. Boscana identified the sacrificed bird as a condor, as also did Kroeber (1908a, 1908c, 1925), Sparkman (1908), Curtis (19260, Strong (1929), Johnston (1962), and Hudson and Underhay (1978). The system was also described by Harrington

(1934: 34–40) and Harris (1941: 33). It was held in honor of the founder of this religion, who was believed to have changed a bird into a young woman, held to be the same bird sacrificed which was repeatedly reborn. This system involved the building of ceremonial enclosures, followed by ceremonial dancing and sacrifice and skinning of the bird, followed by its burial in the enclosure accompanied by mourning, and more celebrations and festivities (Simons 1983).

Several groups within the Toloache system practised dances which included condors. One of these was the Huhuna/Holhol dance of the Monache, Tubatulabal, Yokuts and Chumash (Gayton 1930, 1948a, 1948b; Driver 1937; Aginsky 1943; Latta 1949, 1977; Hudson and Underhay 1978; Spier in Heizer 1978: 435, 480). This dance was performed by male dancers in condor feather cloaks. It included a search for hidden money, which was located by the dancers by pointing sticks and was shared when found, among the Shamans and dancers and accompanying singers. The Holhol dance of the Chumash was similar, but appears to have been led by a single dancer in the condor cloak (Hudson and Underhay 1978). There was also the Tatahuila or whirling dance, of the Gabrielino, Luiseno-Juanero, Serrano, Cahuilla, Cupeno and IpaiTipai (Dubois 1908; Kroeber 1908a, 1908b, 1908c, 1925; Waterman 1910; Spier 1923; Benedict 1924; Curtis 1926; Strong 1929; Boscana 1933; Harrington 1933, 1942; Drucker 1937; Johnston 1962; Bean and Shipek in Heizer 1978: 556; Bean and Smith in Heizer 1978: 573; Luomala in Heizer 1978: 605). Here dancers wore feathered headdresses and eagle-feather skirts, which were raised in the whirling dance. Some accounts describe condor feathers as part of the costumes (Kroeber 1908a, 1908c, 1925; Curtis 1926; Boscana 1933; Harrington 1933; Koford 1953; Hudson and Underhay 1978).

Another system was the World Renewal System. Here, the only role of the California Condor was the use of its feathers, within ceremonies involving a network of priests officiating at an annual cycle of ceremonies, the purpose being the harmony and stability of nature and human health (Kroeber 1925; Kroeber and Gifford 1949; Bean and Vane in Heizer 1978: 663–665).

Other aspects of the vulture relations with people in prehistoric Western North America involved human roles in vulture ecology. Simons (1983) gives evidence of 'condor cafeterias' located at the Five Mile Rapids site, on the Columbia River east of the Dalles, Oregon. Here condor remains comprise 10% of over 9,000 mostly predatory bird bones dated 7300 to 10,000 years BP (Miller 1957; Cressman 1960). Salmon remains, partly as waste from human activities are a possible reason for the attraction of condors to this location (Miller 1957). Fishing has been mentioned as an important activity for Native American peoples in the prehistoric era (Powers 1877; Mckern 1922; Kroeber 1932; Barrett and Gifford 1933; Beals and Hester 1960; Kroeber

1925; Baumhoff 1963, in Heizer 1978: 16–24; Swezey 1975; Schulz and Simons 1973; Swezey and Heizer 1977).

Simons (1983) suggests that such fishing activities were common over most of prehistoric California, and this production of protein could provide an important source of food for condors and other vultures and scavengers. It is possible that the food sources explain the large population of condors that existed at this time, despite the exploitation of condors by people. 'One may therefore conclude that within the prehistoric Far West, condor populations probably remained in overall equilibrium through time, with local conditions largely determined by interplay between the two principal types of condor-human relationships' namely food provision and ritual killing, and may have been a factor for the ancient belief in the rebirth of the same birds in perpetuity (Simons 1983: 489).

Snyder and Snyder (2000) describe possible impacts of historical cultural activities on California condors. They divided the number of square miles of the historic range of the condor by the number of square miles occupied by the average tribe of natives. They concluded that up to seven hundred native tribes could have occupied the range of the condor and if only one tenth of these tribes regularly killed condors, there would be a significant population impact. This may have reduced the population of the condor across its range before the advent of Europeans. Therefore, prior to native killings, condors may have been much more abundant than any historical records indicate (Snyder and Synder 2000).

In ancient Meso-America, vultures were evidently an important part of the culture. Van Dooren (2011) notes that people associated vultures with agriculture, rain and fire, a likely reason being the attraction of vultures to farmland under fire clearance, immediately preceding the rainy season. An example is the classic Vera Cruz culture, in the city of el Tajin in southern Mexico, towards the end of the first millennium BCE. This included vulture figures in stone wall panels, and figures of humans in vulture costumes, raising the possibility of a vulture cult in the city (Kampen 1972, 1978). Vulture images have been found on ancient knives from the contemporary country of Peru. There is also an Inca knife decorated with the images of two vultures and a larger condor devouring a human body (Benson 1997). Another example is a ceramic vessel from the Aztec culture of Mexico with what appears to be a king vulture head (Heilbrunn Timeline of History 2006).

9.2 Current Public Attitudes to New World Vultures

The current public attitudes to New World Vultures stem from their predation on livestock, pollution and nuisance in urban areas, and sympathy for the declining populations of the condors. As noted by Gross (2006)

Americans once persecuted vultures, because they believed vultures transmitted disease, when actually vultures help control brucellosis, anthrax, and other livestock diseases by removing infected carcasses. Koford (1953) also records a time in the 1930s when the California condor was viewed as an enemy of progress and a threat to American values, because condor conservation was used as a reason to halt the building of a road by the United States Forest Service. However, as noted in Chapter 7 of this book, the Black vulture has acclimatized to urban areas, contributing to property damage, roosting pollution due to defecation, negative opinions on their danger to livestock and people, and aircraft collisions (Buckley 1999; Lowney 1999; Avery 2004; Avery and Cummings 2004; Blackwell and Wright 2006; Novaes and Cintra 2013).

An important barometer of public opinion of New World vultures is the legislation related to their presence. Vultures are still killed by people who believe them to be troublesome, but laws have been implemented for their protection. Millsap et al. (2007) note that as most national conservation laws are modeled on international or regional laws and treaties, the components are similar. Examples include the Convention on the International Trade in Endangered Species of Wild Fauna and Flora (CITES), signed in 1973 in Washington, D.C., USA, and implemented by 169 countries as of July 2006; this convention gives a 'uniform system of control on the international movement of CITES-listed species, including raptors' (Millsap 2007: 437). Other relevant agreements are the Convention on Wetlands (Ramsar), the Convention on Biological Diversity and the Convention on the Conservation of Migratory Species of Wild Animals.

In the United States, the Federal Government's involvement with raptors prior to 1900 was mostly for predator control (Millsap 2007). After this, conservation and science actors promoted the benefits of raptor presence and conservation in support of new legislation. Currently, important regulations are in the Code of Federal Regulations Title 50 (50 C.F.R.), Migratory Bird Treaty Act (Parts 10 and 21), Bald and Golden Eagle Protection Act (Part 22), and Endangered Species Act (Parts 17 and 23).

The first relevant legal act that might affect vultures through their status as wildlife was the Lacey Act of 1900 (16 U.S.C. §§ 3371–3378) which was named after its sponsor Representative John F. Lacey, an Iowa Republican. It was signed into law by President William McKinley on May 25, 1900. The Lacey Act protects wildlife and plants by creating civil and criminal penalties, and prohibits trade in wildlife, fish, and plants that have been illegally captured, transported or sold. This Act is still effective, although there have been several amendments (Fisher and Cleva 2000). The next environmentally related law was the Weeks-McLean Act (effective 4 March 1913). This latter act was named after its sponsors, Republican Representative John W. Weeks of Massachusetts and Senator George P.

McLean (R) of Connecticut. It prohibited the spring hunting and marketing of migratory birds and the importation of wild bird feathers for women's fashion. The Secretary of Agriculture was given power to regulate and set hunting seasons across the United States. This was replaced by the Migratory Bird Treaty Act of 1918.

In 1913, the Congress enacted the Migratory Bird Act (MBA; 37 Stat. 878, ch. 45). This protected all migratory game and insectivorous birds and prohibited hunting of these species except as allowed by federal court regulations. This act was however challenged successfully in a federal because the property clause of the Constitution granted states primary management authority over wildlife in their jurisdiction (Bean 1983).

The Migratory Bird Treaty Act (MBTA; 16 U.S.C. 703–711), signed in March 1916, and implemented in 1918 was for the protection of migratory birds between the United States and Canada (Great Britain was acting on behalf of Canada). The constitutionality of the act was upheld by Supreme Court in 1920. Subsequent, similar treaties were enacted between the United States and Mexico, and with Japan and Russia. Although the original treaties did not provide specifically for raptors, these were added to the treaty with Mexico in a 1972 amendment (Bond 1974). The list of birds in the MBTA now includes all Falconiformes and Strigiformes that occur (not accidentally) within the U.S.

The act made it illegal, without a waiver, to pursue, hunt, take, capture, kill or sell birds mentioned in the Act ('migratory birds'). This included live and dead birds and bird parts such as feathers, eggs and nests. There are 800 species currently on the list. Exceptions are federal regulations (50 C.F.R. 22), which control both possession and transportation of bald eagles and golden eagles, and their 'parts, nests, and eggs' for 'scientific, educational, and depredation control purposes; for the religious purposes of American Indian tribes; and to protect other interests in a particular locality.' Members of federally recognized tribes are given the right to apply for an eagle permit for use in 'bona fide tribal religious ceremonies.' The U.S. Fish and Wildlife Service has also issued exemptions, bird-killing permits to public and private entities, to these provisions. These are usually disclosed by Freedom of Information Act requests. The Migratory Bird Treaty Act was enacted in an era when many bird species were threatened by the commercial trade in birds and bird feathers. Later, other laws were added to the MBTA based on conventions between the United States and Mexico (1936), Japan (1972) and the Soviet Union (now with Russia 1976). Although the MBTA allows birds to rest in private properties, and prohibits the removal of the birds from such properties, a federal permit may be granted for the relocation of listed species (also, in some states a state permit is needed as well as a federal permit). The criteria for such a permit are listed in Title 50, Code of Federal Regulations, 21.27, Special Purpose Permits.

Another relevant act was the Endangered Species Preservation Act (ESPA; P.L. 89–669), which was passed by Congress in 1966. The initial ESPA empowered the Secretary of the Interior to ensure the conservation, protection, restoration and propagation of declining species of wildlife and fish. This scope was broadened in 1969, when the Endangered Species Conservation Act (ESCA; P.L. 91–135), was passed, as this further action empowered the Secretary of the Interior to list wildlife species under global extinction threat and prohibit the importation of these species into the U.S. The act also directed the Secretaries of State and Interior to convene an international ministerial meeting on the conservation of endangered species (Bean 1983). The international meeting of 3 March 1973 resulted in the creation of CITES (the Convention on International Trade in Endangered Species of Wild Fauna and Flora, also known as the Washington Convention).

As the ESCA did not provide detailed methods for the conservation of most native endangered species from federal actions and left the taking of endangered species to the states, in 1973 the United Congress enacted the Endangered Species Act (ESA; 16 V.S.C. 1513–1543). The ESA implements the provisions of CITES. It also notes that species may include subspecies, and stipulates the criteria for listing species as endangered or threatened (Section 4). The Secretaries of the Interior and Agriculture are directed to establish and implement a land conservation for the listed, threatened species (Section 5). The Secretary of the Interior is also directed to work with the different states through management, cooperative and financial agreements with state level agencies for the conservation of listed species (Section 6). Very importantly it also directs that all federal agencies should ensure their actions do not pose a risk to any listed species, and it defines a consultation process for assessing impacts (Section 7). Sections 9, 10 and 11, also prohibit the importation, export, taking, possession, transport, sale, and trade of any listed species, normalize the exemptions and prescribe civil and criminal penalties for violations of the Act (U.S. Congress 1983).

In Canada, there is the Migratory Birds Convention Act (also MBCA) passed in 1917 and updated in June 1994. This act protects migratory birds, their eggs, and their nests from hunting, trafficking and commercialization and defines the grounds for a permit to carry out any of these activities. In the general law, there is also the Criminal Code of Canada, Section 446, Cruelty to Animals (from 1892), which outlaws actions 'causing unnecessary suffering.' It directs 'Everyone commits an offence who willfully causes or, being the owner, willfully permits to be caused unnecessary pain, suffering or injury to an animal or bird . . .'

However, as noted by Millsap et al. (2007: 441) 'in Canada, raptors are not protected by any overarching federal legislation, as they are in the U.S.

Rather, basic legal protection from disturbance and harassment is provided by provincial and territorial legislation. Raptors were not included in the Migratory Bird Convention with the U.S. in 1916, the enabling Canadian legislation in 1918, nor in any subsequent amendments to that Act.' As a result, each provincial and territorial government issues the relevant permits for actions related to raptors. However, the federal government enacted the Species at Risk Act (SARA) in 2003; this act is designed to protect the nationally listed raptors and determines the requirements for permits for research and conservation activities. Environment Canada issues the permits, except in National Parks, where such permits are issued by Parks Canada Agency.

International and inter-provincial transport of raptors is controlled by the Wild Animal and Plant Protection and Regulation of International and Inter-provincial Trade Act (WAPRITTA). Any projects that may handle raptors, and potentially harm them are also subject to the Canadian Council on Animal Care Committee. Projects on crown land, whether federal, provincial or territorial must be approved by the relevant authorities. There is also the Canada Wildlife Act (CWA), which although it does not mention raptors, provides regulations for activities on National Wildlife Refuges and Migratory Bird Sanctuaries, including those that affect raptors and hence would require permits under the CWA. There is also the Canadian Council on Animal Care (CCAC), founded in 1968, the mandate of which is 'to work for the improvement of animal care and use on a Canadawide basis.'

9.3 Historic Public Attitudes to Old World Vultures

Vultures have long been associated with spiritual powers, the afterlife and heavenly contacts. One of the oldest temples in the world, Gobekli Tepe of Neolithic Anatolia of the ninth millennium BCE in southeast Turkey, includes stone pillars with carvings depicting many birds including vultures, a vulture figure sitting with a ball (or the sun) in its hands, and numerous limestone vulture figurines. There is a possibility that sun figures represent the ancient, regional view that vultures and other soaring scavenging birds carried the bodies of the dead to heaven (Curry 2008; Van Dooren 2011). In another site, the ancient town of Catal Huyuk of the seventh millennium BCE, vulture skulls are stuck in wall features, and paintings depict vultures in association with headless human figures, probably corpses. Another picture shows a person wearing a sling, beside a corpse, with vultures approaching, representing either a person warding vultures off a corpse, or beckoning them to a feast (Mellaart 1967; Peters and Schmidt 2004). It is possible that exposure of the dead to vultures was practiced at this site, because there is evidence that bones were placed on the ground after the flesh had been removed (MacQueen 1978).

Van Dooren (2011: 97) notes that 'among the peoples of the ancient world, however, none held the vulture in higher regard than the Egyptians', as the vulture appeared as a 'protective and mothering figure' and a scavenger of carrion on the battlefield.' In ancient Egyptian mythology, the vulture was associated with several goddesses, especially Nekhbet and Mut (te Velde 2008). Nekhbet was associated with both birth and death, related to the double perspective on the vulture. There was also a belief in the 'spell of a golden vulture' (Faulkner 2008; Van Dooren 2011). A few words would be inscribed in a golden vulture that would be positioned around the neck of a dead person, and the same words would be uttered over the body for the protection of the spirit of the deceased. There is also evidence that such rites were performed for the living, as a medical document, the Edwin Smith Surgical Papyrus, dated around 1700 BCE, describes words to be uttered over two vulture feathers, to be borne by people to prevent sickness (Mackinney 1942). Possible reasons for the status of the vulture was the belief in their female identity (possibly due to the minimum sexual dimorphism of vultures) (Winter and Winter 1995; Houston 2001), and a perceived commitment to breeding and nesting (including, as noted by Horapollo, the notion that a vulture mother would even give her blood to her starving young).

One of the important factors in the relationship between people and vultures in the Old World was the practice of leaving the dead in particular places for vultures to eat (Pain et al. 2003; see also Schüz and König 1983). This followed the teachings of the prophet Zoroaster (Zoroastrianism), which probably started around the beginning of the first millennium BCE in ancient Persia. In this Persian culture, the dead were sometimes left in open stone towers for the vultures to strip them of their flesh. This practice declined, as did Zoroastrianism in Persia with the advent of Islam. The practice moved to India, with the Parsi descendants of Zoroastrianism, and communities in Mumbai and Gujarat, among others. This ancient practice, termed 'dokhimenishini', has influenced cultural attitudes to vultures for centuries (Modi 1979; Williams 1997; Hinnells 2005; Van Dooren 2011). The Parsees are described as believing that fire, earth, and water are sacred and not to be contaminated with human corpses. The result is the construction of 'towers of silence' allowing only avian scavengers access to corpses. In Mumbai these towers were built 400 years ago. With an average of three bodies a day added to the towers, vultures could consume an entire human corpse in 30 minutes leaving only bones by nightfall (Pain et al. 2003). With the collapse of the vulture populations in India, the avian visitors are only kites and crows, which are unable to consume bodies with the same speed as vultures.

There is a broadly similar culture, among the Tibetan Buddhists of the Himalayas termed 'Sky Burials', which largely came to Western attention through the writings of Friar Odoric in the 14th century. Arguably due to the relative absence of timber resources for cremation in the barren, rocky and ice-covered Himalayas, corpses are offered to vultures, after religious ceremonies (Martin 1996; Bruno 2005). Even the bones would be ground down and mixed with blood or barley for the consumption of the vultures. The consumption by vultures of all the flesh and bones was of vital significance, as it was 'well known that a man must have been wicked in life whose body is rejected by the vultures and wolves' (Wollaston 1922: 12). Van Dooren (2011: 70) notes that both the Indian and Chinese systems are based on religious doctrines and also the idea of charity, or giving the birds alms.

Other cultures portray a more violent relationship between people and vultures. Examples may be found in ancient Greek mythology. The Hesiods Theogony describes the titan Prometheus as punished by Zeus, after his theft of fire from the gods for people, by being bound to a rock with a shaft through his waist, and liver (regenerated each night) consumed daily by an eagle (Evelyn-White 1920). Later literature replaces the eagle with a vulture (Bullfinch 2000).

Another common theme is the portrayal of vultures devouring the bodies of the losing side in battles. In the ancient Sumerian city of Girsu, a stone carving called the Steele of Vultures has been dated to about 2400 BCE, and depicts the defeat of the city-state of Umma by the neighboring King Eannatum of Lagash. The bodies of the losers appear to be trampled by the victors and devoured by vultures (Winter 1985). Another example is the Egyptian Battlefield Palette of the Predynastic period, which shows vultures, other avian scavengers and lions devouring the dead (Mundy et al. 1992). Another is example from ancient Assyria, from the period of Tiglath-Pileser, is a stone carved picture of two armored horsemen riding with a vulture flying above them, carrying what appears to be entrails (Van Dooren 2011).

Aristotle in his 'History of Animals' noted that large numbers of vultures followed armies before battles. This influenced many later Western writers. The Etymologies of Isidore of Seville of the 7th century CE described vultures as being similar to eagles, in that they could detect corpses over great distances, possibly even beyond the seas (Barney 2005). This belief in long distance sensing was echoed by Bartholomaeus Anglicus in the 13th century (Steele 2004), and arguably relates to the writings of John Milton, who notes that vultures arrive at battlefields just at the right time for scavenging (Milton 1968).

The Romans also had opinions of vultures. In the Life of Romulus, Plutarch writes that the founders of Rome, Romulus and Remus used vultures to determine the location of the future city. Hence, 'Romans, in their divinations from birds, chiefly regard the vulture'.

Later European thinkers wrote ideas based on ancient Egyptian and other roots, and perceived vultures within medical ideas. 'Vultures have often been connected to human health and wellbeing', and unlike the 'symbolic and religious' focus of the ancient Egyptians, 'the European Epistula Vultaris calls for the use vultures' body parts to achieve medical and magical ends' (Van Dooren 2011: 108). Although there are several versions of the Epistula Vultaris, some versions promote the use of vulture parts for curing physical ailments such as migraines and epilepsy and also spiritual ailments such as sorcery and demon possession (Mackinney 1943). These ideas, including some from Pliny the Elders Natural History of the 1st century CE, influenced later medical work, for example the *Parnassus Medicinalis Illustratus* of the 15th century, the Dictionnaire Universel des Drogues Simples and Universal-Lexicon of the 18th century, and the Historie des Medicaments of the late 19th century (Zedler 1732–54; Mackinney 1942).

Later writers in Europe scorned vultures. Van Dooren (2011: 50) quotes Goldsmith (1816) in his comparison of vultures and eagles, as representative of a dominant mindset of the period:

'The first rank in the description has been given to the eagle; not because it is stronger or larger than the vulture, but because it is more generous and bold. The eagle, unless pressed by famine, will not stoop to carrion; and never devours but what he has earned by his own pursuit. The vulture, on the contrary, is indelicately voracious; and seldom attacks living animals, when it can be supplied with the dead. The eagle meets and singly opposes his enemy; the vulture, if it expects resistance, calls in the aid of its kind, and basely overpowers its prey by a cowardly combination. Putrefaction and stench, instead of deterring, only serve to allure them.'

During the 19th century, people were portrayed as vultures in political cartoons; these implied a person, e.g., a banker, politician, media person or business person who preyed on others in dishonest fashion. Examples are 'vulture fund', 'vulture investor' and 'vulture capitalist' quoted from the 2006 edition of the *Oxford English Dictionary* (Van Dooren 2011: 53).

In southern Africa, there were various perspectives on vultures. The Zulus saw vultures as possessing prophetic vision, predicting for example a death in the home by calling outside the house. They could predict rain and even locate missing livestock (Mander et al. 2007). Possibly, these beliefs promoted the associated trade in vulture body parts, which were believed to confer clairvoyant powers and greater intelligence on the user (Cunningham 1990). To obtain such parts, poisoning, shooting and trapping of vultures is common, with poisoning of carcasses to kill vultures being of serious conservation concern (Mundy et al. 1992; McKean 2007; Boshoff and Anderson 2008).

Vultures have also been associated with kings and regal power (Van Dooren 2011). For example, there is the vulture king Jatayu from the epic

Ramayana of the Hindu religion in India. The epic story concerns a prince Rama and his brothers, who are incarnations of the God Vishnu. Born to destroy the demon king Ravana, a conflict begins when Rama's wife Sita is kidnapped by Ravana, and the vulture God Jatayu, who is killed in combat with Ravana, aids Rama by telling him the location and future of his quest to find his kidnapped wife. This legend apparently influences positive feelings towards vultures in modern India and Nepal (Griffith 1870–74; Baral et al. 2007). A similar myth is found in ancient China where the God Shaohao, is connected to birds, and started a kingdom were phoenixes appeared, but possibly was a vulture king and his subjects other birds (Yang and An 2005).

9.4 Current Public Attitudes to Old World Vultures

Public attitudes to Old World vultures appear to be similar to those concerning New World vultures; fear, respect, admiration, repulsiveness and conservation sympathy. As noted by Gross (2006) Europeans, like Americans, used to persecute vultures believing them to be the transmitters of disease, when in fact the vultures controlled brucellosis, anthrax, and other livestock diseases by eliminating the carcasses that spread infection.

Vultures in recent times and currently are still killed by people in Europe. However, increasingly, legislation has been created and implemented to protect the vultures, other raptors and birds and wildlife in general. Europe has more legislation for the protection of raptors and other birds than Africa or Asia. The two main international treaties for the conservation of birds of prey in Europe are: the 1979 EEC Council Directive 79/409/EEC on the Conservation of Wild Birds (commonly known as the 'Birds Directive') and the 1979 Convention on the Conservation of European Wildlife and Natural Habitats (popularly known as the 'Berne Convention') (Stroud 2003). The Birds Directive was adopted by the European Community in 1979, for the protection of all the resident and other naturally occurring wild birds in the territory of the EC Member States. Parts of the law, especially species lists were modified when new states joined the EU. The Birds Directive is broad and considers both animal and human issues (Temple-Lang 1982). Article 2 requires Member States to 'maintain the population of the species referred to in Article 1 [i.e., all wild birds] at a level which corresponds in particular to ecological, scientific and cultural requirements, while taking account of economic and recreational requirements, or to adapt the population of these species to that level.' Article 3 requires Member States to preserve, maintain or re-establish a sufficient diversity and area of habitats to meet the obligations in Article 2. Vultures added in 1979 were the Bearded vulture, Egyptian vulture, Griffon vulture and the Cinereous vulture. Article 4 also requires member states to classify the most suitable landcover as special protection areas (SPAs), for the rare or vulnerable species listed in Annex I (Article 4.1), and for regularly occurring migratory species (Article 4.2).

Annex I lists species which 'shall be the subject of special conservation measures concerning their habitat in order to ensure their survival and reproduction in their area of distribution'. This means that for European raptors, Member States are required to classify SPAs under Article 4. EU member states are however allowed a limited degree of discretion in defining 'the most suitable territories'. The limitations of this discretion was defined in the European Court of Justice case-law, e.g., ECJ Case C-3/96—Commission of the European Communities v. The Kingdom of The Netherlands supported by the Federal Republic of Germany (Judgement of the Court). OJ, 25 July 1998, C234/8. Also included in the Birds Directive are details of species protection (Article 5), banning of trade in live or dead birds (Article 6), limitations of hunting (Article 7), and capture or killing (Article 8), especially for the species listed in Annex IV(a). It also allows exceptions in the interests of public health and safety, and air safety, to prevent serious damage to crops, livestock, forests, fisheries and water, for the protection of flora and fauna, for research and teaching, of re-population, of re-introduction and for the breeding necessary for these purposes and also to permit, under strictly supervised conditions and on a selective basis, the capture, keeping or other judicious use of, certain birds in small numbers (Article 9). Research in support of conservation is also encouraged (Article 10).

The Berne Convention, similar to the Birds Directive provides a broad framework for the conservation of fauna and flora within the signatory countries of the Council of Europe. As noted by Stroud (2003: 58) 'As they relate to birds of prey, the species protection regime (including derogation measures) of the Berne Convention is virtually identical to that of the Birds Directive.' Other important legislative pieces in Europe relevant to vultures are commonly called the Habitats Directive, the Convention on Trade in Endangered Species (CITES) and the Convention on the Conservation of Migratory Species (Bonn Convention). The 'Habitats Directive' (Council Directive 93/43/EEC on the conservation of natural habitats and of wild fauna and flora) is important as raptors benefit indirectly from the habitat protection measures (including the classification of a European network of Special Areas of Conservation). Article 6 of the Directive is also directly relevant to the management and conservation of SPAs identified under Article 4 of the Birds Directive. The Convention on Trade in Endangered Species (CITES) (1975) regulates the international trade in specimens, live, dead or derivatives of wild fauna and flora. It is based on a system of conditional permits. Raptors occurring in Europe are listed in Appendix II of CITES, except for the White-tailed eagle *Haliaeetus albicilla*, Imperial eagle *Aquila heliaca*, Spanish imperial eagle (*A. adalberti*), Peregrine (*Falco peregrinus*) and Gyrfalcon (*F. rusticolus*), which are listed in Appendix I. Although the EU is not yet a Party to CITES (Stroud 2003), most European countries are signatories to CITES. The EU fully implements the CITES from 1984 through EU Regulations. An example is the Council Regulation (EC) No. 338/97 on the Protection of Species of Wild Fauna and Flora by regulating Trade therein. This Regulation actually goes beyond CITES by adding non-CITES listed species in its Annexes.

The Convention on the Conservation of Migratory Species (Bonn Convention) (1983) provides a framework within which the contracting parties may act and co-operate for conservation purposes. In particular, these include the adoption of: protection measures for endangered migratory species; agreements for the conservation and management of threatened migratory species; and the undertaking of joint research activities. Appendix 1 lists species that are in danger of extinction throughout all or a significant proportion of their range. Appendix II lists migratory species whose conservation status (not necessarily endangered) would benefit from conservation or related actions. All European diurnal raptors are listed in Appendix II of the Bonn Convention, with the white-tailed eagle additionally listed in Appendix I.

International action plans also exist, these being initiated by the Council of Europe and the European Commission. For example, such plans have been published for the Cinereous Vulture (Heredia et al. 1996; Stroud 2003). EU governments have also united to promote raptor conservation that are 'among the strongest for any group of species anywhere in the world, and apply over a huge extent of Europe, from North Africa to high Arctic Svalbard, and from the Iberian Peninsula to the shores of the Caspian Sea' (Stroud 2003: 73). Nevertheless, Stroud (2003) notes that despite these actions, many birds of prey are still in conservation trouble; in fact the main problems (persecution, including shooting, trapping, direct and indirect poisoning, nest destruction; disturbance at nest sites; egg-robbing; illegal trade in live birds for falconry; habitat change, loss and fragmentation) 'are exactly those that the Convention and Directive explicitly aims to control' (ibid. 79). Factors for these failures are the wide interpretation of international treaty provisions, inadequate enforcement measures, and inadequate environmental education and public awareness programmes. Some laws had negative effects. For example, the EU Regulation 1774/2002 designed to stop the spread of BSE had the unintended effect of starving scavengers, many of which are protected species. Another fundamental problem is the impact of landcover change on raptor nesting, foraging and presence. The main landcover changes are based on agricultural policies that transform extensive and low-intensity farming systems to high-intensity and/or monocultural forms of agriculture, which may affect the entire ecosystems upon which raptors depend (Beaufoy et al. 1994; Tucker and Heath 1994; Donázar et al. 1997; Hagemeijer and Blair 1997; Pain and Pienkowski 1997; Heath and Evans 2000; Stroud 2003).

9.5 Conclusions

Vultures have gone through several changes in their relations with people; however it is the people, rather than the vultures, who changed the relationship. The relation was based on both fiction and a prioritization of fact. In some regions and time periods people prioritized some vultures as negative, while in some other areas or time periods the same vultures were seen as positive. For example, the idea of vultures devouring human corpses was seen as negative by some people, and positive by other cultures. This polarization in time and space also exists within modern polities, as some people for example complain that vultures collide with aeroplanes, pollute urban contexts and kill vulnerable livestock, while other people defend vultures for their aesthetic beauty, cleaning role and even right to exist. From the evidence presented in this chapter there are clear similarities and differences on these points between the continents and between the New and Old Worlds. In North America, only the California Condor is threatened, and the conservation methods for this species (captive breeding and reintroductions) are similar to those for the Old World Cinereous and Griffon vultures. The Black and Turkey vultures are rather increasing, to numbers similar to the Griffons in India, Pakistan and Nepal before the diclofenac disaster. The incorporation of vultures into a quasi-religious culture, which existed in India, ceased to exist in North America with European colonial dominance. In Africa, only South Africa has a serious conservation policy, while in other regions of the continent vultures are declining due to a variety of factors. Throughout Asia, Europe and Africa, ungulate decline is an important factor for the rarity of vultures. Legislative actions for conservation in Europe and North America are similar, but the protection of the vultures is only one issue in a larger scenario of habitat change, prey extinction and urbanisation, from which vultures receive little protection from such laws. The question of future vulture survival must therefore draw together all the threads examined so far, and explore how, based on the limitations of vulture behavior, these birds can cope with the changes, the sustainability of the changes and the possibility of changes in these changes.

10

The Future of Vultures: Conclusions and Summary

10 INTRODUCTION

The future of vultures rests on the trends in the factors examined in this book: conservation policies, landcover change, ecological competition and social attitudes, the sustainability and flexibility of these trends, and the likelihood of new events that may either enhance or degrade the possibilities for vulture survival. Particular points to be considered are changes in the factors for the death of vultures; poisons, electrocution and line collisions, wind turbines; shootings; habitat loss; food losses; and air collisions. There are also changes in the factors for the conservation of vultures; public attitudes, conservation policy, technological changes and landscape issues. The evidence presented in this book shows the strength for the argument that vultures are relics from the past; birds that once fed on mastodons and competed with the huge cats may have no real sustainable place in the modern world, except as recipients of handouts in vulture 'restaurants' and foragers for scraps in increasingly clean cities. There is also another strand; are vultures adaptive, hence their ability to forage in urban areas, or is this just a phenomenon of the smaller vultures, which behave like corvids and seagulls? Can their huge relatives adapt? Also, as the world modernizes, and sanitation becomes more effective, the urban adaptation may fade, leading to a retreat based on livestock attacks until the sources are exhausted. Shortly, will vultures go entirely extinct in the wild, or is there another future, where they reclaim part of their heritage as nature's cleaners, contributors to ecosystems and free rather than shackled to human generosity?

This last chapter is an assessment of the future of both New and Old World Vultures, based on the arguments, analyses and information presented in the first nine chapters. The summary of vulture issues are: (1) vultures are large, specialised birds and their slow breeding creates difficulties for their conservation; (2) vultures appear to have evolved from different species, therefore despite the similarities in appearance, they are actually different birds, with different habits and needs; (3) these differences emerge when comparing generalists and specialists, occasional fresh meat eaters, total scavengers and occasional plant eaters, ground, tree and cliff nesters, human tolerant and human intolerant species; (4) despite their size, vultures are not relics from the past, but part of the modern environment as long as their needs, which are actually quite simple, are met; (5) vultures are important for ecosystems and merely creating a cleaner world does not negate this role; and (6) creating physical components for vulture ecology, while avoiding factors for their extinction does not unduly burden human life.

10.1 Changes in the Poisons that Kill Vultures

The ban on diclofenac in 2006, although not fully implemented, was a major milestone for the survival of vultures. As has been cited in this book, recovery of vultures is already beginning for some species. There are two main issues: (1) whether an effective ban will help the vultures species recover; and (2) if other related issues such the decline in cattle carcass dumping, sky burials, increased urban sanitation and incineration of carcasses, increased populations of competitors such as feral dogs and corvids, and habitat losses will prevent a full recovery, even if the drug is totally banned. Few studies have yet examined these issues, as the full evidence of a vulture recovery across Asia is not yet documented. If the vulture decline led or leads to permanent socio-environmental change, the prospects of a full vulture recovery may be conditional on the adaptation of the birds to the new reality. An important issue is that neither the current vultures nor the ecosystems may have had to deal with such a situation before, where vultures are virtually non-existent, facultative scavengers do most of the scavenging work, and carcasses of animals and people must be disposed of in another fashion.

Increasing attention to sanitation, not only in the affected Asian nations but globally is a related concern. In a so-called civilized society, it is unacceptable for carrion to be lying in the urban to rural environment. As health and development policies demand cleaner environments, to promote human health and well being, the issue is 'what about the vultures'? A clean environment is an intrinsic aspect of human quality of life, enumerated as physical, material, social and emotional well-being, none of which are possible with dead, decaying flesh in close proximity. With alternatives to vultures being found what is the future for these birds?

Another poison that kills vultures, namely lead from gunshots may be solved by the ban in lead in guns. Such a ban may be possible in the United States with improved technology, and possibly in Europe. If such bans are in effect, a global ban may still be decades away. Associated with this is the practice of hunting (which may be a declining activity), and the disposal of some of the remains from hunting. Hunting is viewed by many among the science community and the general public as wrong. If either hunting declines or carcasses are disposed of in a more sanitary manner, the result for vultures is negative. The question then is what will replace the source of meat. So far the only replacements have been the vulture restaurants, but these still require surplus meat, may be expansive to run and being artificial, reduce the natural ecological behavior of vultures.

Pesticides also affect vultures. Despite the ban on the agricultural use of DDT in the United States in 1972, and the global ban in the Stockholm Convention (1972), it is arguably still used, and other dangerous chemical compounds have taken its place. The organochlorine pesticides used in South African agriculture have been detected in the blood of vultures, and are yet to be banned. The question concerns the future for the use of these chemicals. The only future solution is to find a compound that does not threaten vultures. This is possible, given that not all animals are threatened by the same compounds. For example, Turkey Vultures were found to be unaffected by diclofenac. Future avoidance of the diclofenac issue would require Vulture, or Bird Impact Assessments to be exhaustively conducted before any pesticide could be marketed. This, of course would be a major undertaking, as even testing the chemicals on birds would require many test animals from already endangered species.

The use of poisons, such as Furadan in carcasses to kill predators in African countries and some other countries is a major problem that may not disappear soon. With increased human populations and increased pressure for food supplies, livestock keepers may become more reluctant to listen to the policies of conservationists and tolerate carnivores killing their livestock. A further problem in Africa and Asia is desertification, which reduces grazing land for both livestock and wild ungulates and by encouraging competition for grazing land, attracts large carnivores to close proximity. The killing of 'problem' carnivores and the protection of livestock is likely to hamper food access for vultures. For this problem, there is no clear solution in sight, because solutions link to issues outside the reach of environmental policy; these include the rising human populations in the Global South, the increasingly variable weather that encourages drought and famine and the lack of economic development in many contexts.

The killing with guns is illegal in the North America and within the European Union. Few studies have been able to document the effectiveness of the vulture protection laws in this regard, because this would require the location of any shot birds. In the tropical countries, the shooting of vultures, like the poisoning of these birds, is even more difficult to document. The future for vultures in this regard can only be bright if guns are banned and somehow all removed from the hands of all relevant people. However, this is unlikely with the world rather moving in the opposite direction. Increased proliferation of guns means that the protection of vultures will depend more on human desire for their survival than on the possibility of unrealistic anti-gun legislation.

10.2 Electrocution and Power Line Collisions

The use of pylons and other above ground electrical equipment has been found to be a major factor behind the loss of raptor populations. The obvious solution to this, which has been already implemented in some European countries is the location of electrical lines underground. However, there are several problems with this approach. Power lines may be more difficult to locate for repairs when underground especially under concrete, laying of lines underground creates dangers for anyone who digs without knowledge of the location of the line, and this alternative has not spread to the majority of countries with vulture populations. Rural electrification is a major aspect of development in most developing nations; in such instances the 'green' environmental issues and the 'brown' human issues may be ranged against each other.

How this problem may be solved is now uncertain. Like the pesticide issues, human population growth is a major factor. As urban areas expand and roads are built, power lines have become ubiquitous features of the landscape. Solutions such as building lines away from areas of vulture foraging are not realistic especially in developing countries. Alternative designs have made little impact on electrocutions and possibly less on collisions. Rapidly electrifying countries in the Global South are ready candidates for the expansion of such lines. The future of vultures for this issue rests on alternative designs that discourage perching, lower the risk of birds connecting with two live wires, more visible structures, and possibly gradual acceptance of at least some wires underground. This cannot be anything more than a gradual process, considering the socio-economic contexts under which it would be attempted.

10.3 Wind Turbines

Problems with wind turbines are similar, as they are also an aspect of development; cleaner, greener energy sources are increasingly seen as manifestation of the twenty-first century's approach to human development.

Fortunately, or unfortunately, depending on one's viewpoint, these developments are currently mostly limited to North America and Europe. The prospects of huge numbers of wind turbines being built in the savannas of Africa or South America, or the densely populated plains of south east Asia are at present bleak. Nevertheless, wind turbines are presently being developed in China, Japan, India, the Philippines and South Korea. The beginnings for Africa are in South Africa, Morocco and Kenya and for South America, Brazil and Chile.

Currently, there is no clear solution for bird collisions. Several possibilities can be mooted, e.g., changes in the design and/or size of the turbines, surrounding the turbines with high fencing, driving raptors from the location (e.g., more sanitary surroundings) and locating turbines far from possible raptor habitats (e.g., in the sea). Currently, some wind turbines are located in the sea, therefore distant from vulture presence. However, many are located in precisely the type of area vultures might soar over; open windy, thermal source regions. More monitoring of the location and impact on birds may be needed, but an immediate solution is plainly not currently possible.

10.4 Air Collisions

Air collisions are even more serious than turbine collisions, as aeroplanes are moving bodies. Airports are also usually built on flat land near cities and towns, areas that produce strong winds and thermals where vultures and other birds practice soaring flight. Therefore, to solve this problem in the vulture's favor there must be acknowledgment that the birds are worth saving, this justifying the costly exercise of redesigning airports or creating landcover developments that would reduce the problem, rather than simply killing the birds.

Some researchers have suggested expensive methods, to avoid collisions: removing the environmental factors that attract birds to the airport zone, such as refuse, carcasses, food markets, farms, livestock rearing and even human settlements (for vultures that are attracted to human settlements such as Black and Turkey vultures in the New World, and Hooded vultures in the Old World). Other possibilities would be to attract vultures to distant areas, away from the airports, and also from the main approach the aeroplanes use for connecting to the airport.

The future for ameliorate actions is linked to the increasing size and speed of commercial aeroplanes, the increased number and size of airports, and the increased use of small supersonic military aircraft. These developments mean that matters may be getting worse for vultures. Although none of the larger vulture species have been documented as strongly attracted to urban settlements, the problems of shared air space (with aircraft) in high thermals and orographic lift over mountains, preferences for foraging over highlands and also flat open land, and the long hours spent soaring (one vulture could pass twenty or more aeroplanes, miles from an airport in one day). Therefore, there are natural reasons from the perspective of vulture ecology and also from aeronautical developments and landscape design that may ensure that the problems increase rather than decline. This might especially be the case if vulture numbers increase. Possible externalities that may produce a brighter future include the redesign of the airport context to discourage vulture presence, by the methods suggested above or possibly vultures may adapt to the presence of aeroplanes.

While these options may be practicable in North America, the application of such measures in the Global South is less certain. In these countries there are fewer aeroplanes and airports, a situation which may change with increased populations and transport needs. Consideration of avian needs when designing new airports will depend on the awareness of conservation problems. Otherwise a simple solution might be mooted; kill the birds so aeroplanes are safer. In Asia, the question would be whether the vulture sympathy that developed around the diclofenac crisis would translate into more permanent perspectives that would resolve collision problems in the vultures' favor. Africa might face more intractable problems, as there may be less knowledge of the declining status of vultures, there is greater difficulty in assessing the factors for vulture decline and the organization of conservation is weaker (with the possible exception of South Africa). The greater socio-economic and environmental upheavals in Africa may also preclude conservation priorities for this problem.

10.5 Habitat and Food Losses

Habitat loss is complicated for vultures. While the presence of food is the key issue, habitat preferences are diverse; some vultures prefer mountains, others live in forest and a few in cities, and the majority in savannas and meadows. Any type of terrain may lack the carcasses that are necessary for the future of vultures. Different species of vultures appear to be coping with habitat change in markedly different ways; this ensures that an assessment of their futures is different, in some cases even polarized. Changes that decimate some species (for example reforestation) may even encourage other species, or be neutral for others.

Concerning habitats, the future for New World vultures in North America appears to be reasonably good. Even the California condor has a brightening future, due to the captive breeding programme and the moderate success of the reintroduction programmes. Turkey and Black vultures are expanding northwards, without a real change in the landscapes concerned. Studies quoted in this book have recorded a northward expansion in the Turkey and Black Vulture populations since the 1920s, with a temporary decline during the 1970s when DDT was used. This is partly attributed to the ban in DDT which caused eggshell thinning. The future for this expansion may depend on the temporary or permanent nature of the factors responsible. These were cited as a general warming of the climate, increases in the population of deer and road-killed animals, reduced use of pesticides and lowered human persecution and an increase in the number of landfills. Climate change is cited as possibly the most important, due to the similar northward expansion in other species.

Cathartid adaptation to environmental change is an important factor for expansion. However, this is not the whole story. Increased deer numbers and road kills may not necessarily be permanent, as hunting regulations may change and improved sanitary services may remove roadkills earlier before putrefaction. Road kills, which increase with increased mileages of roads, larger numbers of mammals, more cars and possibly forest growth near roads, may not be a permanent phenomenon. Deer conservation is also an important environmental issue. Efficient environmental management may reduce roadkills by building underpasses under roads, overpasses or more fences near roads. Deer hunting may also be modified, so that hunters do not leave parts of carcasses in the woods, as these may attract coyotes and other facultative scavengers deemed troublesome. Burning, burying or other disposal of remains, deemed more hygienic, may reduce vulture feeding and the presence of facultative scavengers, while enhancing human quality of life.

The same arguments may apply to landfills. Where landfills contain organic matter that may be eaten by vultures, they also attract facultative scavengers and possible agents for human diseases. These landfills may not be permanent, as more hygienic methods will ensure that meat is wrapped in plastic bags to prevent scents, or in large bins which prevent access by scavengers. Landfills may also be periodically covered with layers of soil beyond the capacity of vultures to uncover. All these methods encourage increased hygiene, cleanliness and enhanced human quality of life.

The decreased use of pesticides may be permanent, unless agricultural pests increase markedly. Many researchers advocate the use of 'greener' methods; these include integrated pest management, use of natural pest predators and removal of areas suitable for pest breeding. Whether these methods enable a permanent decline in pesticide use remains to be seen. Such bans may not necessarily be continent-wide, and considering the migratory habits of some New World Vultures, problems may still arise when the birds cross borders.

Laws of varying permanence are an important factor for the decline in the persecution of vultures in Europe and North America. While in the past vultures were frequently shot, such incidents appear to be on the wane. However, with increased vulture numbers, and increased incidents of, for example Black Vultures attacking young livestock in North America, and possible similar occurrences (real or imagined) for vultures in Europe, the question is whether human persecution is abolished for good, or will it return when the vulture populations cause problems beyond the tolerance of the relevant stakeholders.

Several studies quoted in the earlier chapters argue that the decline or local extinction of vultures contributes to the increase of facultative scavengers that create problems for people. However with increased sanitation services, this problem might be ameliorated, at least for large carcasses such as deer, which may be easier to detect and destroy. The idea of increased sanitary services performing the work of scavengers is already important in North America and Europe. These arguments are complicated by the increasing human populations in North America, which require new settlements in previously natural habitats.

Therefore, for vultures in North America, the future for the vultures, appears to be actually be reliant on direct human intervention, either to stop actions in the name of hygiene and quality of life that may impede vulture food access, or actions to create vulture opportunities such as conservation of wildlife. These actions depend on the commitment of those concerned about vulture welfare and the scenarios they are able to negotiate with those who are concerned with sanitation and human quality of life.

In Africa, the future for vultures based on habitat and food losses appears to be more bleak. Even the most adaptive species, the Hooded Vulture has declined in its favored urban foraging grounds. In Africa, habitat issues include desertification, large wild ungulate declines and local extinction outside reserves and increased urbanisation. None of these issues appear to solvable in the near future. Larger vultures are more affected than the Hooded vulture, as they are usually less tolerant of human presence and landcover change.

Human population increase in Africa is forecast by many commentators to continue at current levels at least well into the twenty first century. This will be negative for vultures, despite the increased cattle and small livestock raising, savannaisation as a result of deforestation for agriculture and rural settlements, and urbanisation with abattoirs and food production areas. Vultures will be more dependent on unsanitary human activities, which may eventually be resolved. Also, as already noted, most African vultures are not urban denizens like the small Hooded Vulture.

Other social problems in Africa include hunger, disease and conflict, all of which are combated by increased food supply, increased medical and sanitation services, birth control and conflict resolution. Increased social and economic development, may take decades to resolve, considering the current trends towards substate and tribal conflict (e.g., South Sudan, Mali, Central African Republic, Chad, Somalia, Eithiopia and northern Kenya). These incidents create the need for more effective management, the result of which may eventually reduce the food sources of vultures. Effective conservation wild ungulates and suitable habitat will be balanced against the needs of growing human populations, and similar to the evolving situation in South Africa vulture habitats may be largely within protected areas.

Africa differs from Asia in that there are no clear, over-riding factors for the decline of vultures. Possible factors include deliberate killing for food, local religions, livestock protection; local declines in wild ungulates, rural to urban migration for people, and consequent declines in cattle herding lifestyles and urban spread. The lack of single factor, such as diclofenac use in Asia, hampers government action and impedes the formulation of policies that may ameliorate impacts on vultures. It is much simpler to deal with one problem factor than several. Considering the problems listed above, the possibility of vulture protection in the future is uncertain even with increased awareness of the vulture issues. The commitment of governments and local people is difficult to acquire.

In Asia, the habitats and food supply of vultures still exist, however the catastrophic decline may have caused some problems (increased numbers of facultative scavengers and human diseases). It remains to be seen if alternative methods of carcass disposal could reduce vulture food supply, even with a full recovery of vulture numbers. Human population growth is also high in Asian nations, and sanitation is a prominent issue. The ban on diclofenac shows that policy makers are aware of vulture issues. Another issue that remains to be seen is whether the vulture awareness created by the diclofenac debacle will be temporary or will lead to long-term conservation concerns, which will translate into effective policy on the ground. The evidence shows that people in local contexts are aware of the impact of the vulture declines, and may also be aware that increased sanitation to replace vultures may be more costly in time and money than simply letting the birds do the sanitation work. Participationary methods between local people and the government may therefore yield dividends, but it is now too early for a judgement on the result for long term vulture survival.

10.6 Academic Research

To conclude, we must also look at the systems of thought that have articulated and suggested some solutions to problems concerning vultures. So far, academic and scientific research is the main tool by which problems are identified, factors are analysed and solutions are mooted, from which may emerge suggestions, advice and pressure for policy makers and information for changes in public attitudes. The academic revolution that occurred in the 1960s, 70s and 80s, allowed interdisciplinary and multidisciplinary fields to emerge; the better to understand and analyze the complex issues underpinning general and vulture conservation. The complexity of the issues discussed in this book, which were severely simplified for lack of space in one volume, could not be investigated or solved by single academic discipline. For example, the diclofenac problem was uncovered by toxicologists, veterinary scientists, ornithologists, zoologists, ecologists and biogeographers. Solutions and further understanding of the externalities have emerged through the work of sociologists, cultural scientists, economists and geographers. Amateur bird-watchers, policy makers and advocates and stakeholders in landuse activities also play a role. In Europe and North America legal scholars and political scientists are also important. If one looks at the journals from which the articles cited in this book were derived it will be seen that some are from the hard sciences such as toxicology and veterinary medicine, while others are from the ecological and social sciences.

The background from which such interdisciplinary work could emerge has taken time to develop. The emergence of 'environmentalism' as a contemporary intellectual and political debate was a major contributory factor behind the revised, integrated perceptions that occurred during the 1970s. New terms such as sustainable development became popular 'battlecries' and value representations for a new politically and ecologically based awareness, within the theme of sustainable development. Sustainability and environmentalism have acquired increasingly broad perspectives, giving 'a fuller understanding of the multiplicity of ways of comprehending the extraordinarily complex nexus of development-environment relations' (Peet and Watts 1996: 38). The Stockholm Conference of 1972 may be seen as an expression of the international environmental and wildlife concerns developed over the previous decade. By 1974, more than one hundred nations had set up environmental departments, agencies and/or committees and laws, and non-governmental organizations had developed.

Older studies were less likely to attempt to cross-disciplinary borders in a manner suited to the solution of integrated problems. Important recent fields upon which such studies may rest include political ecology, human ecology, cultural ecology, the cultural ecology of development and environmental sociology. These disciplines and their associated fields, are based on the knowledge that the environment is not static but chaotic. As aptly described by Holling (1993: 553–554), the 'knowledge of the system we deal with is always incomplete. Surprise is inevitable. Not only is science incomplete, the system itself is a moving target' and therefore 'there is an inherent unknowability, as well as unpredictability concerning these evolved managed ecosystems and the societies with which they are linked'.

Another relevant position, from which complex problems may be understood, concerns the work of Norgard (1994), which postulates the truism that society and nature are in permanent coevolution; society changes, and consequent social changes affect nature, with changes in nature also affecting society and so on forever. This describes a coevolutionary cosmology, which comprises some basic principles, as being the basis of a coevolutionary framework. These principles enumerate an enabling structure and guide for the study of complex interfaces between society and nature.

The first principle holds that people are part of the larger environment in which their social spaces are situated, and should be seen as such when socioenvironmental relations are appraised. The second principle argues that 'how we reason affects the social and environmental systems in which we evolve' (Norgaard 1994: 94) and 'understandings must necessarily keep evolving, replacing themselves, to keep up with the cosmos that our understandings are changing' (*ibid*. 95). The third principle is of conceptual pluralism, rooted in the tenets of the second principle, and holds that there are different ways of knowing and that participants in such situations must be aware and tolerant of different conceptual frameworks, paradoxes and of the position that specific knowledges may have limited rather than universal applications. The fourth principle describes the development of scientific and other knowledge is a socially derived process, which is affected by social needs and perceptions, which shows the 'collective nature of understanding' (Norgaard 1994: 97). Changing conditions (the fifth principle) and context (the sixth principle) are related to history. Norgaard (1984: 525) argues that coevolution provides the 'linkage between economic and ecological paradigms' and that the 'perspective emphasises how man's agricultural activities modify the ecosystem, and how the ecosystem's responses provide cause for subsequent individual action and social organisation' (Norgaard 1994: 26). Ever more complex relations are created between the socio-economic system and the ecosystem, as production must be increased to feed expanding populations and this enhances the 'material transformation of nature' as the 'creation of human activity' (Woodgate and Redclift 1998: 6).

Consideration of multiple issues within societies and cultures across continents is also relevant. It is acknowledged that we must study different structures within these cultures, as a society is dominated by structures, which are 'any recurring pattern of social behavior' and the 'ordered interrelationship between the elements of a social system or society'; examples being 'kinship, religious, economic, political and other institutions as well as associated norms, values and social roles' (Marshall 1994: 517).

Structures may 'constrain action or even determine it' (Craib 1992: 34), and are the foundation of the human-vulture relations described in this book.

Participation of governments in conservation is important, but participation of the local people in a context is even more important. For example, will local people listen to a government order banning diclofenac? Therefore, academic research on participation is important. True participation has been termed 'socio-economic empowerment' (Zimmerer 1994: 118), implying total support and motivation from the local people. Scholars have researched social participation, why sometimes it does not achieve its objectives and what must be done to enable effectiveness. This is a particularly relevant issue, considering the entwined nature of the vulture issues between local people's activities and ecology.

One study articulated seven forms of participation as being in common usage. The first four, described as manipulative, passive, consultative and material incentive based, are in reality 'non-participation' and have 'no positive, lasting effect on people's lives' (Pretty 1995: 1252). In manipulative participation 'participation is simply a pretence' as local people or their representatives have no real power: passive participation is also unilateral, as local people are not consulted (*ibid.*). Consultative participation involves an agenda set, defined and controlled by professionals, without the devolution of real consultative power to local respondents. Incentive-based participation occurs when rewards are given to participants. However, when the rewards or external pressure ceases, so does the co-operation and vice-versa (Pretty 1995; Pretty and Shah 1997). The last three, arguably better forms of participation advocated by Pretty (1995) are: functional participation, which involves local people in meeting objectives, and may or may not involve them fully in decision making; interactive participation, where local people participate in joint analysis and development; and finally self-mobilisation which refers to participation where people take independent action to solve local problems and exercise control over decision-making, even though they may be supported by requested aid agencies. This, similar to the socio-economic empowerment mentioned above, is probably the only form of participation that would benefit local vulture conservation. Local people might still require information on the conservation of vultures and their food sources, the effects of local activities on their ecology and the value of these birds to local ecosystems.

10.7 Conclusions

These approaches to the study of the environmental and social bases from which issues such as vulture ecology emerge, point to the next stage of the solution of vulture problems. Most of the cited articles called for more research, more research, more research, then more suggestions and solutions, followed by more action, more action, more action. Whether this approach will yield dividends remains to be seen, as the issues are becoming more and more complex and the tools of research, policy and implementation, and monitoring and evaluation must expand to incorporate these complexities. Ten, twenty and thirty years down the road, new studies may evaluate the status quo and assess the trajectories towards sustainable vulture futures or extinctions.

References

Abuladze, A. 1998. The Bearded Vulture *Gypaetus barbatus* in Caucasia. pp. 177–182. *In*: R.D. Chancellor, B.-U. Meyburg and J.J. Ferrero (eds.). Holarctic Birds of Prey. ADENEX-WWGBP, Mérida and Berlin.

Acharya, R., R. Cuthbert, H.S. Baral and K.B. Shah. 2009. Rapid population declines of Himalayan Griffon *Gyps himalayensis* in Upper Mustang, Nepal. Bird Conservation International 19: 99–107.

Adamian, M.S. and D. Klem, Jr. 1999. Handbook of the Birds of Armenia. American University of Armenia, an affiliate of the University of California, Oakland-Yerevan.

Adams, L.G., S.D. Farley, C.A. Stricker, D.J. Demma, G.H. Roffler and D.C. Miller. 2010. Are inland wolf-ungulate systems influenced by marine subsidies of the Pacific salmon? Ecological Applications 20: 251–262.

AEE (Asociación Eólica Española). 2009. Observatorio Eólico 2009. Asociación Empresarial Eólica, Madrid, Spain.

Aginsky, B.W. 1943. Culture element distributions, XXIV: Central Sierra. University of California Anthropological Records 8: 393–468.

Agostini, N., G. Premuda, U. Mellone, M. Panuccio, D. Logozzo, E. Bassi and L. Cocchi. 2004. Crossing the sea en route to Africa: Autumn migration of some Accipitriformes over two Central Mediterranean islands. Ring 26(2): 71–78.

Agudo, R., C. Rico, C. Vilà, F. Hiraldo and J. Donázar. 2010. The role of humans in the diversification of a threatened island raptor. BMC Evolutionary Biology 10: 384.

Ahrens, C.D. 2007. Meteorology Today. An Introduction to Weather, Climate, and the Environment. 8th ed. Thomson Books/Cole, Belmont.

Åkesson, S. and A. Hedenström. 2000. Wind selectivity of migratory flight departures in birds. Behavioral Ecology Sociobiology 47: 140–144.

Akyeampong, S., E.J.D. Belford, F. Akuffo-Lartey and C. Anyinam. 2009. Distribution pattern of nests of the hooded vulture (*Necrocytes monachus*) on KNUST Campus, Kumasi. 26th Biennial Conference of the Ghana Science Association, Cape Coast, August 4–9.

Albuja, L. and T.J. De Vries. 1977. Aves colectadas y observadas alrededor de la Cueva de los Tayos, Morona-Santiago, Ecuador. Rev. Universidad Cat61ica, Quito, Afio 5(16): 199–215.

Alerstam, 1. 1990. Bird Migration. Cambridge University Press, Cambridge.

Alexander, K.A. and M.J.G. Appel. 1994a. African wild dogs (*Lycaon pictus*) endangered by a canine distemper epizootic among domestic dogs near the Masai Mara National Reserve, Kenya. Journal of Wildlife Diseases 30: 481–485.

Alexander, K.A., P.W. Kat, R.K. Wayne and T.K. Fuller. 1994b. Serologic survey of selected canine pathogens among free-ranging jackals in Kenya. Journal of Wildlife Diseases 30: 486–491.

Alexander, R.D. 1974. The evolution of social behavior. Annual Review of Ecological Systems 5: 325–383.

Alexander, R.M. 2003. Principles of Animal Location. Princeton University Press, Princeton.

Ali, S. 1993. The Book of Indian Birds. Bombay Natural History Society, Bombay.

Ali, S. 1996. The Book of Indian Birds (12th edn.). Bombay Natural History Society, Bombay.

Ali, S. and R.B. Grubh. 1984. Ecological Study of Bird Hazards at Indian Aerodromes Phase II. First Annual Report (1982–83). Delhi, Bombay, Hindan. Bombay Natural History Society, Bombay.

Ali, S. and S.D. Ripley. 1987. Compact Handbook of the Birds of India and Pakistan together with those of Bangladesh, Nepal, Bhutan and Sri Lanka. Oxford University Press, Delhi.

Ali, S. and S.D. Ripley. 1978. Handbook of the Birds of India and Pakistan, Vol. 1 (2 edn.). Oxford University Press, Oxford.

Ali, S. and S.D. Ripley. 1968–1998. Handbook of the Birds of India and Pakistan, together with those of Nepal, Sikkim, Bhutan and Ceylon. Oxford University Press, Bombay.

Ali, S. 1996. The Book of Indian Birds. Bombay Natural History Society, Bombay.

Allaby, M. 1992. The Concise Oxford Dictionary of Zoology. Oxford University Press, Oxford.

Allen, P.E., L.J. Goodrich and K.L. Bildstein. 1996. Within- and among-year effects of cold fronts on migrating raptors at Hawk Mountain, Pennsylvania, 1934–1991. Auk 113: 329–338.

Alonso, J.C., J.A. Alonso and R. Muñoz-Pulido. 1993. Marking Electric Power Lines for Protection of Birds (in Spanish). REE, Madrid.

Alström, P. 1997. Field identification of Asian Gyps vultures. Oriental Bird Club Bulletin 25: 32–49.

Alvarenga, H.M.F. and S.L. Olson. 2004. A new genus of tiny condor from the Pleistocene of Brazil (Aves: Vulturidae). Proceedings of the Biological Society of Washington 117: 1–9.

Alvarez, L.W., W. Alvarez, F. Asaro and H.V. Michel. 1980. Extraterrestrial cause for the Cretaceous–Tertiary extinction. Science 208(4448): 1095–1110.

Amadon, D. and J. Bull. 1988. Hawks and Owls of the World. Proceedings of the Western Foundation of Vertebrate Zoology 3: 295–357.

Amadon, D. 1964. The evolution of low reproductive rates in birds. Evolution 18: 105–110.

Amadon, D. 1977. Notes on the Taxonomy of Vultures. Condor (Cooper Ornithological Society) 79(4): 413–416.

Amadon, D. 1977. Review on the taxonomy of vultures. Condor 79: 413–416.

Amadon, D. and J. Bull. 1988. Hawks and owls of the world: a distributional and taxonomic list. Proceedings of the Western Foundation of Vertebrate Zoology 3: 295–357.

Amason, U., A. Gullberg and B. Widegren. 1991. The complete nucleotide sequence of the mitochondrial DNA of the fin whale, *Balaenoptera physalus*. Journal of Molecular Evolution 33: 556–568.

American Ornithologists Union. 1998. The AOU Check-list of North American Birds, 7th ed. The American Ornithologists' Union.

American Ornithologists' Union. 2009. Check-list of North American Birds, Tinamiformes to Falconiformes 7th edition. AOU.

American Ornithologists' Union. 1983. Check-list of North American birds, 6th edition. American Ornithologists' Union, Washington, D.C.

American Ornithologists' Union. 1998. Check-list of North American birds. 7th edition. American Ornithologists' Union, Washington, D.C.

Amores, F. 1979. Estructura de una comunidad de rapaces en el ecosistema meditarreneo de Sierra Morena durante el periodo reproductor. Unpublished PhD Thesis, Universidad Complutense de Madrid, Madrid.

Amuno, J.B. 2001. The ecology and behavior of vulture populations in Kampala. Unpublished BSc diseratation. Makerere University, Kampala.

Anderson, M.D. 1999. Africa's Hooded Vulture: a dichotomy of lifestyle. Vulture News 41: 3–5.

Anderson, S., A.T. Bankier and B.G. Barrell. 1981. Sequence and organisation of the human mito chondrial genome. Nature 290: 457–465.

Anderson, D.L., D.A. Wiedenfeld, M.J. Bechard and S.J. Novak. 2004. Avian diversity in the Moskitia region of Honduras. Ornitologia Neotropical 15: 467–482.

Anderson, D.J. and R.J. Horwitz. 2008. Competitive interactions among vultures and their avian competitors. 121(4): 505–509.

Anderson, M.D. 2002. Karoo Large Terrestrial Bird Powerline Project Report No. 1. Johannesburg: Eskom (Unpublished report).

Anderson, M.D., A.W.A. Maritz and E. Oosthuysen. 1999. Raptors drowning in farm reservoirs in South Africa. Ostrich 70(2): 139–144.

Anderson, M.D. 1999. Africa's Hooded Vulture: a dichotomy of lifestyle. Vulture News 41: 3–5.

Anderson, M.D. 2000. African white-backed vulture. *In*: K.N. Barnes (ed.). The Eskom Red Data Book of Birds of South Africa, Lesotho and Swaziland, 77. Johannesburg: BirdLife South Africa.

Anderson, M.D. 2000. Cape vulture. *In*: K.N. Barnes (ed.). The Eskom Red Data Book of Birds of South Africa, Lesotho and Swaziland. BirdLife South Africa, Johannesburg.

Anderson, M.D. 2000. Lappetfaced vulture *Torgos tracheliotus*. *In*: K.N. Barnes (ed.). The Eskom Red Data Book of Birds of South Africa, Lesotho and Swaziland. BirdLife South Africa, Johannesburg.

Anderson, M.D. 2003. Northern Cape Province—Annual Report 2002. Vulture Study Group Workshop, Aandstêr, Namibrand Nature Reserve, Namibia.

Anderson, M.D. 2004a. The problem with drugs. Earth Year 2004(2): 72–73.

Anderson, M.D. 2004b. Vulture crises in South Asia and West Africa .... and monitoring, or the lack thereof, in Africa. Vulture News 52: 3–4.

Anderson, M.D. and P. Hohne. 2007. African white-backed vultures nesting on electricity pylons in the Kimberley area, Northern Cape and Free State provinces, South Africa. Vulture News 57: 44–50.

Anderson, M.D., A.W.A. Maritz and E. Oosthuysen. 1999. Raptors drowning in farm reservoirs in South Africa. Ostrich 70(2): 139–144.

Anderson, M.D., S.E. Piper and G.E. Swan. 2005. Non-steroidal anti-inflammatory drug use in South Africa and possible effects on vultures. South African Journal of Science 101(3-4): 112–114.

Andersson, Ch.J. 1872. Notes on the Birds of Damara Land and the Adjacent Countries of South West Africa. Van Voorst, London.

Andersson, K., D. Norman and L. Werdelin. 2011. Sabretoothed Carnivores and the killing of Large Prey. PLoS ONE 6(10): e24971.

Andrle, R.F. 1968. Raptors and other North American migrants in Mexico. Condor 70: 393–395.

Andryushchenko, Y.A., M.M. Beskaravayny and I.S. Stadnichenko. 2002. Demise of Great Bustards and other bird species because of their collision with power lines on the wintering grounds (in Russian with English). Branta 5: 97–112.

Angelov, I. 2009. Egyptian Vultures *Neophron percnopterus* exposed to toxic substances. BirdLife Europe e-News 3(2): 7.

Angelov, I., T. Yotsova, M. Sarrouf and M.J. McGrady. 2013. Large increase of the Egyptian Vulture Neophron percnopterus population on Masirah Island, Oman. Sandgrouse 35: 140–152.

Angelov, I., I. Hashim and S. Oppel. 2011. Persistent electrocution mortality of Egyptian Vultures Neophron percnopterus over 28 years in East Africa. Unpublished Report.

Animal News May 6, 2013 Woman Eaten By Vultures After Fall From Cliff ^ http://www.inquisitr.com/650400/woman-eaten-by-vultures-after-fall-from-cliff/. Retrieved 25/12/2013.

Anon. 2003. Vulture death mystery explained? The Babbler: BirdLife in Indochina 2(2): 7.

Anon. 1998. California condor restoration. *In*: The Peregrine Fund. Annual Report 1998. Boise, USA: The Peregrine Fund.

Anon. 2001. Lead kills condors in the Grand Canyon National Park. Oryx 35: 108.

Anon. 2006. Namibia's Cape Griffon Vultures receive boost. Bulletin of the African Bird Club 13: 16.

Anon. 2007. Red-headed Vulture breeding programme launched. Vulture News 79–80.

Anon. 2007. Species under the wing: what BSPB is doing for globally threatened species. Neophron 2–3.

Anon. 2008. Local increase in vultures thanks to diclofenac campaign in Nepal. Buceros 13(2): 5.

Anon. 2008. Traditional medicine demand threatens vultures in southern Africa. Vulture News 77–79.

Anon. 2009. Second California Condor shot in three weeks—Wildlife Extra News. Available at: #http://www.wildlifeextra.com/go/news/condor-shot009.html#.

Anon. 2011. Cape Vulture Breeding and Reintroduction. Website. Available at: http://www.vultureconservation.co.za/?p=527 (Accessed: 20/03/2012).

Antal, M. 2010. Policy measures to address bird interactions with power lines—a comparative case study of four countries. Ostrich: Journal of African Ornithology.

Anton, M., A. Galobart and a. Turner. 2004. Co-existence of scimitar-toothed cats, lions and hominins in the European Pleistocene. Implications of the post-cranial anatomy of Homotherium latidens (Owen) for comparative palaeoecology. Quaternary Science Reviews 24: 1287–1301.

Apanius, V., S.A. Temple and M. Bale. 1983. Serum proteins of wild turkey vultures (*Cathartes aura*). Comparative Biochemistry and Physiology B 76: 907–913.10.

APLIC (Avian Power Line Interaction Committee). 2006. Suggested practices for avian protection on power lines: The state of the art in 2006. Edison Electric Institute, Washington, D.C.

Applegate, R.D. 1990. Can vultures smell? Turkey Vulture caught in cage trap. North American Bird Bander 15: 141–142.

Ararat, K., O. Fadhil, R.F. Porter and M. Salim. 2011. Breeding birds in Iraq: important new discoveries. Sandgrouse 33(1): 12–33.

Archer, G. and E.M. Godman (eds.) 1937. The Birds of British Somaliland and the Gulf of Aden. Gurney and Jackson, London.

Arctander, P. 1988. Comparative studies of avian DNA by restriction fragment length polymorphism analysis. Journal of Ornithology 129: 205–216.

Arctander, P., C. Johansen and M.-A. Coutellec-Vreto. 1999. Phylogeography of three closely related African bovids (Tribe Alcelaphini). Molecular Biology and Evolution 16: 1724–1739.

Arnett, E.B. 2005. Relationships between bats and wind turbines in Pennsylvania and West Virginia: an assessment of bat fatality search protocols, patterns of fatality, and behavioral interactions with wind turbines. A final report submitted to the Bats and Wind Energy Cooperative. Bat Conservation International, Austin.

Arrington, D.P. 1994. US Airforce Bird Aircraft Strike Hazard (BASH) Survey Report for 1989–1993. Proc. BSCE 22. Vienna, 29 August–2 September WP, 29, pp. 201–208.

Arroyo-Cabrales, J. and E. Johnson. 2003. Catálogo de los ejemplares tipo procedentes de la Cueva de San Josecito, Nuevo León. Revista Mexicana de Ciencias Geológicas 20(1): 79–93.

Arshad, M., M.J.I. Chaudhary and M. Wink. 2009. High mortality and sex ratio imbalance in a critically declining Oriental White-backed Vulture population (*Gyps bengalensis*). Journal of Ornithology 150(2): 495–503.

Arshad, M., J. Gonzalez, A.A. El-Sayed, T. Osborne and M. Wink. 2009. Phylogeny and physio-geography of critically endangered Gyps species based on nuclear and mitochondrial markers. Journal of Ornithology 150: 419–430.

Arthur, C. and V. Zenoni, refered to in A. Margalida, D. Campion and J. Donazar. 2014. Vultures vs livestock: conservation relationships in an emerging conflict between humans and wildlife. Fauna & Flora International, Oryx 1–5 doi:10.1017/S0030605312000889.

Arun, P.R. and P.A. Azeez. 2004. Vulture population decline, diclofenac and avian gout. Current Science 87: 565–568.

Ascanio, D. and R. Rincón. 2006. A first record of the Pleistocene saber-toothed cat Smilodon populator Lund, 1842 (Carnivora: Felidae: Machairodontinae) from Venezuela. Asociación Paleontologica Argentina 43(2).

Ash, J. and J. Atkins. 2009. Birds of Ethiopia and Eritrea: An Atlas of Distribution. Christopher Helm, London.

Ash, J.S. and J.E. Miskell. 1998. Birds of Somalia. Pica Press, Robertsbridge.

Atkinson, C.T., K.L. Woods, R.J. Dusek, L.S. Sileo and W.M. Iko. 1995. Wildlife disease and conservation in Hawaii: pathogenicity of avian malaria (*Plasmodium relictum*) in experimentally infected Iiwi (Vestiaria coccinea). Parasitology 111: S59–S69.

Attwell, R.J. 1963. Some observations on feeding habits, behavior and inter-relationships of Northern Rhodesian vultures. Ostrich 34: 235–247.

Augusti, J. 2002. Mammoths, Sabertooths and Hominids 65 Million Years of Mammalian Evolution in Europe, Columbia University Press, Columbia.

Augustí, J. 2002. Mammoths, Sabertooths and Hominids: 65 Million Years of Mammalian Evolution in Europe. Columbia University Press, New York.

Austin, W., K. Day, A. Gatto, J. Humphrey, C. Parish, J. Rodgers, R. Sieg, B. Smith, K. Sullivan and J. Young. 2012. A Review of the Third Five Years of the California Condor Reintroduction Program in the Southwest (2007–2011). Arizona Ecological Services Office, Phoenix.

Avery, M. 2004. Managing Depredation and Nuisance Problems Caused by Vultures. United States Department of Agriculture. Washington.

Avery, M.L. 2004. Trends in North American vulture populations. pp. 116–121. *In*: R.M. Timm and W.P. Gorenzel (eds.). Proceedings of the 21st Vertebrate Pest Conference; 1–4 March 2004, Visalia, CA. University of California, Davis, CA.

Avery, M.L. and J.L. Cummings. 2004. Livestock depredations by Black vultures and Golden eagles. Sheep & Goat Research Journal 19(1): 62–67.

Avery, M.L., J.S. Humphrey, E.A. Tillman, K.O. Phares and J.E. Hatcher. 2002. Dispersal of vulture roosts on communication towers. Journal of Raptor Research 36: 44–49.

Avery, M.L. 2004. Trends in North American vulture populations. *In*: R.M. Timm and W.P. Gorenzel (eds.). Proceedings of the 2lst Vertebrate Pest Conference, 116–121. University of California, Davis.

Avery, M.L., J.S. Humphrey, E.A. Tillman, K.O. Phares and J.E. Hatcher. 2002. Dispersing vulture roosts on communication towers. Journal of Raptor Research 36: 45–50.

Avian Power Line Interaction Committee (APLIC). 1996. Suggested practices for raptor protection on power lines—the state of the art in 1996. Edison Electric Institute and Raptor Research Foundation, Washington, DC.

Avise, J.C., W.S. Nelson and C.G. Sibley. 1994. DNA sequences support for a close phylogenetic relationship between some storks and New World vultures. Proceedings of the National Academy of Sciences 91: 5173–5177.

Avise, J.C. 1994. Molecular Markers, Natural History and Evolution. Chapman & Hall, New York, London.

Avise, J.C. and W.S. Nelson. 1995. Reply. Molecular Phylogenetics and Evolution 4: 353–356.

Bahat, O. 1995. Physiological adaptations and foraging ecology of an obligatory carrion eater—the Griffon Vulture *Gyps fulvus*. PhD thesis, Tel-Aviv University, Tel-Aviv.

Bahat, O. 1997. Conservation of threatened raptor populations in Israel. pp. 177–189. *In*: Y. Leshem, A. Froneman, P. Mundy and H. Shamir (eds.). Wings over Africa. Proceedings of the International Seminar on Bird Migration and Flight Safety: Research, Conservation, Education and Flight Safety. Tel Aviv Univerity, Tel Aviv.

Baker, A.J., D.F. Whitacre and O. Aguirre. 1996. Observations of king vultures (*Sarcoramphus papa*) drinking and bathing. Journal of Raptor Research 30(4): 246–47.

Baker, E.C.S. 1932–1935. The nidification of birds of the British Empire 1–4. London: Taylor and Francis.

Baker, J.E. 1997. Trophy hunting as a sustainable use of wildlife resources in southern and eastern Africa. Journal of Sustainable Tourism 5: 306–321.

Baker, N. In BirdLife International 2014. Species factsheet: Trigonoceps occipitalis. Downloaded from http://www.birdlife.org on 03/02/2014.

Baldus, R.D., G.R. Damm and K. Wollscheid. 2008. Best practices in sustainable hunting—a guide to best practices from around the world, CIC—International Council for Game and Wildlife Conservation, Hungary. Downloadable from http://www.cicwildlife.org/index.php?id=412.

Baldus, R.D., G.R. Damm and K. Wollscheid. 2008. Best Practices in Sustainable Hunting—a Guide to Best Practices from Around the World. CIC—International Council for Game and Wildlife Conservation, Budakeszi.

Bamford, A.J., M. Diekmann, A. Monadjem and J. Mendelsohn. 2007. Ranging behaviour of Cape Vultures Gyps coprotheres from an endangered population in Namibia. Bird Conservation International 17(4): 331–339.

Bamford, A.J., A. Monadjem and I.C.W. Hardy. 2009. Nesting habitat preference of the African White-backed Vulture *Gyps africanus* and the effects of anthopogenic disturbance. Ibis 151(1): 51–62.

Bamford, A.J., A. Monadjem and I.C.W. Hardy. 2009a. Nesting habitat preference of the African White-backed Vulture and the effects of anthropogenic disturbance: a statistical model. Ibis 151: 51–62.

Bamford, A.J., A. Monadjem, M.D. Anderson, A. Anthony, W.D. Borello, M. Bridgeford, P. Bridgeford, P. Hancock, B. Howells, J. Wakelin and I. Hardy. 2009b. Regional generality of habitat-association models: a case study on African white-backed and lappet-faced vultures. Journal of Applied Ecology 46: 852–860.

Band, W., M. Madders and D.P. Whitfield. 2007. Developing field and analytical methods to assess avian collision risks at wind farms. *In*: M. de Lucas, G.F.E. Janss and M. Ferrer (eds.). Birds and Wind Farms. Quercus, Madrid.

Bang, B.G. 1972. The nasal organs of the Black and Turkey Vultures: A Comparative Study of the Cathartid Species *Coragyps atratus atratus* and *Cathartes aura septentrionalis* (with notes on Cathartes aura falklandica, Pseudogyps bengalensis, and Neophron percnopterus. Journal of Morphology 115: 153–184.

Bang, B.G. and S. Cobb. 1968. The size of the olfactory bulb in 108 species of birds. Auk 85: 55–61.

Bang, B.G. 1964. The nasal organs of the Black and Turkey vultures: A Comparative Study of the Cathartid Species *Coragyps atratus atratus* and *Cathartes aura septentrionalis* (with notes on *Cathartes aura falklandica, Pseudogyps engalensia* and *Neophronp ercnopterus*).

Banks, R.C., R.T. Chesser, C. Cicero, J.L. Dunn, A.W. Kratter, I.J. Lovette, P.C. Rasmussen, J.V. Remsen, Jr., J.D. Rising and D.F. Stotz. 2007. Forty-eighth supplement to the American Ornithologists' Union Check-list of North American Birds. Auk 124: 1109–1115.

Bannerman, D.A. 1963. Birds of the Atlantic Islands. Vol. I. A History of the Birds of the Canary Islands and of the Salvages. Oliver & Boyd. Edinburgh & London.

Bannerman, D.A. and W.M. Bannerman. 1968. Birds of the Atlantic Islands. Vol. IV. A history of the Birds of the Cape Verde Islands. Oliver & Boyd, Edinburgh & London.

Baral, H.S. 2005. In BirdLife International 2014. Species factsheet: *Aegypius monachus*. Downloaded from http://www.birdlife.org on 12/02/2014.

Baral, H.S. 2010. [Lecture: Vultures in Nepal (2010: 09 December)]. Cited in Harris, R.J. 2013. The conservation of Accipitridae vultures of Nepal: a review. Journal of Threatened Taxa | www.threatenedtaxa.org | 26 February 2013 | 5(2): 3603–3619.

Baral, N. et al. 2007. Conservation Implications of Contingent Valuation of Critically Endangered White Rumped Vulture *Gyps bengalensis* in South Asia. Internation Journal of Biodiversity Science and Management 3: 151.

Baral, N., R. Gautam and B. Tamang. 2005. Population status and breeding ecology of white-rumped vulture *Gyps bengalensis* in Rampur Valley, Nepal. Forktail 21: 87–91.

Baral, N., R. Gautam, N. Timilsina and M.G. Bhat. 2007. Conservation implications of contingent valuation of critically endangered white-rumped vulture *Gyps bengalensis* in South Asia. International Journal of Biosciences 3: 145–156.

Barlow, C. and T. Wacher. 1997. A Field Guide to the Birds of the Gambia and Senegal. Christopher Helm, London.

Barnard, P. 1998. Biological Diversity in Namibia—A Country Study. ABC Press, Cape Town.

Barnes, K.N. 2000. The Eskom Red Data Book of Birds of South Africa, Lesotho and Swaziland. BirdLife South Africa, Johannesburg.

Barnett, R., I. Barnes, M.J. Phillips, L.D. Martin, C.R. Harington, J.A. Leonard and A. Cooper. 2005. Evolution of the extinct Sabretooths and the American cheetah-like cat. Current Biology 15(15): R589–R590.

Barov, B. and M.A. Derhé. 2011. Lammergeier *Gypaetus barbatus* species action plan implementation review. *In*: B. Barov and M.A. Derhé (eds.). Review of the Implementation of Species Action Plans for Threatened Birds in the European Union 2004–2010. Final report. BirdLife International for the European Commission.

Barov, B. and M.A. Derhé. 2011. Review of the Implementation of Species Action Plans for Threatened Birds in the European Union 2004–2010. Final report. BirdLife International for the European Commission.

Barrau, C., M. Clouet and J.L. Goar. 1997. Deux jeunes gypaètes barbus (*Gypaetus barbatus meridionalis*) à l'envol dans une aire des monts du Balé (Éthiopie). Alauda 65: 200–201.

Barrett, S.A. and E.W. Gifford. 1933. Miwok material culture. Bulletin of the Public Museum of the City of Milwaukee 2: 117–376.

Barrientos, R., J.C. Alonso, C. Ponce and C. Palacın. 2011. Meta-analysis of the effectiveness of marked wire in reducing avian collisions with power lines. Conservation Biology 25: 893–903.

Barrios, L. and A. Rodriguez. 2004. Behavioural and environmental correlates of soaring-bird mortality at on-shore wind turbines. Journal of Applied Ecology 41: 72–81.

Barrios Partida, F. 2006. Nomadas del Estrecho de Gibraltar. Guia de la Migracion de aves, los Parques Naturales de Estrecho y Los Alcornocales y el Penon de Gibraltar. Acento 2000.

Barros Valenzuela, R. 1962. [Notes on the Turkey vulture and Black vulture]. Revista Universitaria (Universidad Católica de Chile) 47: 155–166.

Barrows, W.B. 1887. The sense of smell in Cathartes aura. Auk 4: 173–174.

Barzen, J. 1995. Other SIS sightings in northeastern Cambodia. Specialist Group on Storks, Ibises and Spoonbills Newsletter 7: 4–5.

Batbayar, N. 2005. In BirdLife International 2014. Species factsheet: *Aegypius monachus*. Downloaded from http://www.birdlife.org on 12/02/2014.

Batbayar, N., M. Fuller, R.T. Watson and B. Ayurzana. 2006. Overview of the Cinereous vultures *Aegypius monachus* L. ecology research results in Mongolia. *In*: N. Batbayar, P.W. Kee and B. Ayurzana (eds.). Conservation and Research of Natural Heritage. Proceedings of the 2nd International Symposium between Mongolia and Republic of Korea, Ulaanbaatar, Mongolia, in September 30, 2006, 8–15. Wildlife Science and Conservation Centre of Mongolia, Ulaanbaatar.

Bauer, S., Z. Barta, B.J. Ens, G.C. Hays, J.M. McNamara and M. Klaassen. 2009. Animal migration: linking models and data beyond taxonomic limits. Biology Letters 5: 433–435.

Bauer, S., M. van Dinther, K.-A. Høgda, M. Klaassen and J. Madsen. 2008. The consequences of climate-driven stop-over sites changes on migration schedules and fitness of Arctic geese. Journal of Animal Ecology 77: 654–660.

Baumgart, W., M. Kasparek and B. Stephan. 1995. Birds of Syria. Max KAsparekVerlag, Heidelberg.

Bayle, P. 1999. Preventing birds of prey problems at transmission lines in Western Europe. Journal of Raptor Research 33: 43–48.

Baynard, O.E. 1909. Notes from Florida on Catharista urubu. Oologist 26: 191–193.

BBC (British Broadcasting Corporation). 2004. Vet drug 'killing Asian vultures'. BBC News. 2004–02–28.

Beals, R.L. 1933. Ethnology of the Nisenan. University of California Publications in American Archaeology and Ethnology 31: 335–414.

Beals, R.L. and J.A. Hester, Jr. 1960. A new ecological typology of the California Indians. *In*: A.F.C. Wallace (ed.). Men and Cultures. Selected Papers of the Fifth International Congress of Anthropological and Ethnological Sciences. University of Pennsylvania Press, Philadelphia.

Bean, L. 1975. Power and its application in California. Journal of California Anthropology 2: 25–33.

Bean, L.J. 1972. Mukat's People: The Cahuilla Indians of Southern California. University of California Press, Berkeley, Los Angeles, London.

Bean, M.J. 1983. The Evolution of Wildlife Law. Prager Publishers, New York.

Beasley, J.C. and O.E. Rhodes, Jr. 2008. Relationship between raccoon abundance and crop damage. Human-Wildlife Conflicts 2: 36–47.

Beasley, J.C. and O.E. Rhodes, Jr. 2010. Influence of patch- and landscape-level attributes on the movement behavior of raccoons in agriculturally fragmented landscapes. Canadian Journal of Zoology 88: 161–169.

Beasley, J.C., T.L. DeVault and O.E. Rhodes, Jr. 2007. Home range attributes of raccoons in a fragmented agricultural region of northern Indiana. Journal of Wildlife Management 71: 844–850.

Beasley, J.C., T.L. DeVault, M.I. Retamosa and O.E. Rhodes, Jr. 2007. A hierarchical analysis of habitat selection by raccoons in northern Indiana. Journal of Wildlife Management 71: 1125–1133.

Beason, R.C. 2003. Through a Birds Eye: Exploring Avian Sensory Perception. USDA Wildlife Services, Washington, D.C.

Beaufoy, G., D. Baldock and J. Clark. 1994. The Nature of Farming. Low Intensity Farming Systems in Nine European Countries. Institute for European Environmental Policy, London.

Beilis, N. and J. Esterhuizen. 2005. The potential impact on Cape Griffon *Gyps coprotheres* populations due to the trade in traditional medicine in Maseru, Lesotho. Vulture News 15–19.

Bekoff, M. and R.D. Andrews. 1978. Coyotes: Biology, Behavior, and Management. Academic Press, New York.

Belant, J.L. 1997. Gulls in urban environments: landscape-level management to reduce conflict. Landscape and Urban Planning 38: 245–258.

Belcher, C. and G.D. Smooker, 1934. Birds of the Colony of Trinidad and Tobago 76(3): 572–595.

Beletsky, L. 2006. Birds of the World. JHU Press.

Belton, W. 1984. Birds of Rio Grande do Sul, Brazil. Part 1. Rheidae through Furnariidae. Bulletin of the American Museum of Natural History 178: 369–636.

Benedict, R.F. 1924. A Brief Sketch of the Serrano Culture. American Anthropology 26: 366–392.

Benhaiem, S., M. Delon, B. Lourtet, B. Cargnelutti, S. Aulagnier, A.J.M. Hewison, N. Morellet and H. Verheuden. 2008. Hunting increases vigilance levels in roe deer and modifies feeding site selection. Animal Behavior 76: 611–618.

Bennun, L.A. and P. Njoroge (eds.). 1996. Birds to watch in East Africa: a preliminary red data list. Research Reports of the Centre for Biodiversity, National Museums of Kenya: Ornithology 23: 1–16.

Benson, E.P. 1997. Birds and Beasts of Ancient Latin America. University Press of Florida, Gainsville.

Benson, P.C. 1981. Large raptor electrocution and power pole utilization: a study in six western states. Ph.D. Dissertation, Brigham Young University, Provo.

Benson, P.C. 2000. Causes of Cape Vulture *Gyps coprotheres* mortality at the Kransberg colony: a 17-year update. pp. 77–86. *In*: R.D. Chancellor and B.-U. Meyburg (eds.). Raptors at Risk: Proceedings of the 5th World Conference on Birds of Prey and Owls. Hancock House Publishers, Washington, D.C. and World Working Group on Birds of Prey and Owls, Berlin.

Benson, S. and P. Hellander. 2007. Peru. Lonely Planet Publications.

Bent, A.C. 1937. Life histories of North American birds of prey. Part 1. Smithsonian Inst. U.S. Natl. Mus. Bull. 167.

Benton, T.G., J.A. Vickery and J.D. Wilson. 2003. Farmland biodiversity: Is habitat heterogeneity the key? Trends in Ecology and Evolution 18: 182–188.

Berger, K.M. and E.M. Gese. 2007. Does interference competition with wolves limit the distribution and abundance of coyotes? Journal of Animal Ecology 76(6): 1075–1085.

Berger, R., R. Protsch, R. Reynold, C. Rozaire and J.R. Sackett. 1971. New radiocarbon dates based on bone collagen of California Paleoindians. Contributions of the University of California Archaeological Research Facility 12: 43–49.

Bergier, P. and G. Cheyland. 1980. Statut, succes de reproduction et alimentation du vautour percnoptere Neophron percnopterus France mediterranfienne. Alauda 48: 75–97.

Berlanga, M. and P. Wood. 1992. Observations of the king vulture Sarcoramphus papa in Calakmul Biosphere Reserve Campeche Mexico. Vulture News 26: 15–21.

Bernis, F. 1983. Migration of the Common Griffon Vulture in the western Palearctic. pp. 185–196. *In*: S.R. Wilbur and J.A. Jackson (eds.). Vulture Biology and Management. University of California Press, Berkeley.

Bernshausen, F. and J. Kreuziger. 2009. Review of the effectiveness of new developed wire markers based on behaviour observations of overwintering and year round present birds at the Alfsee/Niedersachsen. Planungsgruppe für Natur und Landschaft, Hungen.

Berreman, J.V. 1944. Cheto archaeology: a report of the Lone Ranch Creek shell mound on the coast of southern Oregon. General Series in Anthropology 1: 1–40.

Berta, A. 1985. The status of Smilodon in North and South America. Contributions in Science, Natural History Museum of Los Angeles County 370: 1–15.

Bester, B. 1996. Bush encroachment: a thorny problem. Namibia Environment. 1: 175–177.

Bevanger, K. 1994. Bird interactions with utility structures: collision and electrocution, causes and mitigating measures. Ibis 136(4): 412–425.

Bevanger, K. and H. Brøseth. 2001. Bird collisions with power lines—an experiment with ptarmigan (*Lagopus* spp.). Biological Conservation 99(3): 341–346.

Bezuijen, M.R., J.A. Eaton, Gidean, R.O. Hutchinson and F.E. Rheindt. 2010. Recent and historical bird records for Kalaw, eastern Myanmar (Burma), between 1895 and 2009. Forktail 26: 49–74.

Biggs, D. 2001a. Eagles feast at quelea colony in Kruger. Africa—Birds & Birding 6: 16–17.

Biggs, D. 2001b. Observations of eagle congregations at a Red-billed Quelea colony in the Kruger National Park. Bird Numbers 10: 25–28.

Bildstein, K.L. 1999. Racing with the sun: the forced migration of the Broad-winged Hawk. pp. 79–92. *In*: K.P. Able (ed.). Gatherings of angels. Comstock Press, Ithaca.

Bildstein, K.K., M.J. Bechardi, C. Farmer and L. Newcomb. 2009. Narrow sea crossings present major obstacles to migrating Griffon Vultures *Gyps fulvus*. 151(2): 382–391.

Bildstein, K.L. 2006. Migrating Raptors of the World: Their Ecology and Conservation. Cornell University Press, Ithaca.

Bildstein, K.L. and M. Saborio. 2000. Spring migration counts of raptors and New World vultures in Costa Rica. Ornitologia Neotropical 11: 197–205.

Bildstein, K.L., M.J. Bechard, C. Farmer and L. Newcomb. 2009. Narrow sea crossings present major obstacles to migrating Griffon Vultures Gyps fulvus. Ibis 151(2): 382–391.

BirdLife International. 2007. *Gyps fulvus. In*: IUCN 2007.

BirdLife International. 2012. Bucorvus abyssinicus. IUCN Red List of Threatened Species. Version 2012.1. International Union for Conservation of Nature. Retrieved 28 July 2013.

BirdLife International. 2012. Bucorvus leadbeateri. IUCN Red List of Threatened Species. Version 2012.1. International Union for Conservation of Nature. Retrieved 28 July 2013.

BirdLife International. 2014. Species factsheet: Aegypius monachus. Downloaded from http://www.birdlife.org on 05/02/2014.

BirdLife International. 2014. Species factsheet: Gymnogyps californianus. Downloaded from http://www.birdlife.org on 13/03/2014.

BirdLife International. 2014. Species factsheet: *Gyps rueppellii*. Downloaded from http://www.birdlife.org on 15/01/2014. Recommended citation for factsheets for more than one species: BirdLife International (2014) IUCN Red List for birds. Downloaded from http://www.birdlife.org on 15/01/2014.

BirdLife International. 2014. Species factsheet: *Sarcogyps calvus*. Downloaded from http://www.birdlife.org on 03/02/2014.

BirdLife International. 2014. Species factsheet: *Torgos tracheliotos*. Downloaded from http://www.birdlife.org on 06/02/2014.

BirdLife International. 2014. Species factsheet: *Vultur gryphus*. Downloaded from http://www.birdlife.org on 21/02/2014.

Birdlife International. 2000. Threatened Birds of the World. Lynx Editions, Barcelona and Birdlife International, Cambridge.

Birdlife International. 2001. Threatened Birds of Asia: The Birdlife International Red Data Book Birdlife International, Cambridge.

BirdLife International. 2007. *Gyps fulvus. In*: IUCN 2007. 2007 IUCN red list of threatened species. www.iucnredlist.org.

BirdLife International. 2007. News [www.birdlife.org/news].

BirdLife International. 2007. Lobby governments to outlaw the marketing and sale of diclofenac for veterinary purposes (BirdLife International 2007).

BirdLife International. 2008a. Asian vulture populations have declined precipitously in less than a decade. Presented as part of the BirdLife State of the world's birds website. Available from: http://www.birdlife.org/datazone/sowb/casestudy/113. Checked: 09/03/2014.

BirdLife International. 2008b. Anthropogenic junk ingestion leads to high nestling mortality in vultures and condors. Presented as part of the BirdLife State of the world's birds website. Available from: http://www.birdlife.org/datazone/sowb/casestudy/180. Checked: 09/03/2014.

BirdLife International. 2008c. Anthropogenic junk ingestion leads to high nestling mortality in vultures and condors. Presented as part of the BirdLife State of the world's birds website. Available from: http://www.birdlife.org/datazone/sowb/casestudy/180. Checked: 10/03/2014.

BirdLife International. 2008d. Threatened Birds of the World. Cambridge, UK: BirdLife International.

BirdLife International. 2012. *Aegypius monachus. In*: IUCN 2013. IUCN Red List of Threatened Species. Version 2013.1. <www.iucnredlist.org>. Downloaded on 12 October 2013.

BirdLife International. 2012. *Cathartes aura. In*: IUCN 2013. IUCN Red List of Threatened Species. Version 2013.2. <www.iucnredlist.org>. Downloaded on 16 February 2014.

BirdLife International. 2012. Cathartes burrovianus. *In*: IUCN 2013. IUCN Red List of Threatened Species. Version 2013.1. <www.iucnredlist.org>. Downloaded on 05 September 2013.

BirdLife International. 2012. *Cathartes melambrotus. In*: IUCN 2013. IUCN Red List of Threatened Species. Version 2013.1. <www.iucnredlist.org>. Downloaded on 06 September 2013.

BirdLife International. 2012. *Coragyps atratus. In*: IUCN 2013. IUCN Red List of Threatened Species. Version 2013.1. <www.iucnredlist.org>. Downloaded on 05 September 2013.

BirdLife International. 2012. *Gymnogyps californianus. In*: IUCN 2013. IUCN Red List of Threatened Species. Version 2013.1. <www.iucnredlist.org>. Downloaded on 06 September 2013.

BirdLife International. 2012. *Gypaetus barbatus. In*: IUCN 2013. IUCN Red List of Threatened Species. Version 2013.1. <www.iucnredlist.org>. Downloaded on 05 September 2013.

BirdLife International. 2012. *Gyps africanus. In*: IUCN 2013. IUCN Red List of Threatened Species. Version 2013.1. <www.iucnredlist.org>. Downloaded on 04 September 2013.

BirdLife International. 2012. *Gyps bengalensis. In*: IUCN 2013. IUCN Red List of Threatened Species. Version 2013.1. <www.iucnredlist.org>. Downloaded on 04 September 2013.

BirdLife International. 2012. *Gyps coprotheres. In*: IUCN 2013. IUCN Red List of Threatened Species. Version 2013.1. <www.iucnredlist.org>. Downloaded on 05 September 2013.

BirdLife International. 2012. *Gyps himalayensis. In*: IUCN 2013. IUCN Red List of Threatened Species. Version 2013.1. <www.iucnredlist.org>. Downloaded on 05 September 2013.

BirdLife International. 2012. *Gyps indicus. In*: IUCN 2013. IUCN Red List of Threatened Species. Version 2013.1. <www.iucnredlist.org>. Downloaded on 05 September 2013.

BirdLife International. 2012. *Gyps rueppellii. In*: IUCN 2013. IUCN Red List of Threatened Species. Version 2013.1. <www.iucnredlist.org>. Downloaded on 05 September 2013.

Birdlife International. 2012. IUCN Red List for Birds. <http://www.birdlife.org/>. Downloaded on 12 January 2014.

BirdLife International. 2012. *Necrosyrtes monachus. In*: IUCN 2013. IUCN Red List of Threatened Species. Version 2013.1. <www.iucnredlist.org>. Downloaded on 04 September 2013. BirdLife International. 2001. Threatened birds of Asia: the BirdLife International Red Data Book. BirdLife International, Cambridge.

BirdLife International. 2012. *Necrosyrtes monachus*. *In*: IUCN 2013. IUCN Red List of Threatened Species. Version 2013.2. <www.iucnredlist.org>. Downloaded on 31 January 2014.

BirdLife International. 2012. *Sarcogyps calvus*. *In*: IUCN 2013. IUCN Red List of Threatened Species. Version 2013.1. <www.iucnredlist.org>. Downloaded on 05 September 2013.

BirdLife International. 2012. *Sarcoramphus papa*. *In*: IUCN 2013. IUCN Red List of Threatened Species. Version 2013.1. <www.iucnredlist.org>. Downloaded on 06 September 2013.

BirdLife International. 2012. *Torgos tracheliotos*. *In*: IUCN 2013. IUCN Red List of Threatened Species. Version 2013.1. <www.iucnredlist.org>. Downloaded on 05 September 2013.

BirdLife International. 2012. *Trigonoceps occipitalis*. *In*: IUCN 2013. IUCN Red List of Threatened Species. Version 2013.1. <www.iucnredlist.org>. Downloaded on 04 September 2013.

BirdLife International. 2014. Species factsheet: *Aegypius monachus*. Downloaded from http://www.birdlife.org on 12/02/2014.

BirdLife International. 2014. Species factsheet: *Cathartes aura*. Downloaded from http://www.birdlife.org on 14/03/2014.

BirdLife International. 2014. Species factsheet: *Gyps tenuirostris*. Downloaded from http://www.birdlife.org on 11/01/2014. Recommended citation for factsheets for more than one species: BirdLife International (2014) IUCN Red List for birds. Downloaded from http://www.birdlife.org on 11/01/2014.

BirdLife International. 2014. Species factsheet: *Neophron percnopterus*. Downloaded from http://www.birdlife.org on 25/01/2014.

BirdLife International. 2014. Species factsheet: *Sarcoramphus papa*. Downloaded from http://www.birdlife.org on 20/02/2014.

BirdLife International. 2014. Species factsheet: *Trigonoceps occipitalis*. Downloaded from http://www.birdlife.org on 03/02/2014.

BirdLife International. 2000. Threatened birds of the world. Lynx Editions, Barcelona, and BirdLife International, Cambridge.

Birdlife International. 2001. Threatened Birds of Asia; The BirdLife International Red Data Book. Cambridge.

BirdLife International. 2004. Birds in Europe: population estimates, trends and conservation status. BirdLife International, Cambridge, U.K.

BirdLife International. 2007. Species factsheet: *Gyps africanus*. Available at http://www.birdlife.org [accessed 27 December 2013].

BirdLife International. 2007. Species factsheet: *Gyps africanus*. Available at http://www.birdlife.org [accessed 12 September 2013].

BirdLife International. 2008. Drugs firms told to do more to prevent vulture extinctions. Available at: #http://www.birdlife.org/news/news/2008/08/indian_drug_announcemment.html#.

Birt, T.P., V.L. Birt-Friesen, J.M. Green, W.A. Montevecchi and W.S. Davidson. 1992. Cytochrome b sequence variation among parrots. Hereditas 117: 67–72.

Blackwell, B.F. and S.E. Wright. 2006. Collisions of Red-tailed Hawks (*Buteo jamaicensis*), Turkey Vultures (*Cathartes aura*), and Black Vultures (*Coragyps atratus*) with aircraft: implications for bird strike reduction. Journal of Raptor Research 40(1): 76–80.

Blackwell, B.F., M.L. Avery, B.D. Watts and M.S. Lowney. 2007. Demographics of Black Vultures in North Carolina. USDA National Wildlife Research Center—Staff Publications. Paper 681.

Blake, E.R. 1953. Birds of Mexico: A Guide for Field Identification. University of Chicago Press, Chicago.

Blake, E.R. 1977. Manual of Neotropical Birds. University of Chicago Press, Chicago.

Blanco-Aguiar, J.A., P. González-Jara, M.E. Ferrero, I. Sánchez-Barbudo, E. Virgós, R. Villafuerte and J.A. Dávila. 2008. Assessment of game restocking contributions to anthropogenic hybridization: the case of the Iberian red-legged partridge. Animal Conservation 11: 535–545.

Blanford, W.T. 1895. The Fauna of British India, including Ceylon and Burma. Birds.–Vol. III. Taylor and Francis, London.

Blanford, W.T. 1907. Fauna Section in Imperial Gazetteer of India. The Indian Empire Vol. 1. Oxford University Press, London.

Blench, R. 1993. Ethnographic and linguistic evidence for the prehistory of African ruminant livestock, horses and ponies. pp. 71–103. *In*: T. Shaw, P. Sinclair, B. Andah and A. Okpoko (eds.). The Archeology of Africa. Routledge, London.

Blokpoel, H. 1976. Bird Hazard to Aircraft. Books Canada Inc., Buffalo, New York.

Blumenschine, R. 1986. Early Hominoid Scavenging Opportunities: Implications of Carcass Availability in the Serengeti and Ngorongoro Ecosystems. Series 283Bar International, Oxford.

Boeker, E.L. and P.R. Nickerson. 1975. Raptor electrocutions. Wildlife Society Bulletin 3: 79–81.

Bogliani, G., R. Viterbi and M. Nicolino. 2011. Habitat use by a Reintroduced Population of bearded vultures (*Gypaetus barbatus*) in the Italian Alps. Journal of Raptor Research 45(1): 56–62.

Bohrer, G., D. Brandes, J.T. Mandel, K.L. Bildstein, T.A. Miller, M. Lanzone, T. Katzner, C. Maisonneuve and J.A. Tremblay. 2011. Estimating updraft velocity components over large spatial scales: contrasting migration strategies of golden eagles and turkey vultures. Ecology Letters 15(2): 96–103.

Bombay Natural History Society. 2004. Report of the International South Asian Vulture Recovery Plan Workshop. Buceros 1–48.

Boone, R.B., S.J. Thirgood and J.G.C. Hopcraft. 2006. Serengeti wildebeest migratory patterns modeled from rainfall and new vegetation growth. Ecology 87: 1987–1994.

Borello, W. 2004. Conservation status of vultures in Botswana. pp. 139–143. *In*: A. Monadjem, M.D. Anderson, S.E. Piper and A.F. Boshoff (eds.). The Vultures of Southern Africa—Quo Vadis? Birds of Prey Working Group, Johannesburg.

Borello, W.D. and R.M. Borello. 2002. The breeding status and colony dynamics of Cape Vulture (*Gyps coprotheres*) in Botswana. Bird Conservation International 12: 79–97.

Borello, W.D. 2007. In BirdLife International 2014. Species factsheet: *Gyps coprotheres*. Downloaded from http://www.birdlife.org on 15/03/2014.

Borrow, N. and R. Demey. 2001. The Birds of Western Africa. Christopher Helm, London.

Borrow, N. and R. Demey. 2010. Birds of Ghana. Christopher Helm, London.

Boscana, G. 1933. Chinigchinich: A revised and annotated version of Alfred Robinson's translation of Father Geronimo Boscana's historical account of the belief, usages, customs, extravageences [sic] (!) of the Indians of this Mission of San Juan Capistrano called the Acagchemen tribe. P.T. Hanna (ed.). Fine Arts Press, Santa Ana, California.

Bose, M. and F. Sarrazin. 2007. Competitive behaviour and feeding rate in a reintroduced population of Griffon vultures *Gyps fulvus*. Ibis 149: 490–501.

Boshoff, A. and M.D. Anderson. 2007. Towards a conservation plan for the Cape Griffon *Gyps coprotheres*: identifying priorities for research and conservation. Vulture News 56–59.

Boshoff, A.F. and M.D. Anderson. 2006. Towards a conservation plan for the Cape Griffon Gyps coprotheres: identifying priorities for research and conservation action. Vulture News 77–79.

Boshoff, A., S. Piper and M. Michael. 2009. On the distribution and breeding status of the Cape Griffon *Gyps coprotheres* in the Eastern Cape province, South Africa. Ostrich 80(2): 85–92.

Boshoff, A.F. and C.J. Vernon. 1980. The past and present distribution and status of the Cape Vulture in the Cape Province. Ostrich 51(4): 230–250.

Boshoff, A.F. and M.D. Anderson. 2006. Towards a conservation plan for the Cape Griffon *Gyps coprotheres*: identifying priorities for research and conservation action. Centre for African Conservation Ecology, Nelson Mandela Metropolitan University, Port Elizabeth.

Boshoff, A.F., M.D. Anderson and W.D. Borello (eds.). 1997. Vultures in the 21st Century: Proceedings of a workshop on vulture research and conservation in Southern Africa. Vulture Study Group, Johannesburg.

Boshoff, A.F., J.C. Minnie, C.J. Tambling and M.D. Michael. 2011. The impact of power line-related mortality on the Cape Vulture *Gyps coprotheres* in a part of its range, with an emphasis on electrocution. Bird Conservation International 21: 311–327.

Boshoff, A.F., S.E. Piper and M. Michael. 2009. On the distribution and breeding status of the Cape Griffon *Gyps coprotheres* in the Eastern Cape province, South Africa. Ostrich 80: 85–92.

Boshoff, A.F., C.J. Vernon and R.K. Brooke 1983. Historical atlas of the diurnal raptors of the Cape Province (Aves: Falconiformes). Annals of the Cape Province Museum of Natural History 14: 173–297.

Bosshoff, A. and M. Anderson. 2008. Towards a conservation plan for the Cape Griffon Gyps coprotheres: Identifying priorities for research and conservation. Bulletin of the African Bird Club 15(1): 10.

Botha, A. 2006. A new vulture colour-marking method for southern Africa. Endangered Wildlife 12–14.

Botha, A.J., D.L. Ogada and M.Z. Virani. 2012. Pan-African vulture summit 2012, April 16–20, 2012, Masai Mara, Kenya. Wildlife without Borders, SASOL, Endangered Wildlife Trust, and the Peregrine Fund, Modderfontein, South Africa.

Boudoint, Y. 1976. Techniques de vol et de cassage d'os chez le gypaète barbu, *Gypaetus barbatus*. Alauda 44: 1–21.

Bowden, C. 2008. In BirdLife International 2014. Species factsheet: *Gyps coprotheres*. Downloaded from http://www.birdlife.org on 15/03/2014.

Bowden, C. 2009. The Asian Gyps vulture crisis: the role of captive breeding in India to prevent total extinction. Birding ASIA 12: 121–123.

Brainerd, S. 2007. European charter on hunting and biodiversity. Document of the Standing Committee 27th meeting, Strasbourg, 26–29 November 2007. Convention on the conservation of European wildlife and natural habitats. Downloadable from http://www.cic-wildlife.org/index.php?id=412.

Brandes, D. and D.W. Ombalski. 2004. Modeling raptor migration pathways using a fluid-flow analogy. J. Raptor Res. 38: 195–207.

Braun, M.J., D.W. Finch, M.B. Robbins and B.K. Schmidt. 2007. A Field Checklist of the Birds of Guyana, 2nd Ed. Smithsonian Institution, Washington, D.C.

Brauneis, W., W. Watzlaw and L. Horn. 2003. The behaviour of birds in the proximity of a selected part of the 110 kV power line between Bernburg and Susigke (Bundesland Sachsen-Anhalt). Flight behaviour, collisions, breeding populations (in German with English summary). Ökol. Vögel 25: 69–115.

Brazil, M. 2009. Birds of East Asia: Eastern China, Taiwan, Korea, Japan, Eastern Russia. Christopher Helm, London.

Breen, B.M. and K.L. Bildstein. 2008. Distribution and abundance of the Turkey Vulture Cathartes aura falklandica in the Falkland Islands, summer 2006–07 and autumn-winter 2007. Falklands Conservation. Stanley, Falkland Islands.

Breuer, W. 2007. Stromopfer und Vogelschutz an Energiefreileitungen. Naturschutz und Landschaftsplanung 39: 69–72.

Brewer, R. 1994. The Science of Ecology (2nd ed.). Saunders College Publishing, Orlando, Florida.

Brickle, N., Nguyen Cu, Ha Quy Quynh, Nguyen Thai Tu Cuong and Hoang Van San. 1998. The status and distribution of Green Peafowl Pavo muticus in Dak Lak Province, Vietnam. BirdLife International Vietnam Programme/Institute of Ecology and Biological Resources. (Conservation Report 1), Hanoi.

Bridgeford, P. 2004. Status of vultures in Namibia. pp. 144–147. *In*: A. Monadjem, M.D. Anderson, S.E. Piper, and A.F. Boshoff (eds.). The Vultures of Southern Africa—Quo Vadis? Birds of Prey Working Group, Johannesburg.

Bridgeford, P. 2001. More vulture deaths in Namibia. Vulture News 44: 22–26.

Bridgeford, P. 2002a. Aerial survey of breeding Lappet-faced Vultures in the Namib-Naukluft Park, Namibia. Unpublished Report. VSG workshop, Spioenkop.

Bridgeford, P. 2002b. Recent vulture mortalities in Namibia. Vulture News 46: 38.

Bridgeford, P. 2004. Lappet-faced Vulture Torgos tracheliotus. pp. 28–33. *In*: A. Monadjem, M.D. Anderson, S.E. Piper and A.F. Boschoff (eds.). The Vultures of Southern Africa—Quo Vadis? Birds of Prey Working Group, Johannesburg.

Bridgeford, P. 2009. Monitoring breeding Lappet-faced Vultures in the Namib. African Raptors 2–4.

Bridgeford, P. and M. Bridgeford. 2003. Ten years of monitoring breeding Lappet-faced Vultures Torgos tracheliotos in the Namib-Naukluft Park, Namibia. Vulture News 48: 3–11.

Britton, P.L. 1980. Birds of East Africa: their Habitat, Status and Distribution. East Africa National History Society, Nairobi.

Brock, H.R. 1896. High nesting of turkey vulture. Nidiologist 4: 24–25.

Brock, K.J. 2007. Brock's birds of Indiana, CD-ROM ed. Amos W. Butler, Audubon Society, Indianapolis.

Brodeur, S., R. Decarie, D.M. Bird and M. Fuller. 1996. Complete migration cycle of golden eagles breeding in northern Quebec. The Condor 98: 293–299.

Brodkorb, P. 1964. Catalogue of fossil birds: part 2 (Anseriformes through Galliformes). Bulletin of the Florida State Museum, Biological Sciences 8: 195–335.

Brooke, R.K. 1982. The South African breeding season of the Egyptian Vulture. Vulture News 8: 30–31.

Brookes, I. 2006. The Chambers Dictionary. Chambers, Edinburgh.

Brooks, W.S. 1917. Notes on some Falkland Island birds. Bulletin of the Museum of Comparative Zoology 61: 135–160.

Brouwer, J. 2012. In BirdLife International 2014. Species factsheet: *Trigonoceps occipitalis*. Downloaded from http://www.birdlife.org on 03/02/2014.

Brown, J.W. and D.P. Mindell. 2009. Diurnal birds of prey (Falconiformes) pp. 436–439. *In*: S.B. Hedges and S. Kumar (eds.). The Timetree of Life. Oxford University Press, Oxford.

Brown, L. and D. Amadon. 1968. Eagles, Hawks and Falcons of the World. Vol. 1. Country Life Books. A. Cooper, C. Mourer-Chauvire, G.K. Chambers, A. von Haeseler, A.C. Wilson and S. Piiiibo. 1992. Independent origins of New Zealand moas and kiwis. Proceedings of the National Academy of Sciences USA 89: 8741–8744.

Brown, C.J. 1986. Biology and conservation of the Lappet-faced Vulture in SWA/Namibia. Vulture News 16: 10–20.

Brown, C.J. and I. Plug. 1990. Food choice and diet of the Bearded Vulture *Gypaetus barbatus* in southern Africa. South African Journal of Zoology 25: 169–177.

Brown, I.H. and D. Amadon. 1968. Eagles, Hawks and Falcons of the World. (Vol. 1). Country Life Books, Sidney.

Brown, L. 1971. African Birds of Prey. Houghton Mifflin Company, Boston,

Brown, L. 1976. Birds of Prey: Their Biology and Ecology. Hamlyn Publishing Group, Feltham.

Brown, L. and D. Amadon. 1968. Eagles, Hawks and Falcons of the World Vol. I Country Life Books. Feltham.

Brown, L., E.K. Urban and K. Newman. 1982. Birds of Africa. Vol. 1. Princeton University Press, Princeton.

Brown, L.H. 1977. The status, population structure, and breeding dates of the African Lammergeier *Gypaetus barbatus meridionalis*. Raptor Research 11(3): 49–80.

Brown, L.H. and D. Amadon. 1968. Eagles, Hawks and Falcons of the World. McGraw-Hill, New York.

Brown, L.H., E.K. Urban and K. Newman (eds.). 1982. The Birds of Africa Vol. 1. Academic Press, London.

Bruderer, B., L.G. Underhill and F. Liechti. 1995. Altitude choice by night migrants in a desert area predicted by meteorological factors. Ibis 137: 44–55.

Bruno, E. 2005. Sky Burial. Harriman, New York.

Bruun, B., H. Mendelssohn and J. Bull. 1981. A new subspecies of the Lappet-faced Vulture *Torgos tracheliotus* from the Negev desert, Israel. Bulletin of the British Ornithological Club 101: 244–247.

Bruun, B. 1981. The Lappet-faced Vulture in the Middle East. Sandgrouse 2: 90–95.

Bryan, A.L. 1965. Paleo-America prehistory. Occasional Papers of the Idaho State University Museum 16: 1–247.

Buckley, N.J. 1996. Food finding and the influence of information, local enhancement, and communal roosting on foraging success of North American vultures. Auk 113: 473–488.

Buckley, N.J. 1999. Black vulture (*Coragyps atratus*). Account 411. *In*: A. Poole and F. Gill (eds.). The Birds of North America, Philadelphia, Pennsylvania.

Buckley, N.J. 1997. Experimental tests of the information-center hypothesis with black vultures (*Coragyps atratus*) and turkey vultures (Cathartes aura). Behavioral Ecology and Sociobiology 41(4): 267–279.

Buckley, N.J. 1997. Experimental tests of the information center hypothesis with black vultures (*Coragyps atratus*) and turkey vultures (*Cathartes aura*). Behavioral Ecology and Sociobiology 41: 267–279.

Buckley, N.J. 1998. Interspecific competition between vultures for preferred roost positions. The Wilson Bulletin 110(l): 122–125.

Buckley, N.J. 1999. Black vulture (*Coragyps atratus*). *In*: The Birds of North America, No. 411 in A. Poole and F. Gill (eds.). The Birds of North America, Inc. Philadelphia.

Buehler, D. 2000. Bald eagle (*Haliaeetus leucephalus*). *In*: A. Poole and F. Gill (eds.). The Birds of North America No. 506. The Birds of North America Inc., Philadelphia.

Buij, R., B.M. Croes and J. Komdeur. 2012. Effects of land transformation on raptors in West African savannas: the role of breeding range, migratory behaviour and body size. pp. 143–163. *In*: R. Buij (ed.). Raptors in Changing West African Savannas; The Impact of Anthropogenic Land Transformation on Populations of Palearctic and Afrotropical Raptors in Northern Cameroon. Leiden University, Leiden.

Buij, R., B.M. Croes, G. Gort and J. Komdeur. 2013. The role of breeding range, diet, mobility and body size in associations of raptor communities and land-use in a West African savanna. Biological Conservation 166: 231–246.

Bullfinch, T. 2000. Bullfinch's Greek and Roman Mythology: The Age of Fable [1859] Mineola New York.

Burkepile, D.E., J.D. Parker, C.B. Woodson, H.J. Mills, J. Kubanek and P.A. Sobecky. 2006. Chemically mediated competition between microbes and animals: Microbes as consumers in food webs. Ecology 87: 2821–2831.

Burnett, L.J., K.J. Sorenson, J. Brandt, E.A. Sandhaus and D. Ciani et al. 2013. Eggshell thinning and depressed hatching success of California Condors reintroduced to central California. Condor 115: 477–491.

Burnham, K.P. and D.R. Anderson. 2002. Model selection and multimodel inference: a practical information-theoretical approach. Springer-Verlag, New York.

Bustamante, J. 1996. Population Viability Analysis of Captive and Released Bearded Vulture Populations. Conservation Biology 10(3): 822–831.

Butler, A.L. 1905. A Contribution to the Ornithology of the Egyptian Soudan. Ibis 8(5): 301–409.

Butler, C.J. 2003. The disproportionate effect of global warming on the arrival dates of short-distance migratory birds in North America. Ibis 145: 484–495.

Butler, J.R.A., J.T. du Toit and J. Bingham. 2004. Free-ranging domestic dogs (Canis familiaris) as predators and prey in rural Zimbabwe: threats of competition and disease to large wild carnivores. Biological Conservation 115: 369–378.

Butler, J.R.A. and J.T. du Toit. 2002. Diet of free-ranging domestic dogs (*Canis familiaris*) in rural Zimbabwe: implications for wild scavengers on the periphery of wildlife reserves. Animal Conservation 5: 29–37.

Buuveibaatar, B., J.K. Young, J. Berger, A.E. Fine, B. Lkhagvasuren, P. Zahler and T.K. Fuller. 2012. Factors affecting survival and cause-specific mortality of saiga calves in Mongolia. Journal of Mammalogy 94(1): 127–136.

Cade, T.J. 2007. Exposure of California Condors to lead from spent ammunition. Journal of Wildlife Management 71(7): 2125–2133.

Calder, W.A. and J.R. King. 1974. Thermal and caloric relations of birds. *In*: D.S. Farner and J.R. King (eds.). Avian Biology. Academic Press, New York 4: 259–413.

California Condor Recovery Program. 2012. California Condor *Gymnogyps californianus* Recovery Program Population Size and Distribution November 30, 2012.

Callcott, A.A. and H.L. Collins. 1996. Invasion and range expansion of imported fire ants (Hymenoptera: Formicidae) in North America from 1918 to 1995. The Florida Entomologist 79: 240–251.

Cambodian Wetland Team. 2001. Final report on participatory rural appraisal in Boung Rei village, Sambo District, Kratie Province, Mekong River. Royal Government of Cambodia, Inventory and Management of Cambodian Wetlands Project (MRC–Danida), Phnom Penh.

Camiña Cardenal, Á. 2007. Muladares para el buitre leonado en el Sistema Ibérico. Quercus 261: 22–27.

Camiña Cardenal, Á. and Ch. López Hernández. 2009. Los buitres en España. El Ecologista 60: 49–51.

Camina, A. 1996. Explotación de carronas por el Biutre Leonado Gyps fulvus y otros carrtnìeros en Ui Rioja. 1995–96. Instituto de Estudios Riojanos. C.A. de La Rioja. Informe Inèdito.

Camina, A. 2004. Effect of the Bovine Spongiform Eneephalopathy (BSE) on breeding success and food availability on Spanish vulture populations. pp. 27–43. *In*: R.D. Chancellor and B.-U. Meyburg (eds.). Raptors Worldwide, WWGBP/MME, Berlin and Budapest.

Camina, A. 2004. Griffon Vulture *Gyps fulvus* monitoring in Spain: current research and conservation projects. pp. 45–66. *In*: R.D. Chancellor and B.-U. Meyburg (eds.). Raptors Worldwide: Proceedings of the VI World Conference on Birds of Prey and Owls. WWGBP/ MME, Berlin and Budapest.

Camiña, A. 2005. The Eurasian Griffon Vulture *Gyps fulvus* in Spain: current research and monitoring. pp. 45–66. *In*: R.D. Chancellor and B.-U. Meyburg (eds.). Raptors worldwide: proceedings of the VI World Conference on Birds of Prey and Owls. World Working Group on Birds of Prey and Owls and MME/BirdLife Hungary, Berlin and Budapest.

Camiña, A. 2007. Muladares para el buitre leonado en el Sistema Ibérico: la recogida de cadáveres afecta al éxito reproductor de la especie. Quercus 261: 22–27.

Camiña, Á. and E. Montelío. 2006. Food shortages for the Eurasian Griffon Vulture *Gyps fulvus* in Los Monegros (Ebro Valley, Aragon Region). *In*: D.C. Houston and S.E. Piper (eds.). Proceedings of the International Conference on Conservation and Management of Populations, 163. Thessaloniki, Greece. Natural History Museum of Crete & WWF Greece.

Camiña, A. and E. Montelío. 2006. Griffon vulture *Gyps fulvus* food shortages in the Ebro Valley (NE Spain) caused by regulations against bovine spongiform encephalopathy (BSE). Acta Ornithologica 41: 7–13.

Camina, A., A. Onrubia and A. Sensosiain. 1995. Attacks on Livestock by Eurasian Griffon Gyps fulvus. Journal of Raptor Research 29(3): 214.

Campbell, K.E. and E.P. Tonni. 1983. Size and locomotion in teratorns (Aves: Teratornithidae). The Auk 100: 390–403.

Campbell, K.E. and E. Tonni. 1980. A new genus of teratorn from the Huayquerian Argentina (Aves: Teratornithidae). Natural History Museum of Los Angeles Co. Contributions to Science 330: 59–68.

Campbell, K.E. and E. Tonni. 1981. Preliminary observations on the paleobiology and evolution of teratorns (Aves: Teratornithidae). Journal of Vertebrate Paleontology 1: 265–272.

Campbell, K.E., Jr. and E.P. Tonni. 1983. Size and locomotion in teratorns. Auk 100(2): 390–403.

Campbell, K.E., Jr., E. Scott and K.B. Springer. 1999. A new genus for the Incredible Teratorn (Aves: Teratornithidae). Smithsonian Contributions to Paleobiology 89: 169–175.

Campbell, K.E., E. Scott and K.B. Springer. 1999. A new genus for the Incredible Teratorn (Aves: Teratornithidae). Smithsonian Contributions to Paleobiology 89: 169–175.

Campbell, Kenneth E. Jr. and T. Stenger, Allison. 2002. A New Teratorn (Aves: Teratornithidae) from the Upper Pleistocene of Oregon, USA. pp. 1–11. *In*: Z. Zhou and F. Zhang (eds.). Proceedings of the 5th Symposium of the Society of Avian Paleontology and Evolution Beijing. China Science Press, Beijing.

Campbell, M. 2010. An animal geography of avian foraging competition on the Sussex coast of England. Journal of Coastal Research 26(1): 44–52.

Campbell, M. 1998. Interactions between biogeography and rural livelihoods in the coastal savanna of Ghana. PhD Thesis, University of London, London.

Campbell, M. 2007. An animal geography of avian ecology in Glasgow. Applied Geography 27: 78–88.

Campbell, M. 2008. An animal geography of avian feeding habits in Peterborough. Area 40(4): 472–480.

Campbell, M. 2008. The impact of vegetation, river, and urban features on waterbird ecology in Glasgow, Scotland. Journal of Coastal Research (4): 239–245.

Campbell, M. 2009. Factors for the presence of avian scavengers in Accra and Kumasi, Ghana. Area 41(3): 341–349.

Campbell, R.W., M.I. Preston, L.M. Vand Damme and D. Macrae. 2005. Featured species: Turkey Vulture. Wildlife Afield 2: 96–116.

Campbell, R.W. 2007. Turkey vultures feeding on live Western toad toadlets. Wildlife Afield 4(2): 275–276.

Camphuysen, K., J. Verheij and J. Cremer. 2011. Risk Assessment of Bird Strike Hazards: Gulls Laridae. CSR Consultancy, Oosterend.

Canh, L.X., P.T. Anh, J.W. Duckworth, V.N. Thanh and L. Vuthy. 1997. A survey of large mammals in Dak Lak Province, Vietnam. WWF/IUCN, Hanoi.

Carette, M., J.A. Sanchez-Zapata, J.R. Benıtez, M. Lobon and J.A. Donazar. 2009. Large scale risk-assessment of wind-farms on population viability of a globally endangered long-lived raptor. Biological Conservation 142: 2954–2962.

Carey, J.R. and D.S. Judge. 2000. Longevity Records: Life Spans of Mammals, Birds, Amphibians, Reptiles, and Fish. Monographs on Population Aging. Odense University Press. p. 8.

Caro, T.M. 1994. Cheetahs of the Serengeti Plains. University of Chicago Press, Chicago.

Carpenter, J.W., O.H. Pattee, S.H. Fritts, B.A. Rattner, S.N. Wiemeyer, J.A. Royle and M.N. Smith. 2003. Experimental lead poisoning in Turkey Vultures (*Cathartes aura*). Journal of Wildlife Diseases 39: 96–104.

Carrete, M., J.A. Donázar and A. Margalida. 2006. Density-dependent productivity depression in Pyrenean Bearded Vultures: implications for conservation. Ecological Applications 16: 1674–1682.

Carrete, M., J.M. Grande, J.L. Tella, J.A. Snchez-Zapata, J.A. Donazar, R. Díaz-Delgado and A. Romo. 2007. Habitat, human pressure, and social behavior: Partialling out factors affecting large-scale territory extinction in an endangered vulture. Biological Conservation 136(1): 143–154.

Carrete, M., S.A. Lambertucci, K. Speziale, O. Ceballos, A. Travaini, M. Delibes, F. Hiraldo and J.A. Donázar. 2010. Winners and losers in human-made habitats: interspecific competition outcomes in two Neotropical vultures. Animal Conservation 13: 390–398.

Carrete, M., J.A. Sánchez-Zapata, J.R. Benítez, M. Lobón and J.A. Donázar. 2009. Large-scale risk-assessment of wind-farms on population viability of a globally-endangered long-lived raptor. Biological Conservation 142: 2954–2961.

Carrete, M., J.L. Tella, G. Blanco and M. Bertellotti. 2009. Effects of habitat degradation on the abundance, richness and diversity of raptors across Neotropical biomes. Biological Conservation 142: 2002–2011.

Carriker, M.A. 1910. An annotated list of the birds of Costa Rica, including Cocos Island. Annals of the Carnegie Museum of Natural History 6: 314–915.

Carroll, R.W. 1988. Birds of the Central African Republic. Malimbus 10: 177–200.

Carswell, M. 1986. Birds of the Kampala area. Scopus, Special Supplement No. 2. East Africa Natural History Society, Nairobi.

Carswell, M., D. Pomeroy, J. Reynolds and H. Tushabe. 2005. Bird Atlas of Uganda. British Ornithologists' Club and British Ornithologist' Union, London.

Carvalho-Filho, E.P.M., G. Zorzin and G.V.A. Specht. 2004. Breeding biology of the King Vulture (*Sarcoramphus papa*) in southeastern Brazil. Ornithology Neotropica 15(2): 219–224.

Casas, F., F. Mougeot, J. Viñuela and V. Bretagnolle. 2009. Effects of hunting on the behaviour and spatial distribution of farmland birds: importance of hunting-free refuges in agricultural areas. Animal Conservation 12: 346–354.

Cassin, J. 1845. Description of a new Vulture [*Catheres burrovianus*] in the Museum of the Academy of Natural Sciences of Philadelphia. Proceedings of the Academy of Natural Science of Philadelphia 2(8): 212.

Catry, P., M. Lecoq and I.J. Strange. 2008. Population growth and density, diet and breeding success of striated caracaras Phalcoboenus australis on New Island, Falkland Islands. Polar Biology 31: 1167–1174.

Ceballos, O. and J.A. Donazar. 1989. Factors influencing the breeding density and the nest-site selection of the Egyptian Vulture (*Neophron percnopterus*). Journal of Ornithology 130: 353–359.

Ceballos, O. and J.A. Donazar. 1990. Roost—tree characteristics, food habits and seasonal abundance of roosting Egyptian Vultures in northern Spain. Journal of Raptor Research 24(1-2): 19–25.

Celoria, F. 1992. The Metamorphoses of Antoninus Liberalis: A Translation with a Commentary. Routledge, London and New York.

Chan, S. 2006. In BirdLife International 2014. Species factsheet: *Sarcogyps calvus*. Downloaded from http://www.birdlife.org on 04/02/2014.

Chandler, K.P. 2013. The distribution and status of cinereous vulture (*Aegypius monachus*) at Jorbeer, Bikaner, Rajasthan, India: A study of near threatened monk vulture. Research Journal of Animal, Veterinary and Fishery Sciences 1(1): 17–21.

Chandler, R.L. 1952. Comparative tolerance of West African N'Dama cattle to trypanosomiasis. Annals of Tropical Medicine and Parasitology 46: 127–134.

Chapman, F.M. 1929. My tropical air castle. D. Appleton Century Co., New York.

Chapman, F.M. 1938. Life in an air castle. D. Appleton Century Co., New York.

Chatterjee, S., R.J. Templin and K.E. Campbell. 2007. The aerodynamics of Argentavis, the world's largest flying bird from the Miocene of Argentina. Proceedings of the National Academy of Sciences 104: 12398–12403.

Chaudhary, A., D.B. Chaudhary, H.S. Baral, R. Cuthbert, I. Chaudhari and Y.B. Nepali. 2010. Influence of safe feeding site on vultures and their nest numbers at Vulture Safe Zone, Nawalparasi. Proceedings of the First National Youth Conference on Environment 1–6. Kathmandu.

Chaudhary, A., T.R. Subedi, J.B. Giri, H.S. Baral, B. Bidari, H. Subedi, B. Chaudhary, I. Chaudhari, K. Paudel and R.J. Cuthbert. 2012. Population trends of Critically Endangered Gyps vultures in the lowlands of Nepal. Bird Conservation International 22(03): 270–278.

Chaudhry, M.J.I., D.L. Ogada, R.N. Malik, and M.Z. Virani. In Review. Assessment of productivity, population changes, and mortality in long-billed vultures *Gyps indicus* in Pakistan since the onset of the Asian vulture crisis. Bird Conservation Internation 119.

Chaudhry, M.J.I., D.L. Ogada, R.N. Malik, Z. Virani and D. Giovanni. 2012. First evidence that populations of the critically endangered Long-billed Vulture Gyps indicus have increased following the ban of the toxic verterinary drug diclofenac in south Asia. Bird Conservation International 1–9. doi:10.1017/S0959270912000445.

Chebez, J.C. 1999. BirdLife International 2014. Species factsheet: *Vultur gryphus*. Downloaded from http://www.birdlife.org on 21/02/2014.

Cheke, A. 1972. Where no vultures fly. World of Birds 2(1): 15–22.

Chemonges, J.k. 1991. The role of scavenging birds in clearing up the refuse disposal of in Kampala. MSc thesis, Makerere University, Kampala.

Chemonges, J.K. 1991. The role of scavenging birds in clearing up the refuse disposed of in Kampala. MSc. Thesis, Makerere University, Kampala.

Cheney, C. 2010. Reading vultures. *In*: M. Mitchell (ed.). The African Expedition Magazine.

Cheylan, G.A., A. Ravayrol, J.-M. Cugnasse, J.-M. Billet and C. Joulot. 1996. Dispersal of Juvenile Bonelli's Eagles *Hieraaetus fasciatus* ringed in France (in French). Alauda 64: 413–419.

Chhangani, A.K. 2002. Vultures the most eco-friendly bird. Science Reporter 39(10): 56–59.

Chhangani, A.K. 2003. Predation on vultures, their eggs and chicks by different predators in and around Jodhpur. Newsletter for Birdwatchers 43(3): 38–39.

Chhangani, A.K. 2004. Status of a breeding population of long-billed vulture (Gyps indicus indicus) in and around Jodhpur (Rajasthan), India. Vulture News 50: 15–22.

Chhangani, A.K. 2005. Population ecology of vultures in the western Rajasthan, India. Indian Forester 131(10): 1373–1382.

Chhangani, A.K. 2007a. Sightings and nesting sites of Red-headed Vulture Sarcogyps calvus in Rajasthan, India. Indian Birds 3(6): 218–221.

Chhangani, A.K. 2007b. Study of the cause of vulture population decline in India with special reference to Jodhpur. Unpublished Progress Report No. 3. Council of Scientific and Industrial Research, New Delhi.

Chhangani, A.K. and S.M. Mohnot. 2004. Is Diclofenac the only cause of vulture decline? Current Science 87(11): 1496–1497.

Chhangani, A.K. 2009. Present status of vultures in the Great Indian Thar Desert. pp. 65–83. In: C. Sivaperuman, Q.H. Baqri, G. Ramaswamy and M. Naseema (eds.). Faunal Ecology and Conservation of the Great Indian Desert. Springer, Heidelberg.

Choudhury, A., K. Lahkar and R.W. Risebrough. 2005. New nesting sites of Gyps vultures in Assam. Mistnet 10–11.

Christiansen, P. 2007. Comparative bite forces and canine bending strength in feline and sabretooth felids: implications for predatory ecology. Zoological Journal of the Linnean Society 151(2): 423–437.

Christiansen, P. and J.S. Adolfssen. 2007. Osteology and ecology of Megantereon cultridens SE311 (Mammalia; Felidae; Machairodontinae), a sabrecat from the Late Pliocene—Early Pleistocene of Senéze, France. Zoological Journal of the Linnean Society 151: 833–884.

Christiansen, P. and J.M. Harris. 2005. Body Size of Smilodon (Mammalia: Felidae). Journal of Morphology 266(3): 369–384.

Christiansen, P. and J.M. Harris. 2005. Body Size of Smilodon (Mammalia: Felidae). Journal of Morphology 266(3): 369–384.

Chu, M., W. Stone, K.J. McGowan, A.A. Dhondt, W.M. Hochachka and J.E. Therrien. 2003. West Nile file. Birdscope 17: 10–11.

Church, M.E., R. Gwiazda, R.W. Risebrough, K. Sorenson, C.P. Chamberlain, S. Farry, W. Heinrich, B.A. Rideout et al. 2006. Ammunition is the Principal Source of Lead Accumulated by California Condors Re-Introduced to the Wild. Environmental Science Technology 40(19): 6143–50.

Cisneros-Heredia, D.F. 2006. Notes on breeding, behaviour and distribution of some birds in Ecuador. Bulletin of the British Ornithologists' Club 126(2): 153–164.

Clark, A.J. and A. Scheuhammer. 2003. Lead poisoning in upland-foraging birds of prey in Canada. Ecotoxicology 12: 23–30.

Clark, P.J. and F.C. Evans. 1954. Distance to nearest neighbor as a measure of spatial relationships in populations. Ecology 35: 445–453.

Clark, W.S. 2001. First record of European Griffon Gyps fulvus for Kenya. Bulletin of the African Bird Club 8(1): 59–60.

Clark, W.S. and N.J. Schmitt. 1998. Ageing Egyptian Vultures. Alula 4: 122–127.

Clarke, T. 2006. Birds of the Atlantic Islands. Christopher Helm, London.

Clements, J.F. and N. Shany. 2001. A Field Guide to the Birds of Peru. Ibis Publishing Company, Temecula.

Clements, T., M. Gilbert, H.J. Rainey, R. Cuthbert, J.C. Eames, P. Bunnat, S. Teak, S. Chansocheat and T. Setha. 2012. Vultures in Cambodia: population, threats and conservation. Bird Conservation International 1–18.

Clinton-Eitniear, J. 1986. King vulture conservation and research program. Endangered Species Technical Bulletin. University of Michigan, Ann Arbor 3: 1–2.

Clouet, M. and J.-L. Goar. 2003. L'avifaune de l'Adrar Tirharhar—Adrar des Iforas (Mali). Alauda 71: 469–74.

Coleman, J.S. and J.D. Fraser. 1987. Food habits of Black and Turkey vultures in Pennsylvania and Maryland. Journal of Wildlife Management 51: 733–739.

Coleman, J.S. and J.D.F. Fraser. 1986. Predation on Black and Turkey Vultures. Wilson Bulletin 98: 600–601.

Coleman, J.S., J.D. Fraser and C.A. Pringle. 1985. Salt-eating by Black and Turkey vultures. Condor 87: 291–292.

Coles, V. 1938. Studies in the life history of the Turkey Vulture (*Cathartes aura septentrionalis* Wied.). Ph.D. dissertation, Cornell University, Ithaca, NY.

Collar, N.J. and S.N. Stuart. 1985. Threatened birds of Africa and related islands: the ICBP/IUCN Red Data Book. International Council for Bird Preservation, and International Union for Conservation of Nature and Natural Resources, Cambridge.

Collar, N.J., M.J. Crosby and A.J. Statterfield. 1994. Birds to watch 2: The world list of threatened birds. Conservation series, no. 4. Birdlife International, Cambridge.

Collar, N.J., L.P. Gonzaga, N. Krabbe, A. Madroño Nieto, L.G. Naranjo, T.A. Parker and D.C. Wege. 1992. Threatened birds of the Americas: the ICBP/IUCN Red Data Book. International Council for Bird Preservation, Cambridge.

Coltrain, J.B., J.M. Harris, T.E. Cerling, J.R. Ehleringer, M.-D. Dearing, J. Ward, and J. Allen. 2004. Rancho La Brea stable isotope biogeochemistry and its implications for the palaeoecology of late Pleistocene, coastal southern California. Palaeogeography, Palaeoclimatology, Palaeoecology 205(3–4): 199–219.

Compañia Sevillana de Electricidad. 1995. Analysis of the impacts of power lines on birds of protected areas: manual for risk assessment and mitigation. Iberdrola & REE, Alcobendas.

Cooper, D.S. 2002. Geographic associations of breeding bird distribution in an urban open space. Biological Conservation 104: 205–210.

Cooper, S.M. 1990. The hunting behaviour of spotted hyaenas (*Crocuta crocuta*) in a region containing both sedentary and migratory populations of herbivores. African Journal of Ecology 28: 131–141.

Cooper, S.M. 1991. Optimal hunting group size: the need for lions to defend their kills against loss to spotted hyaenas. African Journal of Ecology 29: 130–136.

Corbacho, C., E. Costillo and A.B. Perales. 2007. La alimentacion del buitre negro. pp. 179–196. *In*: R. Moreno-Opo and F. Guil (eds.). Manual de gestion del habitat y de las poblaciones de buitre negro en Espana. Direccion General para la Biodiversidad, Ministerio de Medio Ambiente, Madrid.

Cortés-Avizanda, A. 2011. Ecological effects of spatial heterogeneity and predictability in the distribution of resources: individuals, populations and guild of scavengers. PhD Thesis, Universidad Autónoma de Madrid.

Cortés-Avizanda, A., M. Carrete and J.A. Donázar. 2010. Managing supplementary feeding for avian scavengers: guidelines for optimal design using ecological criteria. Biological Conservation 143: 1707–1715.

Cortés-Avizanda, A., M. Carrate, D. Serrano and J.A. Donazar. 2009a. Carcasses increase the probability of predation of ground nesting birds: a caveat regarding the conservation value of vulture resturants. Animal Conservation 12: 85–88.

Cortés-Avizanda, A., N. Selva, M. Carrete and J.A. Donazar. 2009b. Effects of carrion resources on herbivore spatial distribution are mediated by facultative scavengers. Basic and Applied Ecology 10: 265–272.

Cortés-Avizanda, A., O. Ceballos and J.A. Donázar. 2009c. Long-term trends in population size and breeding success in the Egyptian Vulture (*Neophron percnopterus*) in northern Spain. Journal of Raptor Research 43(1): 43–49.

Costillo, E., C. Corbacho, R. Moran and A. Villegas. 2007a. Diet plasticity of Cinereous Vulture Aegypius monachus in different colonies in the Extremadura (SW Spain). Ardea 95: 201–211.

Costillo, E., C. Corbacho, R. Morán and A. Villegas. 2007b. The diet of the black vulture Aegypius monachus in response to environmental changes in Extremadura (1970–2000). Ardeola 54(2): 197–204.

Cottam, P. 1957. The pelecaniform characters of the skeleton of the Shoebilled Stork, Balaeniceps rex. British Museum (Natural History) Zoological Bulletin 5: 51–72.

Coues, E. 1883. Hearing of Birds' Ears. Science 2(39): 586–589.

Coves, E. 1903. Key to North American birds, Vol. 2. Fifth ed. Dana Estes, Boston.

Crabtree, R.L. and J.W. Sheldon. 1999a. Coyotes and canid coexistence in Yellowstone. *In*: T.W. Clark, A.P. Curlee, S.C. Minta and P.M. Kareiva (eds.). Carnivores in Ecosystems: The Yellowstone Experience. Yale University Press, New Haven.

Crabtree, R.L. and J.W. Sheldon. 1999b. The ecological role of coyotes on Yellowstone's northern range. Yellowstone Science 7: 15–23.

Cracraft, J. 1981. Toward a phylogenetic classification of the recent birds of the world (class Aves). Auk 98: 681–714.

Cracraft, J., F.K. Barker, M. Braun, J. Harshman, G.J. Dyke, J. Feinstein, S. Stanley, A. Cibois, P. Schikler, P. Beresford, J. García-Moreno, M.D. Sorenson, T. Yuri and D.P. Mindell. 2004. Phylogenetic relationships among modern birds (Neornithes): toward an avian tree of life. pp. 468–489. *In*: J. Cracraft, and M.J. Donoghue (eds.). Assembling the Tree of Life. Oxford University Press, New York.

Craighead, D., B. Bedrosian, R.T. Watson, M. Fuller, M. Pokras and W. Hunt. 2009. A relationship between blood lead levels of common ravens and the hunting season in the southern Yellowstone ecosystem. Ingestion of Lead from Spent Ammunition: Implications for Wildlife and Humans. Peregrine Fund, Boise, Idaho.

Cramp, S. and K.E.L. Simmons. 1980. Handbook of the Birds of Europe, the Middle East and North Africa. Vol. 2. Oxford University Press, Oxford, London.

Cramp, S. and K.E.L. Simmons (eds.). 1980. The Birds of the Western Palearctic Vol. 2. Oxford University Press, Oxford.

Cressman, L.S. 1960. Cultural sequences at the Dalles, Oregon: A Contribution to Pacific Northwest prehistory. Transactions of the American Philosophical Society n.s. 50(10): 1–108.

Cringan, A.T. 2007. An observation of apparent 'fishing' by Turkey Vultures. Colorado Birds 41: 258–259.

Crook, C. 1935. The incubation period of the Black Vulture. The Auk 52(1): 78–79.

Crooks, K.R. and M.E. Soulé. 1999. Mesopredator release and avifaunal extinctions in a fragmented system. Nature 400: 563–566.

Cuesta, M.R. and E.A. Sulbarán. 2000. Andean Condor (*Vultur gryphus*). pp. 16–21. *In*: R.P. Reading and B. Miller (eds.). Endangered Animals: a Reference Guide to Conflicting Issues. Greenwood Press, London.

Cuesta, M.R. and E.A. Sulbaran. 2000. Andean condor (*Vultur gryphus*). pp. 16–21. *In*: B. Miller and R.P. Reading (eds.). Endangered Animals: A Reference Guide to Conflicting Issues. Greenwood Press, Westport.

Cummins, R.A. 1997a. Assessing quality of life. pp. 116–150. *In*: R.J. Brown (ed.). Assessing Quality of Life for People with Psychiatric Disabilities. Stanley Thornes (Publishers), Cheltenham, England.

Cummins, R.A. 1997b. Comprehensive Quality of Life Scale—Adult. School of Psychology. Deakin University, Melbourne.

Cuneo, F. 1968. Notes on breeding the King vulture *Sarcoramphus papa* at Naples Zoo. International Zoo Yearbook 8(1): 156–157.

Cunningham, A.A., V. Prakash, D.J. Pain, G.R. Ghalsasi, G.A.H. Wells, G.N. Kolte, P. Nighot, M.S. Goudar, S. Kshirsagar and A. Rahami. 2003. Indian vultures: victims of an infectious disease epidemic? Animal Conservation. Animal Conservation 6: 187–197.

Cunningham, A.A., V. Prakash, G.R. Ghalsasi and D. Pain. 2001. Investigating the cause of catastrophic declines in Asian Griffon Vultures (Gyps indicus and G. bengalensis). *In*: T. Katzner and J. Parry-Jones (eds.). Reports from the workshop on Indian Gyps vultures, 4th Eurasian Congress on Raptors, Seville, Spain, 10–11. Estación Biológica Donaña, Raptor Research Foundation, Seville.

Cunningham, A.B. 1990. Vultures and the Trade in Traditional Medicines. Vulture News 24: 5.

Cunningham, D.D. 1903. Some Indian friends and acquaintances. J. Murray, London.

Curtis, E.S. 1824a. The North American Indian: Being a series of volumes picturing and describing the Indians of the United States, the Dominion of Canada, and Alaska Vol. 13. Plimpton Press, Norwood, Massachusetts.

Curtis, E.S. 1824b. The North American Indian: Being a series of volumes picturing and describing the Indians of the United States, the Dominion of Canada, and Alaska Vol. 14. Plimpton Press, Norwood, Massachusetts.

Curtis, E.S. 1826. The North American Indian: Being a series of volumes picturing and describing the Indians of the United States, the Dominion of Canada, and Alaska Vol. 15. Plimpton Press, Norwood, Massachuesetts.

Cuthbert, R.J., V. Prakash, M. Saini, S. Upreti, D. Swarup, A.K. Sharma, A. Das, R.E. Green and M. Taggart. 2011a. Are conservation actions reducing the threat to India's vulture populations? Current Science 101: 1480–1481.

Cuthbert, R.J., R. Dave, S.S. Chakraborty, S. Kumar, S. Prakash, S.P. Ranade and V. Prakash. 2011b. Assessing the ongoing threat from veterinary NSAIDs to critically endangered Gyps vultures in India. Oryx 45: 420–426.

Cuthbert, R., R.E. Green, S. Ranade, S. Saravanan, D.J. Pain, V. Prakash and A.A. Cunningham. 2006. Rapid population declines of Egyptian Vulture (*Neophron percnopterus*) and Red-headed Vulture (Sarcogyps calvus) in India. Animal Conservation 9(3): 349–354.

Cuthbert, R., J. Parry-Jones, R.E. Green and D.J. Pain. 2007. NSAIDs and scavenging birds: potential impacts beyond Asia's critically endangered vultures. Biology Letters 3: 90–93.

Cuthbert, R., M.A. Taggart, V. Prakash, M. Saini, D. Swarup, R. Mateo, S.S. Chakraborty, P. Deori and R. Green. 2011. Effectiveness of action in India to reduce exposure of Gyps Vultures to the toxic veterinary drug Diclofenac. PLoS One 6(5): 1–11. e19069. doi:10.1371/journal.pone.0019069.

Cuthbert, R., R.E. Green, S. Ranade, S. Saravanan, D.J. Pain, V. Prakash and A.A. Cunningham. 2006. Rapid population declines of Egyptian Vulture (*Neophron percnopterus*) and Red-headed Vulture (*Sarcogyps calvus*) in India. Animal Conservation 9(3): 349–354.

Dall, S.R.X. 2002. Can information sharing explain recruitment to food from communal roosts? Behavioural Ecology 13: 42–51.

Daneel, A.B. 1984. Breeding of the Hooded Vulture Necrosyrtes monachus in the Kruger National Park. Koedoe 27: 141.

Dänhardt, J. and Å. Lindström. 2001. Optimal departure decisions of songbirds from an experimental stopover site and the significance of weather. Animal Behaviour 62: 235–243.

Darlington, P.J., Jr. 1930. Notes on the senses of vultures. Auk 47: 251–252.

Darwin, C. 1839. Narrative of the surveying voyages of His Majesty's ships Adventurer and Beagle, between the years 1826 and 1836. Coribun, London.

Das, D., R.J. Cuthbert, R.D. Jakati and V. Prakash. 2011. Diclofenac is toxic to the Himalayan Vulture Gyps himalayensis. Bird Conservation International 21: 72–75.

Daszak, P., A.A. Cunningham and A.D. Hyatt. 2000. Review: emerging infectious diseases of wildlife: threats to biodiversity and human health. Science 287: 443–449.

Davies, R. 2006. In BirdLife International 2014. Species factsheet: *Trigonoceps occipitalis*. Downloaded from http://www.birdlife.org on 03/02/2014.

Davis, E.R. 1998. Strategies for alleviating vulture damage in industrial plants. Proceedings of the Vertebrate Pest Conference 18: 71–3.

Dawson, J.W. and R.W. Mannan. 1994. The ecology of Harris' hawks in urban environments. Unpublished Report, Arizona Game and Fish Department, Tucson, Arizona, USA. Available from the U.S. Geological Survey, Richard R. Olendorff Memorial Library, 970 Lusk St., Boise, ID 83706.

De Bonis, L., S. Peigne, H.T. Mackaye, A. Likius, P. Vignaud and M. Brunet. 2010. New sabre-toothed cats in the Late Miocene of Toros Menalla (Chad). Systematic Palaeontology (Vertebrate Palaeontology) Comptes Rendus Palevol 9: 221–227.

de Castro, M.C. and M.C. Langer. 2008. New postcranial remains of Smilodon populator Lund, 1842 from South-Central Brazil. Revista Brasileira de Paleontologia 11(3): 199–206.

De la Puente, J., R. Moreno-Opo and J.C. Del Moral. 2007. El buitre negro en España. Censo Nacional (2006). SEO/BirdLife, Madrid.

De Lucas, M., G.F.E. Janss, D.P. Whitfield and M. Ferrer. 2008. Collision fatality of raptors in wind farms does not depend on raptor abundance. Journal of Applied Ecology 45: 1695–1703.

De Lucas, M., M. Ferrer and G.F.E. Janss. 2012a. Using wind tunnels to predict bird mortality in wind farms: The case of Griffon Vultures. PLoS ONE 7(11): e48092. doi:10.1371/journal.pone.0048092.

De Lucas, M., M. Ferrer, M.J. Bechard and A.R. Muñoz. 2012b. Griffon vulture mortality at wind farms in southern Spain: Distribution of fatalities and active mitigation measures. Biological Conservation 147: 184–189.

De Lucas, M., G.F.E. Janss, D.P. Whitfield and M. Ferrer. 2008. Collision fatality of raptors in wind farms does not depend on raptor abundance. Journal of Applied Ecology 45: 1695–1703.

De Saussure, R. 1956. Remains of the California Condor in Arizona caves. Plateau 29: 44–45.

Dean, W.R.J. 2004. Historical changes in stocking rates: possible effects on scavenging birds. In: A. Monadjem, M.D. Anderson, S.E. Piper and A.F. Boshoff (eds.). The Vultures of Southern Africa—Quo Vadis? Proceedings of a workshop on vulture research and conservation in southern Africa, 156–165. Birds of Prey Working Group, Johannesburg.

Dean, W.R.J., R.I. Yeaton and S.J. Milton. 2006. Foraging sites of Turkey Vultures and common Ravens Corvus corax in central Mexico. Vulture News 54: 30–33.

DeBoer, L.E.M. and R.P. Sinoo. 1984. A karyological study of Accipitridae (Aves: Falconiformes) with karyotypic descriptions of 16 species new to cytology. Genetica 65: 89–107.

Deignan, H.G. 1945. The Birds of Northern Thailand. Bulletin 186. U.S. National Museum.

Deignan, H.G. 1963. Checklist of the birds of Thailand. Bulletin 226. U.S. National Museum, Washington, D.C.

Del Hoyo, J., A. Elliott and J. Sargatal (eds.). 1994. Handbook of birds of the world. Vol. 2. New World vultures to guineafowl. Lynx Edicions, Barcelona, Spain.

Del Moral, J.C. and R. Martí. 2001. El Buitre Leonado en la Península Ibérica. III Censo. Nacional y I Censo Ibérico coordinado, 1999. Monografía no 7. SEO/BirdLife, Madrid.

Del Moral, J.C. (ed.). 2009. El alimoche común en España. Población reproductora en 2008 y método de censo. SEO/Birdlife, Madrid.

Del Moral, J.C. (ed.). 2009. [The Griffon Vulture: breeding population in 2008 and census method]. Seguimiento de Aves no. 30. SEO/BirdLife International, Madrid, Spain (In Spanish with English summary).

Delacour, J. and P. Jabouille. 1931. Les oiseaux de l'Indochine française, 1–4. Exposition, Coloniale Internationale, Paris.

Delgado, G. 1999. Guirre. P. 1814 in Gran Enciclopedia Canaria, VII. Ediciones Canarias, Santa Cruz de Tenerife.

Dellelegn, Y. and B. Abdu. 2010. Ethiopian vulture survey report. Unpublished report to the Royal Society for the Protection of Birds.

DeLuca, W.V., C.E. Studds, R.S. King and P.P. Marra. 2008. Coastal urbanization and the integrity of estuarine waterbird communities: Threshold responses and the importance of scale. Biological Conservation 141: 2669–2678.

Demerdzhiev, D.A. 2010. Mortality rate in wild birds caused by 20kV power lines. Electrocution in six studied protection sites in the Natura 2000 Network in Bulgaria. Bulgarian Society for the Protection of Birds.

Demerdzhiev, D.A., S.A. Stoychev, T.H. Petrov, I.D. Angelov and N.P. Nedyakov. 2009. Impact of power lines on bird mortality in Southern Bulgaria. Acta Zoologica Bulgarica 61(2): 177–185.

Demeter, I. 2004. Medium-voltage power lines and bird mortality in Hungary. MME BirdLife Hungary, Budapest.

Demey, R. and L.D.C. Fishpool. 1991. Additions and annotations to the avifauna of CÔte D'Ivoire. Malimbus 12: 61–86.

Desai, A. and L. Vuthy. 1996. Status and distribution of large mammals in eastern Cambodia: results of the first foot surveys in Mondulkiri and Rattanakiri Provinces. IUCN/FFI/WWF Large Mammal Conservation Project, Phnom Penh.

Deshler, W. 1963. Cattle in Africa: Distribution, types, and problems. Geographical Review 53(1): 52–58.

Desholm, M., A.D. Fox, P.D.L. Beasley and J. Kahlert. 2006. Remote techniques for counting and estimating the number of bird–wind turbine collisions at sea: a review. Ibis 148: 76–89.

Desjardins, P. and R. Morais. 1990. Sequence and gene organisation of chicken mitochondrial genome. Journal Molecular Biology 212: 599–634.

Desowitz, R.S. 1959. Studies on immunity and host parasite relationships. I. The immunological response of resistant and susceptible breeds of cattle to trypanosomal challenge. Annals of Tropical Medicine and Parasitology 53: 293–313.

DeVault, T.L., I.L. Brisbin, Jr. and O.E. Rhodes, Jr. 2004. Factors influencing the acquisition of rodent carrion by vertebrate scavengers and decomposers. Canadian Journal of Zoology 82: 502–509.

DeVault, T.L., O.E. Rhodes, Jr. and J.A. Shivik. 2003. Scavenging by vertebrates: Behavioral, ecological, and evolutionary perspectives on an important energy transfer pathway in terrestrial ecosystems. Oikos 102: 225–234.

DeVault, T.L. and O.E. Rhodes, Jr. 2002. Identification of vertebrate scavengers of small mammal carcasses in a forested landscape. Acta Theriologica 47: 185–192.

DeVault, T.L., Z.H. Olson, J.C. Beasley and O.E. Rhodes, Jr. 2011. Mesopredators dominate competition for carrion in an agricultural landscape. Basic and Applied Ecology 12: 268–274.

Devault, T.L., B.D. Reinhart, I.L. Brisbin, Jr. and O.E. Rhodes, Jr. 2004. Home ranges of sympatric black and turkey vultures in South Carolina. Condor 106: 706–711.

DeVault, T.L., B.D. Reinhart, I.L. Brisbin and O.E. Rhodes. 2005. Flight behavior of black and turkey vultures: implications for reducing bird–aircraft collisions. Journal of Wildlife Management 69: 592–599.

DeVault, T.L., B.D. Reinhart, I.L. Brisbin and O.E. Rhodes. 2010. Flight behavior of black and turkey vultures: implications for reducing bird-aircraft collisions. The Journal of Wildlife Management 69(2): 601–608.

Di Giacomo, A.S. 2005. A´reas importantes para la conservacio´n de las aves en Argentina Sitios prioritarios para la conservacio´n de la biodiversidad. Temas de Naturaleza y Sociedad 5: 1–514.

Di Vittorio, M. 2006. Reintroduction of the Griffon Vulture *Gyps fulvus* in Nebrodi Regional Park, Sicily. *In*: D.C. Houston and S.E. Piper (eds.). Proceedings of the International Conference on Conservation and Management of Populations, 174. Thessaloniki, Greece. Natural History Museum of Crete & WWF Greece.

Dickerman, R.W. 2003. Talon-locking in the red-tailed hawk. Journal of Raptor Research 37: 176.

Dickerman, R.W. 2007. Birds of the Southern Pacific Lowlands of Guatemala. Special Publication of the Museum of Southwestern Biology 7: 1–45.

Dickerson, D.D. 1983. Black vultures kill skunk in Mississippi. Mississippi Kite 13: 2–3.

Dickey, D.R. and A.J. Van Rossem. 1938. The birds of El Salvador. Field Museum of Natural History, Zoological Series No. 23: 1–609.

Dickinson, E.C. (ed.). 2003. The Howard and Moore Complete Checklist of the Birds of the World. Third edition. Princeton University Press, Princeton.

Diekmann, M. 2004. Status of vultures in the Waterberg Region of Namibia. pp. 148–153. *In*: A. Monadjem, M.D. Anderson, S.E. Piper and A.F. Boshoff (eds.). The Vultures of Africa—Quo Vadis? Birds of Prey Working Group, Johannesburg.

Diekmann, M. and A. Strachan. 2006. Saving Namibia's most endangered bird. WAZA Magazine 16–19.

Dingle, H. 1996. Migration: The Biology of Life on the Move. Oxford University Press, New York.

Directorate of Culture and Cultural and Natural Heritage (DCCHN). 2010. Implementation of recommendation No. 110/2004 on minimising adverse effects of above ground electricity transmission facilities (power lines) on birds. Report by the Governments to the 30th Meeting of the Standing Committee of the Bern Convention, Strasbourg. T-PVS/Files (2010) 11. Council of Europe, Strasbourg.

Dirksen, S., A.L. Spaans and J. van der Winden. 2007. Collision risks for diving ducks at semi-offshore wind farms in freshwater lakes: a case study. *In*: M. de Lucas, G.F.E. Janss and M. Ferrer (eds.). Birds and Wind Farms. Quercus, Madrid.

Dirksen, S., A.L. Spaans, J. van der Winden and L.M.J. van den Bergh. 1998. Nachtelijke vliegpatronen en vlieghoogtes van duikeenden in het IJsselmeergebied. Limosa 71: 57–68.

Dirksen, S., J.V.D. Winden and A.L. Spaans. 1998. Nocturnal collision risks of birds with wind turbines in tidal and semi-offshore areas. pp. 99–107. *In*: C.F. Ratto and G. Solari (eds.). Wind Energy and Landscape. Balkema, Rotterdam.

Dobrowolska, A. and M. Melosik. 2008. Bullet-derived lead in tissues of the wild boar (*Sus scrofa*) and red deer (*Cervus elaphus*). Eurasian Journal of Wildland Resources 54: 231–235.

Dobson, A., M. Borner and T. Sinclair. 2010. Road will ruin Serengeti. Nature 467: 272–273.

Dolata, P.T. 2006. The White Stork *Ciconia ciconia* protection in Poland by tradition, customs, law andactive efforts. pp. 477–492. *In*: P. Tryjanowski, T.H. Sparks and L. Jerzak (eds.). The White Stork in Poland: Studies in Biology, Ecology and Conservation. Bogucki Wydawnictwo Naukowe.

Dolbeer, R.A. 1998. Evaluation of shooting and falconry to reduce bird strikes with aircraft at John F. Kennedy International Airport. pp. 145–153. *In*: Proceedings of the IBSC 2, Stara Lesna, Slovakia 14–18 September. WP13.

Doloh, J. 2007. Red-headed Vulture breeding programme launched. Vulture News 57: 79–80.

Dominguez-Rodrigo, M. 2000. A study of carnivore competition in riparian and open habitats of modern savannas and its implications for hominid behavioral modelling. Journal Human Evolution 40: 77–98.

Donázar, J.A. 1993. Los buitres ibéricos. Biología y conservación. J.M. Reyero Editor, Madrid.

Donázar, J.A. 2004. Alimoche Común Neophron percnopterus. pp. 129–131. *In*: A. Madroño, C. González and J.C. Atienza (eds.). Libro Rojo de las Aves de España. Dirección General para la Biodiversidad & SEO/BirdLife, Madrid.

Donazar, J.A., F. Hiraldo and J. Bustamante. 1993. Factors influencing nest site selection, breeding density and breeding success in the bearded vulture (*Gypaetus barbatus*). Journal of Applied Ecology 30: 504–514.

Donazar, J.A. 1993. Los buitres ibericos [The vultures of Spain: biology and conservation]. *In*: J.M. Reyero (ed.). Biologia y conservacion. Madrid.

Donazar, J.A. and O. Ceballos. 1989. Growth rates of nestling Egyptian Vultures Neophron percnopterus in relation to brood size, hatching order and environmental factors. Ardea 77(2): 217–226.

Donazar, J.A., O. Ceballos and J.L. Tella. 1996. Communal roosts of Egyptian vultures (*Neophron percnopterus*): dynamics and implications for the species conservation. *In*: J. Muntaner and J. Mayol (eds.). Biologia y Conservacion de las Rapaces Mediterraneas, 1994 (Monograf ias no 4), 189–201. SEO, Madrid.

Donázar, J.A., A. Cortés-Avizanda and M. Carrete. 2010. Dietary shifts in two vultures after the demise of supplementary feeding stations: consequences of the EU sanitary legislation. European Journal of Wildlife Research 56: 613–621.

Donazar, J.A., A. Margalida, M. Carrete and J.A. Sanchezzapata. 2009a. Too sanitary for vultures. Science 326: 664.

Donazar, J.A., A. Margalida and D. Campion. 2009b. Vultures, Feeding Stations and Sanitary Legislation: A Conflict and its Consequences from the Perspective of Conservation Biology. Sociedad de Ciencias Aranzadi, Donostia, Spain.

Donazar, J.A., M.A. Naveso, J.L. Tella and D. Campion. 1996. Extensive grazing and raptors in Spain. pp. 117–149. *In*: D.J. Pain and M.W. Pienkowski (eds.). Farming and Birds in Europe. Academic Press, Cambridge.

Donázar, J.A., M.A. Naveso, J.L. Tella and D. Campión. 1997. Extensive grazing and raptors in Spain. pp. 117–149. *In*: D.J. Pain and M.W. Pienkowski, (eds.). Farming and Birds in Europe: the Common Agricultural Policy and its Implications for Bird Conservation. Academic Press, London.

Donázar, J.A., J.J. Negro, C.J. Palacios, L. Gangoso, J.A. Godoy, O. Ceballos, F. Hiraldo and N. Capote. 2002a. Description of a new subspecies of the Egyptian Vulture (Accipitridae: *Neophron percnopterus*) from the Canary Islands. Journal of Raptor Research 36(1): 17–23.

Donazar, J.A., A. Travaini, O. Ceballos, A. Rodrıguez, M. Delibes and F. Hiraldo. 1999. Effects of sex-associated competitive asymmetries on foraging group structure and despotic distribution in Andean condors. Behavioral Ecology and Sociobiology 45: 55–65.

Donazar, J.A., C.J. Palacios, L. Gangoso, O. Ceballos, M.J. Gonzalez and F. Hiraldo. 2002. Conservation status and limiting factors in the endangered population of Egyptian vulture (*Neophron percnopterus*) in the Canary Islands. Biological Conservation 107(1): 89–97.

Donazar, J.A. and J.E. Feijoo. 2002. Social structure of Andean Condor roosts: Influence of sex, age, and season. Condor (Cooper Ornithological Society) 104(1): 832–837.

Donnay, T.J. 1990. Status, Nesting and Nest Site Selection of Cape Vultures in Lesotho. Vulture News 24: 11–24.

Donázar, J.A. 1993. Los buitres ibéricos. Biología y conservación. J.M. Reyero Editor, Madrid.

Donzar, J.A., C.J. Palacios, L. Gangoso, O. Ceballos, M.J. Gonzlez and F. Hiraldo. 2002. Conservation status and limiting factors in the endangered population of Egyptian vulture (*Neophron percnopterus*) in the Canary Islands. Biological Conservation 107: 89–97.

Dorin, M., S. Linda and K.S. Smallwood. 2005. Response to public comments on the staff report entitled Assessment of Avian Mortality from Collisions and Electrocutions (CEC-700-2005-015) (Avian White Paper) written in support of the 2005 Environmental Performance Report and the 2005 Integrated Energy Policy Report. California Energy Commission, Sacramento.

Dorin, M., L. Spiegel, J. McKinney, C. Tooker, P. Richins, K. Kennedy, T. O'Brien and S.W. Matthews. 2005. Assessment of Avian Mortality from Collisions and Electrocutions. California Energy Commission, California.

Dorst, J. 1970. Before Nature Dies. Collins, London.

Douglas, I., D. Goode, M. Houck and R. Wang. 2010. The Routledge Handbook of Urban Ecology. Taylor and Francis, Oxford.

Doutressoulle, G. 1947. L'élevage en Afrique occidentale française. Larose, Paris.

Dowsett, R.J. and F. Dowsett-Lemaire. 1980. The Systematic Status of Some Zambian Birds. Gerfaut 70: 151–199.

Dowsett-Lemaire, F. 2006. An Annotated List and Life History of the Birds of Nyika National Park, Malawi-Zambia. Tauraco Research Report 8: 1–6.

Dowsett-Lemaire, F. 2006. In BirdLife International 2014. Species factsheet: Trigonoceps occipitalis. Downloaded from http://www.birdlife.org on 03/02/2014.

Dowsett-Lemaire, F. and R.J. Dowsett. 2006. The Birds of Malawi. Tauraco Press and Aves, Liège.

Dowsett, R.J., D.R. Aspinwall and F. Dowsett-Lemaire. 2008. The Birds of Zambia: an Atlas and Handbook. Tauraco Press and Aves, Liége.

Drake, V.A. and R.A. Farrow. 1988. The influence of atmospheric structure and motions on insect migration. Annual Review of Entomology 33: 183–210.

Drewitt, A.L. and R.H.W. Langston. 2008. Collision effects of wind-power generators and other obstacles on birds. Annals of the New York Academy of Science 1134: 233–266.

Driver, H.E. 1937. Culture element distributions, X: Northwest California. University of California Anthropological Records 1: 53–154.

Driver, H.E. 1939. Culture element distributions, X. Northwest California. University of California Anthropological Records 1: 297–433.

Drucker, P. 1937. Culture elements distributions, V: Southern California. University of California Anthropological Records 1: 1–52.

Du-Bois, C.G. 1905. Religious ceremonies and myths of the Mission Indians. American Anthropology 7: 620–629.

Du-Bois, C.G. 1908. The Religion of the Luseno and Diegueno Indians of Southern California. University of California Publications in American Archaeology and Ethnology 8: 69–186.

Ducatez, M.F., Z. Tarnagda, M.C. Tahita, A. Sow, S. de Landtsheer, B.Z. Londt, I.H. Brown, A.d.M.E. Osterhaus, R.A.M. Fouchier, J.B. Ouedraogo and C.P. Muller. 2007. Genetic characterization of HPAI (H5N1) viruses from poultry and wild vultures, Burkina Faso. Emerging Infectious Diseases 13(4): 611–613.

Duckworth, J.W., P. Davidson, T.D. Evans, P.D. Round and R.J. Timmins. 2002. Bird records from North Laos, principally Xiangkhouang Province and the upper Lao Mekong, in 1998–2000. Forktail 18: 11–44.

Duckworth, J.W., R.E. Salter and K. Khounboline. 1999. Wildlife in Lao PDR: 1999 Status Report. The World Conservation Union, Wildlife Conservation Society, Centre for Protected Areas and Watershed Management, Gland.

Dwyer, J.F. and S.G. Cockwell. 2011. Social hierarchy of scavenging raptors on the Falkland Islands, Malvinas. Journal of Raptor Research 45(3): 229–235.

Dzhamirzoev, G.S. and S.A. Bukreev. 2009. Status of Egyptian Vulture *Neophron percnopterus* in the North Caucasus, Russian Federation. Sandgrouse 31(2): 128–133.

Eagan, T.S. 2009. Disease and predator ecology of white-footed mice in the Upper Wabash River Basin. M.S. thesis. Purdue University, West Lafayette, IN, USA.

Eames, J.C. 2007a. Cambodian national vulture census 2007. The Babbler: BirdLife in Indochina 33–34.

Eames, J.C. 2007b. Mega transect counts vultures across Myanmar. The Babbler: BirdLife in Indochina 30.

Eames, J.C., Nguyen Duc Tu, Le Trong Trai, Dang Ngoc Can, Ngo Van Tri, Hoang Duc Dat, Thai Ngoc Tri and Nguyen Thi Thu He. 2004. Final biodiversity report for Yok Don National Park, Dak Lak Province, Vietnam: Creating protected areas for resource conservation using landscape ecology (PARC) project. FPD/UNOPS/UNDP/Scott Wilson Asia-Pacific Ltd./Environment and Development Group and Forest Renewable Resources Ltd., Hanoi.

Earl, T.M. 1929. On the Scent of Vultures. Wilson Bulletin 41: 103.

East Africa Natural History Society Nest Records, Nature Kenya, Nairobi.

Edwards, S.V., P. Arctander and A.C. Wilson. 1991. Mitochondrial resolution of a deep branch in the genealogical tree for perching birds. Proceedings of the Royal Society of London B243: 99–107.

Eisermann, K. and C. Avendano. 2007. Lista comentada de las aves de Guatemala/Annotated checklist of the birds of Guatemala. Lynx Edicions, Barcelona.

Eitniear, J.C. 1996. Estimating age classes in king vultures (*Sarcoramphus papa*) using plumage coloration. Journal of Raptor Research 30(1): 35–38.

Eliotout, B., P. Lecuyer and O. Duriez. 2007. Premiers résultats sur la biologie de reproduction du Vautour Moine Aegypius monachus en France. Alauda 75(3): 253–264.

Eliotout, B., P. Lecuyer and O. Duriez. 2007. Premiers résultats sur la biologie de reproduction du vautour moine Aegypius monachus en France. Alauda 75: 253–264.

Eliotout, B., P. Lecuyer and O. Duriez. 2007. Premiers résultats sur la biologie de reproduction du Vautour Moine Aegypius monachus en France. Alauda 75(3): 253–264.

Ellis, C. 2004. Of Gyps vultures, gypsies and satellite technology. Peregrine Fund Newslette 35: 14–15.

Ellis, D.H., R.L. Glinski, J.G. Goodwin and W.H. Whaley. 1983. New World vulture counts in Mexico, Central America, and South America. pp. 124–132. *In*: S.R. Wilbur and J.A. Jackson (eds.). Vulture Biology and Management. University of California Press, Berkeley.

Elorriaga, J., L. Garcia, J. Martinez and E. Unamunzaga. 2000. Quality of life of persons with mental retardation in Spain. *In*: K.D. Keith and R.L. Schlock (eds.). Cross-Cultural Perspectives on Quality of Life. American Association on Mental Retardation, Washington, D.C.

Elosegui, I. 1989. Vautour fauve (*Gyps fulvus*), Gypaete barbu (*Gypaetus barbatus*), Percnoptère d'Egypte (*Neophron percnopterus*): Synthèse bibliographique et recherches. Acta Biologica Montana 3. Série Document de Travail.

Elsasser, A.B. and R.F. Heizer. 1963. The archaeology of Bowers Cave, Los Angeles County, California. Report of the University of California Archaeological Survey 59: 1–60.

Emlen, S.T. 1971. Adaptive aspects of coloniality in the Bank Swallow. American Zoology 11: 628.

Emslie, S.D. 1988a. An early condor-like vulture from North America. The Auk 105: 3529–3535.

Emslie, S.D. 1988b. The fossil history and phylogenetic relationships of condors (Ciconiiformes: Vulturidae) in the New World. Journal of Vertebrate Paleontology 8(2): 212–228.

Epstein, H. and I.L. Mason. 1984. Cattle. pp. 6–27. *In*: I.L. Mason (ed.). Evolution of Domesticated Animals. Longman, London.

Epstein, H. 1971. The Origin of the Domestic Animals of Africa, Vol. 1. Africana, New York.

Erickson, W., G. Johnson, D. Young, D. Strickland, R. Good, M. Bourassa and K. Bay. 2002. Synthesis and Comparison of Baseline Avian and Bat Use, Raptor Nesting and Mortality Information from Proposed and Existing Wind Developments. WEST, Inc., Report. Bonneville Power Administration, Portland, Oregon. <http://www.bpa.gov/power/pgc/wind/Avian_and_Bat_Study_12-2002.pdf> (downloaded 03.03.2014).

Erickson, W.P., G. Johnson, D. Young, D. Strickland, R. Good, M. Bourassa, K. Bay and K. Sernka. 2002. Synthesis and comparison of baseline avian and bat use, raptor nesting and mortality information from proposed and existing wind developments. Bonneville Power Administration, Portland.

Ericson, P.G.P., C.L. Anderson, T. Britton, A. Elżanowski, U.S. Johansson, M. Kallersjö, J.I. Ohlson, T.J. Parsons, D. Zuccon and G. Mayr. 2006. Diversification of Neoaves: integration of molecular sequence data and fossils. Biology Letters 22; 2(4): 543–547.

Erni, B., F. Liechti and B. Bruderer. 2003. How does a first year passerine migrant find its way? Simulating migration mechanisms and behavioural adaptations. Oikos 103: 333.

Erni, B., F. Liechti and B. Bruderer. 2005. The role of wind in passerine autumn migration between Europe and Africa. Behavioral Ecology 16: 732–40.

Erwin, R.M. 1978. Coloniality in terns: The role of social feed. Condor 80: 211–215.

Espuno, N., B. Lequette, M.L. Poulle, P. Migot and J.D. Lebreton. 2004. Heterogeneous response to preventive sheep husbandry during wolf recolonization of the French Alps. Wildlife Society Bulletin 32: 1195–1208.

Eu Asac. 2009. Weapons destruction table 1999–2006. EU ASAC (European Union's Assistance on Curbing Small Arms and Light Weapons in the Kingdom of Cambodia) website www.eu-asac.org. Accessed 10 March 2014.

Evans, G.L. 1961. The Friesenhahn Cave. Bulletin of the Texas Memorial Museum 2: 1–22.

Evans, M.I. and S. Al-Mashaqbah. 1996. Did Lappet-faced Vulture Torgos tracheliotos formerly breed in Jordan? Sandgrouse 18: 61.

Evelyn-White, H.G. 1920. Hesiod, the Homeric Hymms, and Homerica. Heinemann, London.

Everaert, J., K. Devos and E. Kuijken. 2002. Windturbines en vogels in Vlaanderen. Voorlopige onderzoeksresultaten en buitenlandse bevindingen. Instituut voor Natuurbehoud, Brussels. http://publicaties.vlaanderen.be/docfolder/12563/Effecten_windturbines_op_de_fauna_Vlaanderen_2008.pdf.

Everett, M. 2008. Lammergeiers and lambs. British Birds 101(4): 215.

Faanes, C.A. 1987. Bird behavior and mortality in relation to power lines in prairie habitats. Fish and Wildlife Technical Report 7, United States Department of the Interior Fish and Wildlife Service, Washington, D.C.

FAB. 2008. Vulture crisis in Aragón (North Spain). Unpublished report.

Fain, M.G. and P. Houde. 2004. Parallel radiations in the primary clades of birds. Evolution 58: 2558–2573.

FAO (Food and Agricultural Organization of the United Nations). 2000. Africover multipurpose land cover databases for Kenya. Rome: FAO. Available at http://www.africover.org [accessed October 2012].

Faulkner, R., J. Wasserman, O. Goelet and E. Von Dassow. 2008. The Egyptian Book of the Dead: The Book of Going Forth By Day. Chronicle Books, San Francisco.

Faulkner, R.O. et al. 2008. The Egyptian Book of the Dead: the Book of Going Forth by Day. San Fransisco. p. 124.

Feduccia, A. 1995. Explosive evolution in tertiary birds and mammals. Science 267: 637–638.

Feduccia, Alan. 1996. The Origin and Evolution of Birds. Yale University Press, Yale.

Feduccia, A. 1999. The Origin and Evolution of Birds. Yale University Press, Yale.

Felce, D. and J. Perry. 1995. Quality of life: its definition and measurement. Research in Developmental Disabilities 16(1): 51–74.

Felce, D. and J. Perry. 1996. Assessment of quality of life. pp. 63–70. In: R.L. Schalock and G.N Siperstein (eds.). Quality of Life Volume I: Conceptualization and Measurement. American Association on Mental Retardation, Washington D.C.

Felsenstein, J. 1985. Confidence limits on phylogenies: an approach using the bootstrap. Evolution 39: 783–791.

Fennec, R.S. 2005. Growth rate and duration of growth in the adult canine of Smilodon gracilis and inferences on diet through stable isotope analysis. Feranec Bull FLMNH 45(4): 369–77.

Fergus, C. 2003. Wildlife of Virginia and Maryland. Stackpole Books, Washington D.C.

Ferguson-Lees, J. and D.A. Christie. 2000. Raptors of the world: identification guides. Christopher Helm, London.

Ferguson-Lees, J. and D.A. Christie, 2001. Raptors of the World. Christopher Helm, London.

Fernandez, J.A. 1975. Distribucio´n y frecuencia de la co´pula del buitre leonado (Gyps fulvus) en el sur de Espana. Donana Acta Vertebr 2: 193–199.

Fernández-Arroyo, F.J.F. 2007a. Bajón en el censo de pollos de buitre leonado de las hoces del Riaza./Preocupación en la reunión de expertos de Plasencia. Quercus 261: 33–34.

Fernández-Arroyo, F.J.F. 2007b. Señales de alarma sobre la situación de los buitres. Conclusiones de las III Jornadas sobre Buitres. Argutorio 20: 29–31.

Fernández-Arroyo, F.J.F. 2009. Declive de buitres en las hoces del Riaza. Letter. Quercus 275: 4.

Fernández-Arroyo, F.J.F. 2012. New vulture censuses in the Montejo Raptor Refuge, Spain. Vulture News 62: 4–24.

Ferrari, N. and J.M. Weber. 1995. Influence of the abundance of food resources on the feeding habits of the red fox, Vulpes vulpes, in western Switzerland. Journal of Zoology, London 236: 117–129.

Ferrer, M. 1993. El Aguila Imperial. Quercus, Madrid.

Ferrer, M. and G.F.E. Janss. 1999. Birds and Power Lines. Quercus, Madrid.

Ferrer, M., M. de la Riva and J. Castroviejo. 1991. Electrocution of raptors on power lines in Southwestern Spain. Journal of Field Ornithology 62: 181–190.

Ferro, M. 2000. Consumption of metal artefacts by Eurasian Griffons at Gamla Nature Reserve, Israel. Vulture News 43: 46–48.

ffrench, R. 1991. A guide to the birds of Trinidad and Tobago. 2nd ed. Cornel University Press, Ithaca.

Finkelstein, M.E., D.F. Doak, D. George, J. Burnett, J. Brandt, M. Church, J. Grantham and D.R. Smith. 2012. Lead poisoning and the deceptive recovery of the critically endangered California condor. Proceedings of the National Academy of Sciences DOI: 10.1073/pnas.1203141109.

Fischer, A.B. 1969. Laboratory experiments on and open-country observations of the visual acuity and behavior of Old World vultures. Zoologische Jahrbücher Systematik 96: 81–132.

Fisher, H.I. 1942. The pterylosis of the Andean Condor. Condor 44(1): 30–32.

Fisher, H.I. 1944. The skulls of the Cathartid vultures. Condor 46(6): 272–296.

Fisher, H.I. 1945. Locomotion in the fossil vulture Teratornis. American Midland Naturalist 33(3): 725–742.

Fisher, H.I. 1955. Some Aspects of the Kinetics in the Jaws of Birds. The Wilson Bulletin 67(3): 175–188.

Fisher, H.L. 1944. The Skulls of the Cathartid vultures. The Condor 46(6): 272–296.

Fisher, I.J., D.J. Pain and V.G. Thomas. 2006. A review of lead poisoning from ammunition sources in terrestrial birds. Biological Conservation 131: 421–432.

Fisher, P. and S. Cleva. 2000. Nation marks Lacey Act centennial, 100 years of federal wildlife law enforcement. US Fish and Wildlife Service. http://www.fws.gov/pacific/news/2000/2000-98.htm Retrieved on March 8, 2014.

Fitzner, R.E. 1978. Behavioral ecology of the Swainson's hawk Buteo swainsoni in southeastern Washington. Ph.D. Thesis. Washington State University, Pullman.

Fjeldså, J. and N. Krabbe. 1990. Birds of the High Andes. Apollo Books.

Flagstaff. 2011. In BirdLife International 2014. Species factsheet: Gymnogyps californianus. Downloaded from http://www.birdlife.org on 13/03/2014.

Flint, V.E., R.L. Boehme, V.V. Kostin and A.A. Kuznetsov. 1984. A Field Guide to Birds of the USSR, including Eastern Europe and Central Asia. Princeton University Press, Princeton.

Flint, V.E., N. Bourso-Leland and A.A. Kuznetsov. 1989. A Field Guide to Birds of Russia and Adjacent Territories. Princeton University Press, Princeton.

Foster, G.M. 1944. A Summary of Yuki culture. University of California Anthropological Records 5: 155–244.

Foy, H.A. 1911. Third report on experimental work on animal Trypanosomiasis. Annals of Tropical Medicine and Parasitology 20: 303–308.

Frank, A. 1987. Helpers at birds' nests: a worldwide survey of cooperative breeding and related behavior. University of Iowa Press, Iowa City.

Frank, L. 2010. Hey presto! We made the lions disappear. Swara 4: 17–21.

Franson, J.C., L. Siloe and J.J. Thomas. 1995. Causes of eagle deaths. pp. 68. *In*: E.T. LaRoe, G.S. Farris, C.E. Puckett and P.D. Doran (eds.). Our Living Resources: A Report to the Nation on the Distribution, Abundance, and Health of U.S. Plants, Animals, and Ecosystems. National Biological Service, Washington, D.C., USA.

Freeland, L.S. and S.M. Broadbent. 1960. Central Sierra Miwok dictionary with texts. University of California Press Publications in Linguistics 23: 1–71.

Fremuth, W. 2005. In BirdLife International 2014. Species factsheet: *Aegypius monachus*. Downloaded from http://www.birdlife.org on 12/02/2014.

Friedman, R. and P.J. Mundy. 1983. The use of 'restaurants' for the survival of vultures in South Africa. pp. 345–355. *In*: S.R. Wilbur and J.A. Jackson (eds.). Vulture Biology and Management. University of California Press, Berkeley.

Friend, M., R.G. McLean and F.J. Dein. 2001. Disease emergence in birds: challenges for the twenty-first century. The Auk 118: 290–303.

Friesen, V.L., W.A. Montevecchi and W.S. Davidson. 1993. Cytochrome b nucleotide sequence variation among the Atlantic Alcidae. Hereditas 119: 245–252.

Frost, D. 2008. The use of 'flight diverters' reduces mute swan Cygnus olor collision with power lines at Abberton Reservoir, Essex, England. Conservation Evidence 5: 83–91.

Fry, D.M. 1983. Techniques for sexing monomorphic vultures. pp. 356–374. *In*: S.R. Wilbur and J.A. Jackson (eds.). Vulture Biology and Management. University of California Press, Berkeley.

Fuller, M.R., W.S. Seegar and L.S. Schueck. 1998. Routes and travel rates of migrating Peregrine Falcons Falco peregrinus and Swainson's Hawks Buteo swainsoni in the Western Hemisphere. Journal of Avian Biology 29: 433–440.

Gallagher, M. 1989. Vultures in Oman, eastern Arabia. Vulture News 22: 4–11.

Galliformes. Bulletin of the Florida State Museum, Biological Sciences 8: 195–335.

Galushin, V.M. 1971. A huge urban population of birds of prey in Delhi, India. Ibis 113: 522.

Gangoso, L. and C.-J. Palacios. 2005. Ground nesting by Egyptian Vultures (*Neophron percnopterus*) in the Canary Islands. Journal of Raptor Research 39: 186–187.

Gangoso, L., P. Álvarez-Lloret, A. Rodríguez-Navarro, R. Mateo, F. Hiraldo and J.A. Donázar. 2009. Long-term effects of lead poisoning on bone mineralization in vultures exposed to ammunition sources. Environmental Pollution 157(2): 569–574.

García-Fernández, A.J., E. Martínez-López, D. Romero, P. María-Mojica, A. Godino and P. Jiménez. 2005. High levels of blood lead in griffon vultures (*Gyps fulvus*) from Cazorla natural park (Southern Spain). Environmental Toxicology 20: 459–463.

García-Ripollés, C., P. López-López and V. Urios. 2010. First description of migration and wintering of adult Egyptian Vultures *Neophron percnopterus* tracked by GPS satellite telemetry. Bird Study 57(2): 261–265.

García-Ripollés, C., P. López-López and V. Urios. 2011. Ranging behaviour of non-breeding Eurasian Griffon Vultures *Gyps fulvus*: A GPS-Telemetry study. Acta Ornithologica 46(2):127–134.

Garcia, N. and E. Virgos. 2007. Evolution of community in several carnivore palaeoguilds from the European Pleistocene: the role of intraspecific competition. Lethaia 40(1): 33–44.

Gard, K., M.S. Groszos, E.C. Brevik and G.W. Lee. 2007. Spatial analysis of bird-aircraft strike hazard for Moody Air Force Base aircraft in the State of Georgia. Georgia Journal of Science 65(4): 160–169.

Garland, M.S. and A.K. Davis. 2002. An examination of Monarch Butterfly (Danaus plexippus) autumn migration in coastal Virginia. American Midland Naturalist 147: 170–174.

Garrett, K. and J. Dunn. 1981. Birds of Southern California: Status and Distribution. Los Angeles Audubon Society, Los Angeles.

Garrido, J.R. and M. Fernández-Cruz. 2003. Effects of power lines on a White Stork Ciconia ciconia population in Central Spain. Ardeola 50: 191–200.

Garrod, A.H. 1873. On certain muscles in the thigh of birds and their value for classification. Proceedings of the Zoological Society of London 1873: 626–644.

Gasaway, W.C., K.T. Mossestad and P.E. Stander. 1991. Food acquisition by spotted hyaenas in Etosha National Park, Namibia: predation versus scavenging. African Journal of Ecology 29: 64–73.

Gautam, R., B. Tamang and N. Baral. 2003. Ecological studies on White-rumped Vulture *Gyps bengalensis* in Rampur valley, Palpa, Nepal. Oriental Bird Club Bulletin 18–19.

Gauthereaux, S.A. 1993. Avian interactions with utility structures: background and milestones. pp. 1-1–1-6. *In*: J.W. Huckabee (ed.). Proceedings: Avian Interactions with Utility Structures, International Workshop. Electric Power Research Institute, Palo Alto, CA, USA.

Gavashelishvili, A. 2005. Vulture movements in the Caucasus. Vulture News 53: 28–29.

Gavashelishvili, A. 2005. Vultures of Georgia and Caucasus. GCCW and Buneba Print Publishing, Tbilisi.

Gavashelishvili, A. and J. McGrady. 2006. Modelling vulture habitats in the Caucasus. *In*: D.C. Houston and S.E. Piper (eds.). Proceedings of the International Conference on Conservation and Management of Populations, 151. Thessaloniki, Greece. Natural History Museum of Crete & WWF Greece.

Gavashelishvili, A. and M.J. McGrady. 2007. Radio-satellite telemetry of a territorial Bearded Vulture *Gypaetus barbatus* in the Caucasus. Vulture News 56: 4–13.

Gavashelishvili, A., M.J. McGrady and Z. Javakhishvili. 2006. Planning the conservation of the breeding population of cinereous vultures (*Aegypius monachus*) in the Republic of Georgia. Oryx 40(1): 76–83.

Gavashelishvili, A. and M.J. McGrady. 2006. Breeding site selection by bearded vulture (*Gypaetus barbatus*) and Eurasian griffon (*Gyps fulvus*) in the Caucasus. Animal Conservation 9(2): 159–170.

Gavashelishvili, A. and M.J. McGrady. 2006. Geographic information system-based modelling of vulture response to carcass appearance in the Caucasus. Journal of Zoology 269(3): 365–372.

Gavashelishvili, A., M. McGrady, M. Ghasabian and K.L. Bildstein. 2012. Movements and habitat use by immature Cinereous Vultures (*Aegypius monachus*) from the Caucasus. Bird Study 59: 449–462.

Gayton, A.H. 1930. Yokuts—Mono chiefs and shamans. University of California Publications in American Archaeology and Ethnology 23: 361–420.

Gayton, A.H. 1945. Yokuta and Western Mono social organisation. American Anthropology 47: 409–426.

Gayton, A.H. 1948a. Yokuts and Western Mono ethnography 1: Tulare Lake, Southern Valley and Central Foothill Yokuts. University of California Anthropological Records 10: 1–142.

Gayton, A.H. 1948b. Yokuts and Western Mono ethnography 2: Northern Foothill Yokuts and Western Mono. University of California Anthropological Records 10: 143–302.

Gbogbo, F. and V.P. Awotwe-Pratt. 2008. Waste management and Hooded Vultures on the Legon Campus of the University of Ghana in Accra, Ghana, West Africa. Vulture News 58: 16–22.

Geeson, N., C.J. Brandt and Thornes. 2002. Mediterranean Desertification: A Mosaic of Processes and Responses. John Wiley & Sons, Chichester.

Geist, H. 2005. The Causes and Progression of Desertification. Ashgate Publishing, Evanston.

Geist, H.J.B. 2005. The Causes and Progression of Desertification. Ashgate Publishing Ltd., Farnham.

Geist, H.J. and E.F. Lambin. 2004. Dynamic causal patterns of desertification. BioScience 54(9): 817–829.

Genero, J. 2005. In BirdLife International 2014. Species factsheet: Trigonoceps occipitalis. Downloaded from http://www.birdlife.org on 03/02/2014.

Gerdzhikov, G.P. and D.A. Demerdzhiev 2009. Data on bird mortality in 'Sakar' IBA (BG021), caused by hazardous power lines. Ecologia Balkanica 1: 67–77.

Gerlach, H. 1994. Chapter 32, Viruses. pp. 862–948. *In*: B.W. Ritchie, G.J. Harrison and L.R. Harrison (eds.). Avian Medicine: Principles and Application. Zoological Education Network, Lake Worth, Florida.

Gese, E.M., R.L. Ruff and R.L. Crabtree. 1996. Foraging ecology of coyotes (*Canis latrans*): the influence of extrinsic factors and a dominance hierarchy. Canadian Journal of Zoology 74: 769–783.

Ghasabian, M. and K. Aghababian. 2005. The situation of the Griffon Vulture in Armenia. pp. 24–26. *In*: L. Slotta-Bachmayr, R. Bögel and A. Camina-Cardenal (eds.). 2004. The Eurasian Griffon Vulture (Gyps fulvus) in Europe and the Mediterranean—Status Report and Action Plan. The European Griffon Vulture Working Group EGVWG, EGVWG, Salzburg and Madrid.

Gibb, G.C., O. Kardailsky, R.T. Kimball, E.L. Braun and D. Penny. 2007. Mitochondrial genomes and avian phylogeny: complex characters and resolvability without explosive radiations. Molecular Biology Evolution 24: 269–280.

Gibbs, J.P. and E.J. Stanton. 2001. Habitat fragmentation and arthropod community change: Carrion beetles, phoretic mites, and flies. Ecological Applications 11: 79–85.

Gifford, E.W. 1916. Miwok moieties. University of California Publications in American Archaeology and Ethnology 12: 139–194.

Gifford, E.W. 1932. The Northfork Mono. University of California Publications in American Archaeology and Ethnology 31: 15–65.

Gifford, E.W. 1955. Central Miwok ceremonies. University of California Publications in American Archaeology and Ethnology. University of California Anthropological Records 14: 261–318.

Gifford, E.W. 1926. Miwok cults. University of California Publications in American Archaeology and Ethnology 18: 391–408.

Gifford, F.W. and A.L. Kroeber. 1939. Culture element distributions IV: Pomo. University of California Publications in American Archaeology and Ethnology 37: 117–254.

Gil del Pozo, M. and J. Roig. 2003. Interaction between BirdLife and Red Electrica's Transmission Facilities: Experience and Solutions. Proceedings of the 4th technical session on Power Lines and the Environment, Madrid.

Gil del Pozo, M. and J. Roig. 2003. Interaction between BirdLife and Red Electrica's transmission facilities: experience and solutions. Proceedings of the 4th technical session on power lines and the environment, Madrid.

Gilbert, M. and S. Chansocheat. 2006. Olfaction in accipitrid vultures. Vulture News 55: 6–7.

Gilbert, M., J.L. Oaks, M.Z. Virani, R.T. Watson, S. Ahmed, J. Chaudhry, M. Arshad, S. Mahmood, A. Ali, R.M. Khattak and A.A. Khan. 2004. The status and decline of vultures in the provinces of Punjab and Sind, Pakistan: a 2003 update. pp. 221–234. *In*: R.C. Chancellor and B.-U. Meyburg (eds.). Raptors Worldwide. Proceedings of the 6th world conference on birds of prey and owls. WWGBP and MME/Birdlife Hungary, Berlin and Budapest.

Gilbert, M., J.L. Oaks, M.Z. Virani, R.T. Watson, A.A. Khan, S. Ahmed, J. Chaudry, M. Arshad, S. Mahmood and A. Ali. 2003. The Asian vulture crash: Investigating mortality of oriental whitebacked vulture *Gyps bengalensis* in Punjab Province, Pakistan. 52nd Annual Wildlife Disease Association Conference. University of Saskatchewan, Saskatoon.

Gilbert, M., M.Z. Virani, R.T. Watson, J.L. Oaks, P.C. Benson, A.A. Khan, S. Ahmed, J. Chaudhry, M. Arshad, S. Mahmood and Q.A. Shah. 2002. Breeding and mortality of Oriental White-backed Vulture *Gyps bengalensis* in Punjab Province, Pakistan. Bird Conservation International 12: 311–326.

Gilbert, M., R.T. Watson, M.Z. Virani, J.L. Oaks, S. Ahmed, M.J.I. Chaudhry, M. Arshad, S. Mahmood, A. Ali and A.A. Khan. 2006. Rapid population declines and mortality clusters in three Oriental White-backed Vulture *Gyps bengalensis* colonies in Pakistan due to diclofenac poisoning. Oryx 40(4): 388–399.

Gilbert, M., R.T. Watson, S. Ahmed, M. Asim and J.A. Johnson. 2007. Vulture restaurants and their role in reducing diclofenac exposure in Asian vultures. Bird Conservation International 17: 63–77.

Gill, R.E., T.L. Tibbitts, D.C. Douglas, C.M. Handel, D.M. Mulcahy, J.C. Gottschalck, N. Warnock, B.J. McCaffery, P.F. Battley and T. Piersma. 2009. Extreme endurance flights by landbirds crossing the Pacific Ocean: ecological corridor rather than barrier? Proceedings of the Royal Society B 276: 447–457.

Gill, V. 2009. New drug threat to Asian vultures, BBC News December 9, 2009.

Gillard, R. 1977. Unnecessary electrocution of owls. Blue Jay 35: 259.

Gilman, E.F. and D.G. Watson. 1993. Albizia julibrissin. Fact Sheet ST-68. University of Florida, Gainesville.

Glazener, W.C. 1964. Note on the feeding habits of the caracara in south Texas. Condor 66: 162.

Glutz von Blotzheim, U.N., K.M. Bauer and E. Bezzel. 1971. Handbuch der Vogel Mitteleuropas. Falconiformes. Akadem. Verlagsgesellschaft, Frankfurt am Main. 4: 620–637.

Godoy, J.A., J.J. Negro, F. Hiraldo and J.A. Donázar. 2004. Phylogeography, genetic structure and diversity in the endangered bearded vulture (*Gypaetus barbatus* L.) as revealed by mitochondrial DNA. Molecular Ecology 13(2): 371–90.

Goes, F. 1999. Notes on selected bird species in Cambodia. Forktail 15: 25–27.

Goldsmith, O. 1816. A History of the Earth and Animated Nature. Wingrave and Collingwood, London.

Gómez de Silva, H. 2004. The nesting season 2003: Mexico region. North American Birds 57: 550–553.

Gomez, L.G., D.C. Houston, P. Cotton and A. Tye. 1994. The role of Greater Yellow-headed Vultures *Cathartes melambrotus* as scavengers in Neotropical forest. Ibis 136: 193–196.

González, L.M. 1994. Cinereous Vulture. pp. 24–25. *In*: G.M. Tucker and M.F. Heath (eds.). Birds in Europe: Their Conservation Status (BirdLife Conservation Series 3). BirdLife International, Cambridge.

González, L.M. and R. Moreno-Opo Díaz-Meco. 2008. Impacto de la falta de alimento en las aves necrófagas amenazadas. Ambienta 73: 48–55.

González, L.M., A. Margalida, R. Sánchez and J. Oria. 2006. Supplementary feeding as an effective tool for improving breeding success in the Spanish imperial eagle (*Aquila adalberti*). Biological Conservation 129: 477–486.

Gore, M.E.J. 1990. Birds of the Gambia (2nd edn.). BOU Check-list 3. British Ornithologists' Union, London.

Gortázar, C., P. Acevedo, F. Ruiz-Fons and J. Vicente. 2006. Disease risks and overabundance of game species. European Journal of Wildland Resources 52: 81–87.

Graham, C. 2006. Californian Condors eat cold lead. World Birdwatch 28: 22–23.

Grant, C. 1965. The Rock Paintings of the Chumsh: A Study of a California Indian Culture. University of California Press, Berkeley and Los Angeles.

Graves, G.R. 1992. Greater yellow-headed vulture (*Cathartes melambrotus*) locates food by olfaction. Journal of Raptor Research 26: 38–39.

Gray, G.R. 1844. The Genera of Birds, Volume 1. R. & J. Taylor, London.

Grayson, D. 2006. The Late Quaternary biogeographic histories of some Great Basin mammals (western USA). Journal of Biogeography 25(21): 2964–2991.

Grayson, D.K. and D.J. Meltzer. 2002. Clovis hunting and large mammal extinction: a critical review of the evidence. Journal of World Prehistory 16: 313–359.

Green, R.E., M.A. Taggart, D. Das, D.J. Pain, C.S. Kumar, A.A. Cunningham and R. Cuthbert. 2006. Collapse of Asian vulture populations: risk of mortality from residues of the veterinary drug diclofenac in carcasses of treated cattle. Journal of Applied Ecology 43: 949–956.

Green, R.E., I. Newton, S. Shultz, A.A. Cunningham, M. Gilbert, D. Pain and V. Prakash. 2004. Diclofenac poisoning as a cause of vulture population decline across the Indian subcontinent. Journal of Applied Ecology 41: 793–800.

Green, R.E., A.M. Taggart, D. Das, D.J. Pain, C.S. Kumar, A.A. Cunningham and R. Cuthbert. 2006. Collapse of Asian vulture populations: risk of mortality from residues of the veterinary drug diclofenac in carcasses of treated cattle. Journal of Applied Ecology 43: 949–956.

Green, R.E., M.A. Taggart, K.R. Senacha, B. Raghavan, D.J. Pain, Y. Jhala and R. Cuthbert. 2007. Rate of decline of the Oriental White-backed Vulture population in India estimated from a survey of diclofenac residues in carcasses of ungulates. PloS ONE 8: 1–10.

Green, R.E., I. Newton, S. Shultz, A.A. Cunningham, M. Gilbert, D.J. Pain and V. Prakash. 2004. Diclofenac poisoning as a cause of vulture population declines across the Indian subcontinent. Journal of Applied Ecology 41: 793–800.

Greider, M. and E.S. Wagner. 1960. Black vulture extends breeding range northward. Wilson Bulletin 72: 291.

Greig-Smith, P. 1964. Quantitative plant ecology (2nd edn.). Butterworth and Co., London.

Griesinger, J. 1996. Autumn migration of griffon vultures (*Gyps f. fulvus*) in Spain. J. Muntaner and J. Mayol (eds.). Biología conservación de las rapaces Mediterráneas, 1994, 401–410. Monograph. Sociedad Española de Ornitologia/BirdLife International, Madrid.

Griffith, R.T.H. 1870–74. Valmiki. The Ramayan of Valmiki. Trübner & Co. London, E.J. Lazarus and Co., Benares.

Griffiths, C.S. 1994. Monophyly of the Falconiformes based on syringeal morphology. Auk 111: 787–805.

Griffiths, C.S., G.F. Barrowclough, J.G. Groth and L.A. Mertz. 2007. Phylogeny, diversity, and classification of the Accipitridae based on DNA sequences of the RAG-1 exon. Journal of Avian Biology 38(5): 587–602.

Grimmett, R., C. Inskipp and T. Inskipp. 1998. Birds of the Indian subcontinent. A & C Black/Christopher Helm, London.

Grimmett, R., C. Inskipp and T. Inskipp. 1998. Birds of the Indian Subcontinent. Christopher Helm, London.

Grimmett, R., C. Inskipp and T. Inskipp. 1999. A Guide to the Birds of India, Pakistan, Nepal, Bangladesh, Bhutan, Sri Lanka and the Maldives. Princeton University Press, Princeton.

Gross, L. 2006. Switching drugs for livestock may help save critically endangered Asian vultures. PLoS Biol. 4(3): e61. doi:10.1371/journal.pbio.0040061.

Grubac, R.B. 1991. Status and biology of the Bearded Vulture *Gypaetus barbatus aureus* in Macedonia. Birds of Prey Bulletin 4: 101–118.

Grubh, R.B. 1978. Field identification of some Indian vultures (*Gyps bengalensis, G. indicus, G. fulvus*, and *Torgos calvus*). Journal of the Bombay Natural History Society 75: 444–449.

Grubh, R.B., G. Narayam and S.M. Satheesan. 1990. Conservation of vultures in (developing) India. pp. 360–363. *In*: J.C. Daniel and J.S. Serrao (eds.). Conservation in Developing Countries. Bombay Natural History Society/Oxford University Press, Bombay.

Grünkorn, T., A. Diederichs, B. Stahl, D. Pöszig and G. Nehls. 2005. Entwicklung einer Methode zur Abschätzung des Kollisionsrisikos von Vögeln an Windenergie-anlagen. Bioconsult

SH, Hockensbüll, Germany. http//:www.umweltdaten.landsh.de/nuis/upool/gesamt/wea/voegel_wea.pdf.

Grussu, M. 2008. Unusual nest-sit of Lammergeier in Sardinia. British Birds 101: 491–495.

Gu, X. and R. Krawczynski. 2012. Scavenging birds and ecosystem services. Experience from Germany. Proceedings of the Conference on Environmental Pollution and Public Health (CEPPH 2012). Scientific Research Publishing, pp. 647–649.

Guangmei, Z. and W. Qishan. 1998. China Red Data Book of Endangered Animals: Aves. Science Press, Beijing.

Gurney, J.H. 1864. A descriptive catalogue of the raptorial birds in the Norfolk and Norwich Museum. Oxford University, Oxford.

Guthrie, D.A. 1980. Analysis of avifaunal and bat remains from midden sites on San Miguel Island. pp. 689–702. *In*: D.M. Power (ed.). The California Islands: Proceedings of a Multidisciplinary Symposiu., Santa Barbara Museum of Naturak History, Sanra Barbara.

Guzmán, J. and J.P. Castaño. 1998. Electrocution of raptors on power lines in Sierra Morena Oriental and Campo de Montiel. Ardeola 45: 161–169.

Haas, D. and M. Nipkow. 2006. Caution: Electrocution! NABU Bundesverband, Bonn.

Haas, D. 2011. Electrocution of birds. Some further aspects of international high significance. Presentation at International Conference on Power Lines and Bird Mortality in Europe, Budapest.

Haas, D., M. Nipkow, G. Fiedler, R. Schneider, W. Haas and B. Schürenberg. 2005. Protecting birds from powerlines. Nature and Environment, No. 140. Council of Europe Publishing, Strassbourg.

Hackett, S.J., R.T. Kimball, S. Reddy, R.C.K. Bowie, E.L. Braun, M.J. Braun, J.L. Chojnowski, W.A. Cox, K.-L. Han, J. Harshman, C.J. Huddleston, B.D. Marks, K. J. Miglia, W.S. Moore, F.H. Sheldon, D.W. Steadman, C.C. Witt and T. Yuri. 2008. A phylogenomic study of birds reveals their evolutionary history. Science 320(5884): 1763–1768.

Haemig, P.D. 2007. Ecology of Condors. Ecology Online Sweden.

Hagar, J.A. 1988. Swainson's Hawk (migration). pp. 56–64. *In*: R.S. Palmer (ed.). Handbook of North American Birds 5, Diurnal raptors (2). Yale University Press, New Haven.

Hagemeijer, E.J.M. and M.J. Blair (eds.). 1997. The EBCC Atlas of European Breeding Birds: Their Distribution and Abundance. T. & A.D. Poyser, London.

Hall, J.C., A.K. Chhangani, T.A. Waite and I.M. Hamilton. 2011. The impacts of La Niña-induced drought on Indian Vulture *Gyps indicus* populations in Western Rajasthan. Bird Conservation International 0: 1–13. doi:10.1017/S0959270911000232.

Hall, M., J. Grantham, R. Posey, A. Mee, A. Mee and L. Hall. 2007. Lead exposure among reintroduced California condors in southern California. California condors in the 21st century. Series in Ornithology No. 2, 163–184. Nuttall Ornithological Club and American Ornithologists' Union, Cambridge, MA.

Hall, P. 1999. In BirdLife International 2014. Species factsheet: Trigonoceps occipitalis. Downloaded from http://www.birdlife.org on 03/02/2014.

Halleux, D. 1994. Annotated bird list of Macenta Prefecture, Guinea. Malimbus 16: 10–29.

Hamilton, W.J. and F. Heppner. 1967. Radiant solar energy and the function of black homeotherm pigmentation: An hypothesis. Science 155: 196–197.

Hance, I. 2009. Seven White-rumped Vultures found dead in Cambodia. The Babbler: BirdLife in Indochina 14.

Hancock, J.A., J.A. Kushlan, M.P. Kahl, A. Harris and D. Quinn. 1992. Storks, Ibises and Spoonbills of the World. Princeton University Press, Princeton.

Hancock, P. 2008a. Cape vulture. pp. 9–10. *In*: P. Hancock (ed.). The Status of Globally and Nationally Threatened Birds in Botswana. BirdLife Botswana.

Hancock, P. 2008b. The status of globally and nationally threatened birds in Botswana. BirdLife Botswana, Gaborone.

Hancock, P. 2009a. Botswana—major poisoning incidents. African Raptors 10–11.

Hancock, P. 2009b. Poisons devastate vultures throughout Africa. African Raptors Newsletter 2 (November).

Hancock, P. 2010. Vulture poisoning at Khutse. Birds and People 27: 7.

Handelsblatt, 2006-JUN-30: Großer Geier-Einflug über Deutschland.

Handrinos, G. and T. Akriotis. 1997. The Birds of Greece. Christopher Helm, London.

Hardy, E. 1947. The Northern Lappet-faced Vulture in Palestine—a new record for Asia. Auk 64: 471–472.

Hardy, N. 1970. Fatal dinner. Thunder Bay Field Naturalist Club Newsletter 24: 11.

Harness, R. and D.R.S. Gombobaatar. 2008. Raptor electrocutions in the Mongolia Steppe. Winging It 20: 1–6.

Harness, R., D.R.S. Gombobaatar and R. Yosef. 2008. Mongolian distribution of power lines and raptor electrocutions. Rural Electric Power Conference, Charleston, South Caroline.

Harness, R.E. and K.R. Wilson. 2001. Utility structures associated with raptor electrocutions in rural areas. Wildlife Society Bulletin 29: 612–623.

Harrington, J.P. 1933. Annotations of Alfred Robinson's Chinigchinich. *In*: P.T. Hanna (ed.). Chinigchinich: A Revised and Annotated Version of Alfred Robinson's translation of Father Geronimo Boscana's historical account of the belief, usages, customs, and extravagences [sic] (!) of the Indians of this Mission of San Juan Capistrano called the Acagchemen Tribe. Fine Arts Press, Santa Ana, California.

Harrington, J.P. 1934. A new original version of Boscana's historical account of the San Juan Capistrano Indians of southern California. Smithsonian Miscellaneous Collections 92: 1–62.

Harrington, J.P. 1942. Culture element distributions, XIX: Central California. University of California Anthropological Records 7: 1–46.

Harris, H. 1941. The annals of Gymnogyps to 1900. Condor 43: 2–55.

Harris, R.J. 2013. The conservation of Accipitridae vultures of Nepal: a review. Journal of Threatened Taxa 5(2): 3603–3619.

Harrison, J.A., D.G. Allan, L.G. Underhill, M. Herremans, A.J. Tree, V. Parker and C.J. Brown. 1997. The atlas of southern African birds. BirdLife South Africa, Johannesburg.

Harrison, P. 1991. Seabirds: An Identification Guide. Houghton Mifflin, Harcourt.

Harrison, T.M., J.K. Mazet, K.E. Holekamp, E. Dubovi, A.L. Engh, K. Nelson, R.C. Van Horn and L. Munson. 2004. Antibodies to canine and feline viruses in spotted hyena (*Crocuta crocuta*) in the Masai Mara National Reserve. Journal of Wildlife Diseases 40: 1–10.

Hartert, E. 1920. Die Vögel der paläarktischen Fauna. Volume 2. Friendlander & Sohn. Berlin.

Hartley, R. and G. Hulme. 2005. First recorded breeding of three species of vultures on the Save Valley Conservancy. Honeyguide 51: 19–20.

Hartman, J.C., A. Gyimesi and H.A.M. Prinsen. 2010. Are bird flaps effective wire markers in a high tension power line?—Field study of collision victims and flight movements at a marked 150 kV power line Report nr. 10-082, Bureau Waardenburg bv, Culemborg.

Hassanin, A. and E.J.P. Douzery. 2003. Molecular and morphological phylogenies of Ruminantia and the alternative position of the Moschidae. Systematic Biology 52: 206–228.

Hassanin, A. and E.J.P. Douzery. 1999. The tribal radiation of the family Bovidae (Artiodactyla) and the evolution of the mitochondrial cytochrome b gene. Molecular Phylogenetics and Evolution 13: 227–243.

Hatch, D.E. 1970. Energy conserving and heat dissipating mechanisms of the Turkey Vulture. Auk 87: 111–124.

Haverschmidt, F. 1947. The black vulture and the Caracara as vegetarians. Condor 49: 210.

Haverschmidt, F. and G.F. Mees. 1994. Birds of Suriname. VACO N.V., Paramaribo.

Heath, M. and M.I. Evans (eds.). 2000. Important Bird Areas in Europe: Priority Sites for Conservation. 2 volumes. BirdLife Conservation Series No. 8. BirdLife International, Cambridge.

Heck, H. 1963. Successful breeding of the King Vultures (*Sarcoramphus papa* L.) in captivity. Zoologische Garten 27: 295–297.

Hedges, S.B. and C.G. Sibley. 1994. Molecules vs. morphology in avian evolution. The case of the pelecaniform birds. Proceedings of the National Academy of Sciences USA 91: 9861–9865.

Heidrich, P. and M. Wink. 1994. Tawny owl (*Strix aluco*) and Hume's tawny owl (*Strix butleri*) are distinct species. Evidence from nucleotide sequences of the cytochrome b gene. Z. Naturforsch. 49c: 230–234.

Heidrich, P., C. Koenig and M. Wink. 1995. Molecular phylogeny of the South American Screech Owls of the *Otus atricapillus* complex (Aves, Strigidae) inferred from nucleotide sequences of the mitochondrial cytochrome b gene. Z. Naturforsch. SOc 294–302.

Heilbrunn Timeline of History. 2006. Vulture Vessel [Mexico; Aztec] (1981. 297) New York 2006).

Heim de Balsac, H. and N. Mayaud. 1962. Les oiseaux du Nord-Ouest de l'Afrique. Paul Lechevalier, Paris.

Heizer, R.F. 1964. The western coast of North America. pp. 117–148. *In*: J.D. Jennings and E. Norbeck (eds.). Prehistoric Man in the New World. University of Chicago Press, Chicago.

Heizer, R.F. and C.W. Clewlow, Jr. 1973. Prehistroric Rock Art of California, 2 Vols. Ballena Press, Ramona, California.

Hellmayr, C.E. and B. Conover. 1949. Catalogue of Birds of the Americas Vol. 8, Part 1, No. 4. Cathartidae-Accipitridae-Pandionidae. Field Museum of Natural History, Chicago.

Helm-Bychowski, K. and J. Cracraft 1993. Recovering a phylogenetic signal from DNA sequences. Relationships within the Corvine assemblage (Class Aves) as inferred from complete sequence of the mitochondrial DNA cytochrome-b-gene. Molecular Biology and Evolution 10: 1196–1214.

Hemmer, H. 2002. Die Feliden aus dem Epivillafranchium von Untermassfeld. *In*: R.-D. Kahlke (ed.). Das Pleistozäʻn von Untermassfeld Bei Meiningen (Thuringen). Romisch-Germanisches Zentralmuseum Bande 40(3): 699–782.

Heredia, B. 1996. Action plan for the Cinereous Vulture (*Aegypius monachus*) in Europe. pp. 147–158. *In*: B. Heredia, L. Rose and M. Painter (eds.). Globally Threatened Birds in Europe: Action Plans Council of Europe, and BirdLife International, Strasbourg.

Heredia, B., M. Yarar and S.J. Parr. 1997. A baseline survey of Cinereous Vultures *Aegypius monachus* in Western Turkey.

Heredia, R. 1991. Alimentación suplementaria. pp. 101–108. *In*: R. Heredia and B. Heredia (eds.). El Quebrantahuesos (*Gypaetus barbatus*) en los Pirineos Colección Técnica. Madrid: Instituto para la Conservación de la Naturaleza.

Herholdt, J.J. and M.D. Anderson. 2006. Observations on the population and breeding status of the African White-backed Vulture, the Black-chested Snake Eagle, and the Secretarybird in the Kgalagadi Transfrontier Park. Ostrich 77: 127–135.

Herklots, G.A.C. 1961. The Birds of Trinidad and Tobago. Collins, London.

Hernández, A.E. and A. Margalida. 2009. Poison-related mortality effects in the endangered Egyptian Vulture (*Neophron percnopterus*) population in Spain. European Journal of Wildlife Research 55: 415–423.

Hernández, D.A. and J.R. Zook. 1993. Northward migration of peregrine falcons along the Caribbean coast of Costa Rica. J. Raptor Resources 27: 123–125.

Hernandez, M. and A. Margalida. 2008. Pesticide abuse in Europe: effects on the cinereous vulture (*Aegypius monachus*) population in Spain. Ecotoxicology 17: 264–272.

Hernandez, M. and A. Margalida. 2009. Poison-related mortality effects in the endangered Egyptian vulture (*Neophron percnopterus*) population in Spain: conservation measures. European Journal of Wildlife Research 55: 415–423.

Herremans, M., A.J. Tree, V. Parker and C.J. Brown (eds.). The Atlas of Southern African Birds, Vol. 1: Non-passerines. BirdLife South Africa, Johannesburg, South Africa.

Herremans, M. and D. Herremans-Tonnoeyr. 2000. Land use and conservation status of raptors in Botswana. Biological Conservation 94: 31–41.

Hertel, F. 1995. Ecomorphological indicators of feeding behavior in recent and fossil raptors. Auk 112(4): 890–903.

Hertel, Fritz. 1994. Diversity in body size and feeding morphology within past and present vulture assemblages. Ecology 75(4): 1074–1084.

Heuglin, T. von. 1869. Ornithologie Nordost-Afrikas, der Nilquellen- und Küsten-Gebiete des Rothen Meers und des nördlichen Somal-Landes. Cassel: Theodor Fischer.

Hevly, R.H., M.L. Heuett and S.J. Olsen. 1978. Paleoecological reconstruction from an upland Patayan rock shelter, Arizona. Journal of the Arizona Academy of Sciences 13: 67–78.

Hidalgo, C., J. Sánchez, C. Sánchez and M.T. Saborío. 1995. Migración de Falconiformes en Costa Rica. Hawk Migration Association of North America, Hawk Migration Stud. 21(1): 10–13.

Hiebl, I., R.E. Weber, D. Schneeganss, J. Kosters and G. Braunitzer. 1988. Structural adaptations in the major and minor hemoglobin components of adult Rüppell's griffon (*Gyps ruepellii*, Aegypiinae): a new molecular pattern for hypoxia tolerance. Biological Chemistry Hoppe–Seyler 369: 217–232.

Hill, J.R. and P.S. Neto. 1991. Black vultures nesting on skyscrapers in southern Brazil. Journal of Field Ornithology 62(2): 173–176.

Hillis, D.M. and J.P. Huelsenbeck. 1992. Signal, noise and reliability in molecular phylogenetic analyses. J. Heredity 83: 189–195.

Hillis, D.M. and C. Moritz. 1990. Molecular Systematics. Sinauer Press, Sundaland.

Hilton-Taylor, C. 2000. 2000 IUCN Red List of Threatened Species. World Conservation Union (IUCN), Gland.

Hilty, S. 2003. Birds of Venezuela. 2nd edn. Princeton University Press, Princeton, New Jersey.

Hilty, S.L. and W.L. Brown. 1986. A Guide to the Birds of Colombia. Princeton University Press, Princeton.

Hilty, S.L. 1977. A Guide to the Birds of Colombia. Princeton University Press, Princeton.

Hilty, S.L. and W.L. Brown. 1989. A Guide to the Birds of Colombia. Princeton University Press, Princeton.

Hinnells, J.R. 2005. The Zoroastrian Diaspora: Religion and Migration. Oxford University Press, Oxford.

Hinton, A.C. and W.D. Lang. 1939. S.A. Neave (ed.). Nomenclator Zoologicus 2. Zoological Society of London, London.

Hiraldo, F. 1976. Diet of the black vulture *Aegypius monachus* in the Iberian Peninsula.—Donana, Acta Vertebrata 3: 19–31.

Hiraldo, F. 1976. El Buitre Negro (*Aegypius monachus*) en la Peninsula Iberica. Poblacion, biologia general, uso de recursos e interacciones con otra aves. Unpublished PhD Thesis, Universidad de Sevilla, Sevilla.

Hiraldo, F. and J.A. Donázar. 1990. DonaForaging time in the Cinereous Vulture *Aegypius monachus*: seasonal and local variations and influence of weather. Bird Study 37: 128–132.

Hiraldo, F., M. Delibes and J. Calderon. 1984. Comments on the taxonomy of the Bearded Vulture Gypaetus barbatus (Linnaeus 1758). Bonner Zoologische Beiträge 35: 91–95.

Hiraldo, F., M. Delibes and J.A. Donazar. 1991. Comparison of diets of Turkey Vultures in three regions of northern Mexico. Journal of Field Ornithology 62(3): 319–324.

Hirzel, A.H., B. Posse, P.A. Oggier, Y. Crettenand, C. Glenz and R. Arlettaz. 2004. Ecological requirements of reintroduced species and the implications for release policy: the case of the bearded vulture. Journal of Applied Ecology 41(60): 1103–1116.

Hla, H., N.M. Shwe, T.W. Htun, S.M. Zaw, S. Mahood, J.C. Eames and J.D. Pilgrim. 2011. Historical and current status of vultures in Myanmar. Bird Conservation International 21: 376–387.

Hla, H., N.M. Shwe, T.W. Htun, S.M. Zaw, S. Mahood, J.C. Eames and J.D. Pilgrim. 2010. Historical and current status of vultures in Myanmar. Bird Conservation International 1–12. doi:10.1017/S0959270910000560.

Hla, H., N.M. Shwe, T.W. Htun, S.M. Zaw, S. Mahood, J.C. Eames and J.D. Pilgrim. 2011. Historical and current status of vultures in Myanmar. Bird Conservation International 21: 376–387.

Hockey, P.A.R., W.R.J. Dean and P.G. Ryan. 2005. Roberts Birds of Southern Africa. 7th Edition. The Trustees of the John Voelcker Bird Book Fund, Cape Town.

Hoerschelmann, H. von, A. Haack and F. Wohlgemuth. 1988. Bird casualties and bird behaviour at a 380-kV-power line (in German with English summary). İkologie der Vogel 10: 85–103.

Hoerschelmann, H. von, A. Haack and F. Wohlgemuth. 1988. Bird casualties and bird behaviour at a 380-kV-power line. İkologie der Vogel 10: 85–103.

Holling, C.S. 1993. Investing in Research for Sustainability. Ecological Applications 3: 552–555.

Holt, M. 1998. Black vulture fatalities: Botetourt County. Report of investigation. U.S. Department of the Interior, Fish and Wildlife Service, Richmond.

Hooper, L. 1920. The Chuilla Indians. University of California Publications in American Archaeology and Ethnology 16: 315–380.

Hopkins, C.L. 1888. Notes relative to the sense of smell in the Turkey Buzzard (*Cathartes aura*). Auk 5: 248–250.

Hopwood, C. 1912. A list of birds from Arakhan. Journal of the Bombay Natural History Society 21: 1196–1221.

Horapollo. The Hieroglyphics of Horapollo Nilous. Translated by A.T. Cory. Kessinger Publishing, Whitefish, Montana.

Horvath, M., K. Nagy, I. Demeter, A. Kovacs, J. Bagyura, P. Toth, S. Solt and G. Halmos. 2011. Birds and power lines in Hungary: Mitigation planning, monitoring and research. Presentation at International Conference on Power Lines and Bird Mortality in Europe, Budapest.

Horvath, M., K. Nagy, F. Papp, A. Kovacs, I. Demeter, K. Szugyi and G. Halmos. 2008. The evaluation of the Hungarian medium-voltage electricity network from a bird conservation perspective (in Hungarian). Magyar Madártani és Természetvédelmi Egyesület, Budapest.

Hosner, P.A. and D.J. Lebbin. 2006. Observations of plumage pigment aberrations of birds in Ecuador, including Ramphastidae. Boletín de la Sociedad Antioqueña de Ornitología 16(1): 30–42.

Hötker, H., K.-M. Thomsen and H. Köster. 2004. Auswirkungen regenerativer Energiegewinnung auf die biologische Vielfalt am Beispiel der Vögel und der Fliedermäuse—Fakten, Wissenslücken, Anforderungen an die Forschung, ornithologische Kriterien zum Ausbau von regenerativen Energiegewinnungsformen. Nabu / Bundesamt für Naturschutz, Bonn, Germany. Förd. Nr Z1.3-684 11-5/03. http://bergenhusen.nabu.de/bericht/VoegelRegEnergien.pdf.

Hou, L., M. Martin, Z. Zhou and A. Feduccia. 1996. Early adaptive radiation of birds: Evidence from fossils from Northeastern China. Science 274(5290): 1164–1211.

Houde, P. 1988. Paleognathous Birds from the early Tertiary of the Northern Hemisphere. Publications of the Nuttall Ornithological Club 22: 34–35.

Houston, D.C. 1976. Breeding of the white-backed and Rűppell's Griffon Vultures, *Gyps africanus* and *Gyps rueppellii*. Ibis 118: 14–40.

Houston, C. 1986. Scavenging efficiency of Turkey Vultures in Tropical Forest. Condor 88: 318–323.

Houston, C.S. 2006. Saskatchewan Turkey Vulture Nests, 1896–2002. Blue Jay 64: 209–211.

Houston, C.S., B. Terry, M. Blom and M.J. Stoffel. 2007. Turkey Vulture nest success in abandoned houses in Saskatchewan. Wilson Journal of Ornithology 199: 742–747.

Houston, C. 1988. Competition for food between Neotropical vultures in forest. Ibis 130: 402–417.

Houston, C.S., B. Terry, M. Blom and M.J. Stoffel. 2007. Turkey nest success in abandoned houses in Saskatchewan. Wilson Journal of Ornithology 119(4): 742–747.

Houston, D. 2010. Competition for food between Neotropical vultures in forest. Ibis 130(3): 402–417.

Houston, D.C. 1974. The role of griffon vultures *Gyps africanus* and *Gyps ruppellii* as scavengers. Journal of Zoological Society of London 172: 35–46.

Houston, D.C. 1983. The adaptive radiation of the griffon vultures. pp. 135–152. *In*: S.R. Wilbur and J.A. Jackson (eds.). Vulture Biology and Management. University of California Press, Berkeley.

Houston, D.C. 1985. Indian white-backed vulture (Gyps bengalensis). pp. 456–466. *In*: I. Newton and R.D. Chancellor (eds.). Conservation Studies of Raptors. International Council for Bird Preservation, Cambridge.

Houston, D.C. 1994. Cathartidae (New World Vultures). pp. 24–41. *In*: J. del Hoyo, A. Elliott and J. Sargatal (eds.). Handbook of the Birds of the World. Lynx Edicions, Barcelona, Spain.

Houston, D.C. and J.E. Cooper. 1975. The digestive tract of the whitebacked griffon vulture and its role in disease transmission among wild ungultates. Journal of Wildlife Diseases 11: 306–313.

Houston, D.C., A. Mee and M. McGrady. 2007. Why do condors and vultures eat junk? The implications for conservation. Journal of Raptor Reserach 41: 235–238.

Houston, D.C. 1974a. Food searching in griffon vultures. East African Wildlife Journal 12: 63–77.

Houston, D.C. 1974b. The role of Griffon vultures *Gyps africanus* and *Gyps rupellii* as scavengers. Journal of Zoology, London 172: 35–46.

Houston, D.C. 1975. Ecological isolation of African scavenging birds. Ardea 63: 55–64.

Houston, D.C. 1976. Breeding of the White-backed and Rüppell's Griffon Vultures, *Gyps africanus* and *Gyps rueppellii*. Ibis 118: 14–40.

Houston, D.C. 1979. The adaptation of scavengers. pp. 360–363. *In*: A.R.E. Sinclair and N. Griffiths (eds.). Serengeti, Dynamics of an Ecosystem. University of Chicago Press, Chicago.

Houston, D.C. 1983. The adaptive radiation of the griffon vulture. pp. 135–152. *In*: S.R. Wilbur and J.A. Jackson (eds.). Vulture Biology and Management. University of California Press, Berkeley.

Houston, D.C. 1984a. A comparison of the food supply of African and South American vultures. Proceedings of the fifth Pan-African Ornithological Congress, 249–262.

Houston, D.C. 1984b. Does the King Vulture *Sarcoramphus papa* use a sense of smell to locate food? Ibis 126: 67–69.

Houston, D.C. 1985. Evolutionary ecology of Afrotropical and neotropical vultures in forests. pp. 6–864. *In*: P. Buckley, M. Foster, E. Morton, R. Ridgley and F. Buckley (eds.). Neotropical Ornithology, American Ornithologists' Union, Washington D.C.

Houston, D.C. 1985. Indian white-backed vulture (*G. bengalensis*). pp. 465–466. *In*: I. Newton and R.D. Chancellor (eds.). Conservation Studies on Raptors Technical publication 5. International Council for Bird Preservation, Cambridge.

Houston, D.C. 1986. Scavenging efficiency of Turkey Vultures in tropical forest. The Condor 88: 318–323.

Houston, D.C. 1987. The effect of reduced mammal numbers on Cathartes vultures in Neotropical forests. Biological Conservation 129: 91–98.

Houston, D.C. 1988. Competition for food between Neotropical vultures in forest. Ibis 130: 402–417.

Houston, D.C. 1990. A change in the breeding season of Rüppell's Griffon Vultures *Gyps rueppellii* in the Serengeti in response to changes in ungulate populations. Ibis 132(1): 36–41.

Houston, D.C. 1990. Forest vultures depend on smell. pp. 94. *In*: I. Newton (ed.). Birds of Prey. Facts on File, New York.

Houston, D.C. 1994. Family Cathartidae (New World vultures). *In*: J. del Hoyo, A. Elliott and J. Sargatal (eds.). Handbook of the Birds of the World. Vol. 2. New World vultures to guineafowl, 24–41 Lynx Edicions, Barcelona.

Houston, D.C. 1994. Observations on Greater Yellow-headed Vultures *Cathartes melambrotus* and other Carthartes species as scavengers in forest in Venezuela. pp. 265–268. *In*: B.-U. Meyburg and R.D. Chancellor (eds.). Raptor Conservation Today. World Working Group on Birds of Prey and Pica Press, Berlin and London.

Houston, D. C. 1986. Scavenging efficiency of Turkey Vultures in tropical forest. Condor 88: 318–323.

Houston, D.C. 2001. Condors and Vultures. Voyageur Press, Stillwater.

Houston, D.C. 2006. Reintroduction programmes for vulture species. pp. 87–97. *In*: D.C. Houston and S.E. Piper (eds.). Proceedings of the International Conference on Conservation and Management of Vulture Populations. 14–16 November 2005, Thessaloniki, Greece. Natural History Museum of Crete & WWF Greece.

Houston, D.C. and J.A. Copsey. 1994. Bone digestion and intestinal morphology of the Bearded Vulture. The Journal of Raptor Research 28(2): 73–78.

Houton, D.C. 1979. The adaptations of scavengers. pp. 263–286. *In*: A.R.E. Sinclair and M. Norton Griffiths (eds.). Serengeti: Dynamics of an Ecosystem, Chicago University Press, Chicago.

Howard, H. 1932. Eagles and eagle-like vultures of the Pleistocene of Rancho La Brea. Washington, Carnegie Institution of Washington.

Howard, H. 1952. The prehistoric avifauna of Smith Creek Cave, Nevada, with a description of a new gigantic raptor. Bulletin of the South California Academy of Science 51: 50–54.

Howard, H. 1962a. Birds remains from a prehistoric cave deposit in Grant County, New Mexico. Condor 64: 241–242.

Howard, H. 1962b. A comparison of avian assemblages from individual pits at Rancho La Brea, California. Los Angeles County Museum Contributions in Science 58: 1–24.

Howard, H. 1966. Two fossil birds from the Lower Miocene of South Dakota. Los Angeles County Museum of Natural History, Contributions to Science 107: 1–8.

Howard, H. 1968. Limb measurements of the extinct vulture, Coragyps occidentalis. Papers of the Archaeological Society of New Mexico 1: 115–127.

Howard, H. 1972. The incredible teratorn again. Condor 74: 341–344.

Howard, H. 1974. Postcranial elements of the extinct condor Breagyps clarki L. Miller. Contributions in Science 256: 1–24.

Howard, H. and A.H. Miller. 1933. Bird remains from cave deposits in New Mexico. Condor 35: 15–18.

Howard, H. and A.H. Miller. 1939. The avifauna associated with human remains at Rancho La Brea, California. Carnegie Institution of Washington Publications 514: 39–48.

Howell, J.A. 1997. Bird mortality at rotor swept area equivalents, Altamont Pass and Montezuma Hills, California. Transactions of the Western Section of the Wildlife Society 33: 24–29.

Howell, J.A. 1997. Avian Use and Mortality at the Sacramento Municipal Utility District Wind Energy Development Site, Montezuma Hills, Solano County, California. Sacramento Municipal Utility District, California.

Howell, S.N.G. and S. Webb. 1995. A Guide to the Birds of Mexico and Northern Central America. Oxford University Press, Oxford.

Howell, T.R. 1972. Birds of the lowland pine savanna of northeastern Nicaragua. Condor 74: 316–340.

Hoxie, W. 1887. Sense of smell in the Black Vulture. Ornithologist and Oologist 12: 132.

Htin Hla, T. 2003. Preliminary investigation of White-rumped Vultures (*Gyps bengalensis*) in Southern Shan states Myanmar. OBC Report, July 2003.

Hudson, T. and E. Underhay. 1978. Crystals in the sky: an intellectual odyssey involving Chumash astronomy, cosmology and rock art. Ballena Press Anthropology Papers 10: 1–163.

Hughes, C. and B. Hwang. 1996. Attempts to conceptualize and measure quality of life. pp. 51–61. *In*: R.L. Schalock (ed.). Quality of Life. Vol. I: Conceptualization and Measurement. American Association on Mental Retardation, Washington, D.C.

Hume, A.O. 1873. Contributions to the Ornithology of India, Sindh. Stray Feathers 1: 91–289.

Humphrey, J.S., M.L. Avery and A.P. McGrane. 2000. Evaluating relocation as a vulture management tool in north Florida. pp. 81–83. *In*: T.P. Salmon and A.C. Crabb (eds.). Proceedings of the Nineteenth Vertebrate Pest Conference. University of California, Davis.

Humphrey, J.S., E.A. Tillman and M.L. Avery. 2004. Vulture-livestock interactions at a central Florida cattle ranch. pp. 122–125. *In*: R.M. Timm and W.P. Gorenzel (eds.). Proceedings of the 21st Vertebrate Pest Conference. University of California, Davis.

Hunt, W., C. Parish, S. Farry, T. Lord, R. Sieg, A. Mee and L. Hall. 2007. Movements of introduced California condors in Arizona in relation to lead exposure. California condors in the 21st century. Series in Ornithology No. 2, Nuttall Ornithological Club and American Ornithologists' Union, Cambridge, MA, pp. 79–96.

Hunt, W.G., C.N. Parish, K. Orr and R.F. Aguilar. 2009. Lead poisoning and the reintroduction of the California condor in Northern Arizona. Journal of Avian Medicine and Surgery 23: 145–150.

Hunter, J.S., S.M. Durant and T.M. Caro. 2006. Patterns of scavenger arrival at cheetah kills in Serenegeti National Park, Tanzania. African Journal of Ecology 45: 275–281.

Hunting, K. 2002. A Roadmap for PIER Research on Avian Power Line Electrocution in California. California Energy Commission, Sacramento.

Hüppop, O. and K. Hüppop. 2003. North Atlantic Oscillation and Timing of Spring Migration in Birds. Proceedings of the Royal Society B 270: 233–240.

Husein, K.Z. and S.U. Sarkar. 1971. Notes on a collection of birds from Pabnar (Bangladesh). Journal of the Asiatic Society of Bangladesh 16: 259–288.

Hustler, K. and W.W. Howells. 1988. Breeding biology of the hooded and lappet-faced vultures in the Hwange National Park. Honeyguide 34: 109–115.

Huston, M.A. 1997. Hidden treatments in ecological experiments: Re-evaluating the ecosystem function of biodiversity. Oecologia 110: 449–460.

IDAE (Instituto para la Diversificación y el Ahorro de la Energía). 2010. Plan de AcciónNacional de Energías Renovables 2010–2020. Madrid, Spain.

Iezekiel, S. and H. Nicolaou. 2006. Griffon vulture *Gyps fulvus* artificial reproduction. *In*: D.C. Houston and S.E. Piper (eds.). Proceedings of the International Conference on Conservation and Management of Populations, 164. Thessaloniki, Greece. Natural History Museum of Crete & WWF Greece.

Igl, L.D. and S.L. Peterson. 2010. Repeated Use of an Abandoned Vehicle by Nesting Turkey Vultures (Cathartes aura). Journal of Raptor Research 44: 1, 73–75.

Imberti, S. 2003. In BirdLife International 2014. Species factsheet: *Gymnogyps californianus*. Downloaded from http://www.birdlife.org on 13/03/2014.

Imberti, S. 2003. In BirdLife International 2014. Species factsheet: *Vultur gryphus*. Downloaded from http://www.birdlife.org on 21/02/2014.

Infante, S., J. Neves, J. Ministro and R. Brandão. 2005. Impact of distribution and transmission power lines on birds in Portugal. Quercus, ICN and SPEA, Castelo Branco. Unpublished Report.

Iñigo, A., B. Barov, C. Orhun and U. Gallo-Orsi. 2008. Action plan for the Egyptian Vulture *Neophron percnopterus* in the European Union. BirdLife International for the European Commission.

Inigo Elias, E.E. 1987. Feeding habits and ingestion of synthetic products in a black vulture population from Chispas, Mexico. Acta Zoologia: Mexico 22: 1–15.

Inigo, A. and J.C. Atienza. 2007. Efectos del Reglamento 1774/2002 y las decisions adoptadas por la Comisión Europea en 2003 y 2005 sobre las aves necrófagas en la peninsula Ibérica y sus posibles soluciones. Official report to the European Comission. SEO/BirdLife 15-06-2007.

Inskipp, C. and T.P. Inskipp. 1991. A Guide to the Birds of Nepal. Christopher Helm, London.

Inskipp, C. and T.P. Inskipp. 1993. Birds recorded during a visit to Bhutan in spring. Forktail 9: 121–142.

Irwin, D.M., T.D. Kocher and A.C. Wilson. 1991. Evolution of the cytochrome b gene of mammals. Journal of Molecular Evolution 32: 128–144.

Irwin, M.P.S. 1981. The Birds of Zimbabwe. Quest Publishing, Bulawayo.

Isenmann, P., T. Gaultier, A. El Hili, H. Azafzaf, H. Dlensi and M. Smart. 2005. Birds of Tunisia. SEOF, Paris.

Isidore of Seville Etymologiae. Trans. Stephen A. Barney New York, 2005, Book XII, vii12.

IUCN. 2013. IUCN Red List of Threatened Species. Version 2013.1. <www.iucnredlist.org>. Downloaded on 05 September 2013.

IUCN. 2012. IUCN Red List of Threatened Species (ver. 2012.1). Available at: http://www. iucnredlist.org. (Accessed: 19 June 2012).

Jackman, S. 2008. The pscl Package. Political Science Computational Laboratory. Stanford University, Stanford.

Jackson, J.A. 1975. Reguritative feeding of young Black Vultures in December. Auk 92: 802–803.

Jackson, J.A. 1983. Nesting phenology, nest site selection, and reproductive success of Black and Turkey Vultures. pp. 245–270. *In*: S.R. Wilbur and J.A. Jackson (eds.). Vulture Biology and Management. University of California Press, Berkeley.

Jackson, J.A. 1988a. American black vulture. pp. 11–24. *In*: R.S. Palmer (ed.). Handbook of North American Birds. Yale University Press, New Haven.

Jackson, J.A. 1988b. Turkey vulture. pp. 25–42. *In*: R.S. Palmer (ed.). Handbook of North American Birds. Yale University Press, New Haven.

Jackson, J.A., I.D. Prather, R.N. Connor and S.P. Gaby. 1978. Fishing behavior of black and turkey vultures. Wilson Bulletin 90: 141–143.

Jacob, J. 1983. Zur systematischen Stellung von Vultur gryphus (Cathartiformes). Journal für Ornithologie 124: 83–86.

Jacobs, B.F., J.D. Kingston and L.L. Jacobs. 1999. The origin of grass-dominated ecosystems. Annals of the Missouri Botanical Garden 86: 590–643.

James, D.A. and J.C. Neale. 1986. Arkansas birds: their distribution and abundance. University of Arkansas Press, Fayetteville.

Janss, G.F.E. 2000. Avian mortality from power lines: a morphologic approach of a species-specific mortality. Biological Conservation 95: 353–359.

Janss, G.F.E. and M. Ferrer. 1999a. Avian electrocution on power poles: European experiences. pp. 145–164. *In*: M. Ferrer and G.F.E. Janss (eds.). Birds and Power Lines: Collision, Electrocution, and Breeding. Quercus, Madrid.

Janss, G.F.E. and M. Ferrer. 1999b. Mitigation of raptor electrocution on steel power poles. Wildlife Society Bulletin 27: 263–273.

Janss, G.F.E., A. Lazo and M. Ferrer. 1999. Use of raptor models to reduce avian collisions with powerlines. Journal of Raptor Research 33: 154–159.

Janss, G.F.E. and M. Ferrer. 2000. Common Crane and Great Bustard collision with power lines: collision rate and risk exposure. Wildlife Society Bulletin 28: 675–680.

Janss, G.F.E. and M. Ferrer. 1998. Rate of bird collision with power lines: effects of conductor-marking and static wire-marking. Journal of Field Ornithology 69(1): 8–17.

Janzen, D.H. 1976. The depression of reptile biomass by large herbivores. American Naturalist 110: 371–400.

Janzen, D.H. 1977. Why fruits rot, seeds mold, and meat spoils. American Naturalist 111: 691–713.

Jaramillo, A. 2003. Birds of Chile. Princeton University Press, Princeton.

Jarvis, A., A.J. Robertson, C.J. Brown and R.E. Simmons. 2001. Namibian Avifaunal Database. National Biodiversity Programme, Ministry of Environment & Tourism, Windhoek.

Jenkins, A. 2010. Cape vulture. Africa—Birds & Birding 15(2): 62–63.

Jenkins, A. 2007. Electric Eagle Project. Unpublished report to Eskom, Sunninghill.

Jenkins, A.R., J.J. Smallie and M. Diamond. 2010. Avian collisions with power lines: a global review of causes and mitigation with a South African perspective. Bird Conservation International 20: 263–278.

Jennelle, C.S., M.D. Samuel, C.A. Nolden and E.A. Berkley. 2009. Deer carcass decomposition and potential scavenger exposure to chronic wasting disease. Journal of Wildlife Management 73: 655–662.

Jennings, M.C. 2010. Atlas of the Breeding Birds of Arabia. Fauna of Arabia, Vol. 25.

Jennings, M.C. 1996. Summary report of ABBA Survey 19 to south central Saudi Arabia; March and April 1996, Phoenix 13: 21–23.

Jennings, M.C. and R.N. Fryer. 1984. The occurrence of the Lappet-faced Vulture Torgos tracheliotus (J.R. Forster) in the Arabian peninsula, with new breeding records from Saudi Arabia. Fauna of Saudi Arabia 6: 534–545.

Jerdon, T.C. 1871. Supplementary notes to The Birds of India. Ibis 1: 234–247.

Johnsgard, P.A. 1988. North American Owls: Biology and Natural History. Smithsonian Institute Press, Washington, DC.

Johnsgard, P.A. 1990. Hawks, Eagles, and Falcons of North America. Smithsonian Institute Press, Washington, DC.

Johnson, G.D., W.P. Erickson, M.D. Strickland, M.F. Shepherd, D.A. Shepherd and S.A. Sarappo. 2002. Collision mortality of local and migrant birds at a large-scale windpower development on Buffalo Ridge, Minnesota. Wildlife Society Bulletin 30: 879–887.

Johnston, B.E. 1962. California's Gabrieliño Indians. Frederick Webb Hodge Anniversary Publication Fund. Southwest Museum, Los Angeles. 8: 1–198.

Johnson, J.J. 1967. The archaelogy of the Camanche Reservoir locality, California. Sacramento Anthropological Society Paper 6: 1–198.

Johnson, J.A., H.R.L. Lerner, P.C. Rasmussen and D.P. Mindell. 2006. Systematics within Gyps vultures: a clade at risk. BMC Evolutionary Biology 6: 65–77. doi:10.1186/1471-2148-6-65.

Johnson, J.A., M. Gilbert, M.Z. Virani, M. Asim and D.P. Mindell. 2008. Temporal genetic analysis of the critically endangered oriental white-backed vulture in Pakistan. Biological Conservation 141(9): 2403–2409.

Jollie, M. 1976. A contribution to morphology and phylogeny of the Falconiformes. Evolutionary Theory 1: 285–298.

Jollie, M. 1977a. A contribution to morphology and phylogeny of the Falconiformes. Evolutionary Theory 3: 1–141.

Jollie, M. 1977b. A contribution to morphology and phylogeny of the Falconiformes. Evolutionary Theory 2: 115–300.

Jones, H.L. 2003. Birds of Belize. University of Texas Press, Austin.

Jones, H.L. and O. Komar. 2007a. Central America. North American Birds 61: 155–159.

Jones, H.L. and O. Komar. 2007b. Central America. North American Birds 61: 340–344.

Jones, H.L. and O. Komar. 2007c. Central America. North American Birds 61: 521–525.

Jones, H.L. and O. Komar. 2008a. Central America. North American Birds 62: 163–170.

Jones, H.L. and O. Komar. 2008b. Central America. North American Birds 62: 314–318.

Jones, H.L. and O. Komar. 2008c. Central America. North American Birds 62: 487–491.

Jones, H.L. and O. Komar. 2010. Central America. North American Birds 64: 164–168.

Jones, H.L. 2004. Birds of Belize. University of Texas Press, Austin.

Jones, H.P. and S.W. Kress. 2012. A review of the world's active seabird restoration projects. The Journal of Wildlife Management 76(1): 2–9.

Jones, P.J. 1978. A possible function of the 'wing-drying' posture in the Reed Cormorant Phalacrocorax africanus. Ibis 120–542.

Jonzen, N., A. Linden, T. Ergon, E. Knudsen, J.O. Vik, D. Rubolini, D. Piacentini, C. Brinch, F. Spina, L. Karlsson, M. Stervander, A. Andersson, J. Waldenstrom, A. Lehikoinen, E. Edvardsen, R. Solvang and N.C. Stenseth. 2006. Rapid advance of spring arrival dates in long-distance migratory birds. Science 312: 1959–1961.

Junge, G.C. and G.F. Mees. 1961. The Avifauna of Trinidad and Tobago. E.J. Brill, Leiden.

Junge, G.C.A. and G.F. Mees. 1958. The Avifauna of Trinidad and Tobago. Zoologische Verhand. No. 37. E.J. Brill, Leiden.

Junta de Castilla y León. 2003. Libro del Parque Regional de Picos de Europa. Programa Parques Naturales de Castilla y León. Junta de Castilla y León, Valladolid.

Junta de Castilla y León. 2006. Memoria anual de la Reserva Regional de Caza de Riaño, 2005–2006.

Jurek, R.M. 1990. An historical review of California condor recovery programmes. Vulture News 23: 3–7.

Kabouche, B., J. Bayeul, L. Zimmermann and P. Bayle. 2006. Bird mortality in aerial power lines: challenges and prospects in Provence—Alpes-Cote d'Azur (in French). Report DIREN PACA–LPO PACA, Hyères.

Kaczensky, P. 1999. Large carnivore depredation in livestock in Europe. Ursus 11: 59–72.

Kampen, M.E. 1978. Classic Veracruz Grotesques and Sacrificial Iconography. Man 116–126.

Kampen, M.E. 1972. Sculptures of El Tajin, Vera Cruz, Mexico. University of Florida Press, Gainsville.

Kampen, M.E. 1978. Classic Veracruz Grotesques and Sacrificial Iconography. Man, New Series 8: 121.

Karyakin, I.V. and L.M. Novikova. 2006. The Steppe Eagle and power lines in Western Kazakhstan. Does coexistence have any chance? Raptors Conservation 6: 48–57.

Karyakin, I.V., A.V. Kovalenko and L.M. Novikova. 2006. The Imperial Eagle in the Volga-Ural Sands: results of researches in 2006. Raptors Conservation 6: 39–47.

Karyakin, I.V. 2008. Lines-killers continue to harvest the mortal crop in Kazakhstan. Raptors Conservation 11: 14–21.

Karyakin, I., L. Konovalov, M. Grabovskiy and E. Nikolenko. 2009. Carrion-eaters of the Altay-Sayan region. Raptor Conservation 15: 37–65.

Katzner, T. 2004. Vultures on the verge of a biological breakdown. Wildlife Conservation 107: 45–47.

Katzner, T. 2005. In BirdLife International 2014. Species factsheet: *Aegypius monachus*. Downloaded from http://www.birdlife.org on 12/02/2014.

Katzner, T.E., C.H. Lai, J.D. Gardiner, J.M. Foggin, D. Pearson and A.T. Smith. 2004a. Adjacent nesting by Lammergeier *Gypaetus barbatus* and Himalayan griffon *Gyps himalayensis* on the Tibetan Plateau, China. Forktail 20: 94–96.

Katzner, T., A. Gavashelishvili, S. Sklyarenko, M. McGrady, J. Shergalin and K. Bildstein. 2004b. Population and conservation status of Griffon Vultures in the former Soviet Union. pp. 235–240. *In*: R.D. Chancellor and B.-U. Meyburg (eds.). Raptors Worldwide, WWGBP/ MME, Berlin, Budapest.

Kaufman, K. 1996. Lives of North American Birds. Houghton Mifflin Field Guides, Boston.

Kazmierczak, K. 2000. A Field Guide to the Birds of India, Sri Lanka, Pakistan, Nepal, Bhutan, Bangladesh and the Maldives. Pica Press/Christopher Helm, London.

Kelly, T.R. and C.K. Johnson. 2011. Lead exposure in free-flying Turkey Vultures is associated with big game hunting in California. PLoS One 6(4): e15350. doi:10.1371/journal. pone.0015350.

Kemp, A. 2003. Hornbills. pp. 384–389. *In*: C. Perrins (ed.). Firefly Encyclopedia of Birds. Firefly Books, Richmond Hill.

Kemp, A.C. and M.I. Kemp. 1975. Observations on the white-backed Vulture Gyps africanus in the Kruger National Park, with notes on other avian scavengers. Koedoe 18: 51–68.

Kendall, C. 2012. In BirdLife International 2014. Species factsheet: Trigonoceps occipitalis. Downloaded from http://www.birdlife.org on 03/02/2014.

Kendall, C. 2012. Poison empties skies that once were full. SWARA 36(4): 24–29.

Kendall, C., M.Z. Virani, P. Kirui, S. Thomsett and M. Githiru. 2012. Mechanisms of coexistence in vultures: Understanding the patterns of vulture abundance at carcasses in Masai Mara National Reserve, Kenya. Condor 114(3): 523–531.

Kendall, C.J. and M.Z. Virani. 2012. Assessing mortality of African vultures using wing tags and GSM-GPS transmitters. Journal of Raptor Research 46: 135–140.

Kendall, C.J. 2013. Alternative strategies in avian scavengers: how subordinate species foil the despotic distribution. Behavioral Ecology and Sociobiology 67(3): 383–393.

Kendall, C.J. 2014. The early bird gets the carcass: Temporal segregation and its effects on foraging success in avian scavengers. The Auk 131(1): 12–19.

Kendall, C.J., M.Z. Virani, J.G.C. Hopcraft, K.L. Bildstein and D.I. Rubenstein. 2014. African vultures don't follow migratory herds: scavenger habitat use is not mediated by prey abundance. Plos One 9(1): e83470.

Keyes, C.R. and H.S. Williams. 1888. Preliminary annotated catalogue of the birds of Iowa. Proceedings of Davenport Academy of Natural Sciences 5: 113–161.

Khachar, S. and T. Mundkur. 1989. Status and distribution of the King Vulture Sarcogyps calvus in Gujarat-Results of a recent survey. Journal of the Bombay Natural History Society 86: 360–362.

Kiff, L. 2005. BirdLife International. 2014. Species factsheet: *Gymnogyps californianus*. Downloaded from http://www.birdlife.org on 13/03/2014.

Kiff, L.F. 2000. The current status of North American vultures. pp. 175–189. *In*: R.D. Chancellor and B.-U. Meyburg (eds.). Raptors at Risk. WWGBP/HancockHouse. Berlin and Surrey, Canada.

Kiff, L.F., D.B. Peakall and S.R. Wilbur. 1979. Recent changes in California Condor Eggshells. Condor 81(2): 166–172.

Kiff, L.F., D.B. Peakall, M.L. Morrison and S.R. Wilbur. 1983. Eggshell thickness and DDE residue levels in vulture eggs. pp. 440–458. *In*: A. Wilbur and J.A. Jackson (eds.). Vulture Biology and Management. University of California Press, Berkeley.

Kingdon, J. 1997. The Kingdon Field Guide to African Mammals. Academic Press, London.

Kinloch, B.G. 1956. Fishing in Uganda. Uganda Wild Life and Sport 1: 13–21.

Kirazlı, C. and E. Yamaç. 2013. Population size and breeding success of the Cinereous Vulture, *Aegypius monachus*, in a newly found breeding area in western Anatolia (Aves: Falconiformes). Zoology in the Middle East 59(4): 289–296.

Kirk, D.A. and A.G. Gosler. 1994. Body condition varies with migration and competition in migrant and resident South American vultures. The Auk 111(4): 933–944.

Kirk, D.A. and M.J. Mossman. 1998. Turkey vulture (Cathartes aura). *In*: A. Poole and F. Gill (eds.). The Birds of North America, No. 339. The Academy of Natural Sciences, Philadelphia, and the American Ornithologists Union, Washington D.C.

Kleiber, M. 1961. The Fire of Life. An Introduction to Animal Energetics. Wiley and Sons, New York.

Kobierzycki, E. 2011. Le Vautor percnoptere dans les Pyrenees francaises. LPO Mission Rapaces—Pyrenees Vivantes.

Kocher, T.D., W.K. Thomas, A. Meyer, S.V. Edwards, S. Piiiibo, F.X. Villablanqt and A.C. Wilson. 1989. Dynamics of mitochondrial DNA evolution in animals, amplification and sequencing with conserved primers. Proceedings of the National Academy of Sciences USA 86: 6196–6200.

Koening, R. 2006. Vulture research soars as the scavengers' numbers decline. Science 312: 1591–1592.

Koford, C.B. 1953. The California Condor. General Publishing Company, Toronto.

Koford, C.B. 1953. The California Condor. National Audubon Society Research Report 4: 1–154.

Komar, O. and J.P. Dominguez. 2001. Lista de aves de El Salvaor. SalvaNatur, San Salvador, El Salvador.

Komen, L. 2006. In BirdLife International. 2014. Species factsheet: Gyps coprotheres. Downloaded from http://www.birdlife.org on 15/03/2014.

Komen, L. 2009. Namibia—vultures killed deliberately and accidentally. African Raptors 2: 13.

König, C. 1982. Zur systematischen Stellung der der Neuweltgeier (Cathartidae). Journal für Ornithologie (Journal of Ornithology) 123: 259–267.

Konig, C. 1983. Interspecific and intraspecific composition for food among old world vultures. pp. 153–171. *In*: S.R. Wilbur and J.A. Jackson (eds.). Vulture Biology and Management. University of California Press, Berkeley.

Koops, F.B.J. 1987. Collision victims in the Netherlands and the effects of marking (in Dutch). Vereniging van directeuren van electriciteitsbedrijven in Nederland, Arnhem.

Kornegay, J.R., T.H. Kocher, L.A. Williams and A.C. Wilson. 1993. Pathways of lysozyme evolution inferred from the sequences of cytochrome b in birds. Journal of Molecular Evolution 37: 367–379.

Kovacs, A., I. Demeter, I. Fater, J. Bagyura, K. Nagy, T. Szitta, G. Firmanszky and M. Horvath. 2008. Current efforts to monitor and conserve the Eastern Imperial Eagle Aquila heliaca in Hungary. AMBIO: A Journal of the Human Environment 37: 457–459.

Krabbe, N. and J. Fjeldså. 1990. Birds of the High Andes. Apollo Press, Newport News.

Krebs, J.R. 1974. Colonial nesting and social feeding as strategies for exploiting food resources in the Great Blue Heron (Ardea Herodias). Behavior 51: 99–134.

Kreiger, A.D. 1964. Early man in the new world. *In*: J.D. Jennings and E. Norbeck (eds.). Prehistoric Man in the New World. University of Chicago Press, Chicago.

Kretzmann, M.B., N. Capote, B. Gautschi, J.A. Godoy, J.A. Donázar and J.J. Negro. 2003. Genetically distinct island populations of the Egyptian vulture (*Neophron percnopterus*). Conservation Genetics 4(6): 697–706.

Krijgsveld, K.L., K. Akershoek, F. Schenk, F. Dijk and S. Dirksen. 2009. Collision risk of birds with modern large wind turbines. Ardea 97(3): 357–366.

Kroeber, A.L. 1908a. A mission record of the California Indians. University of California Publications in American Archaeology and Ethnology 8: 1–27.

Kroeber, A.L. 1908b. Ethnography of the Cahuilla Indians. University of California Publications in American Archaeology and Ethnology 8: 29–68.

Kroeber, A.L. 1908c. Notes on the Luiseno. University of California Publications in American Archaeology and Ethnology 8: 29–68.

Kroeber, A.L. 1932. The Patwin and their neighbours. University of California Publications in American Archaeology and Ethnology 29: 253–423.

Kroeber, A.L. and E.W. Gifford. 1949. World renewal: A Cult System of Native Northwest California. University of California Anthropological Records 13: 1–156.

Kroeber, A.L. 1925. Handbook of the Indians Bureau of American Ethnology Bulletin. 78: 1–995.

Kroeber, A.L. 1929. The Valley Museum. University of California Publications in American Archaeology and Ethnology 24: 253–290.

Krook, K., W.J. Bond and P.A.R. Hockey. 2007. The effect of grassland shifts on the avifauna of a South African savanna. Ostrich: Journal of African Ornithology 78(2): 271–279.

Kruger, R., A. Maritz and C. van Rooyen. 2004. Vulture electrocutions on vertically configured medium voltage structures in the Northern Cape Province South Africa. pp. 437–441. In: R.D. Chancellor and B.-U. Meyburg (eds.). Raptors Worldwide. WWGBP/MME, Berlin.

Kruger, R. 1999. Towards solving raptor electrocutions on Eskom distribution structures in South Africa. Mini-Thesis. University of the Orange Free State, Bloemfontein.

Krüger, S., S. Piper, I. Rushworth, A. Botha, B. Daly, D. Allan, A. Jenkins, D. Burden and Y. Friedmann (eds.). 2006. Bearded Vulture (*Gypaetus barbatus meridionalis*) Population and Habitat Viability Assessment Workshop Report. Conservation Breeding Specialist Group (SSC/IUCN)/CBSG Southern Africa. Endangered Wildlife Trust, Johannesburg.

Kruger, S.C., D.G. Allan, A.R. Jenkins and A. Amar. 2013. Trends in territory occupancy, distribution and density of the Bearded Vulture *Gypaetus barbatus meridionalis* in southern Africa. Bird Conservation International 24: 162–177.

Kruuk, H. 1967. Competition for food between vultures in East Africa. Ardea 55: 171–193.

Kruuk, H. 1972. The Spotted Hyena: A Study of Predation and Social Behavior. University of Chicago Press, Chicago.

Kumar, S., K. Tamura and M. Nei. 1993. MEGA —Molecular Evolutionary Genetics Analysis. Version 1.0. Pennsylvania State University.

Kurten, B. and L. Werdelinb. 1990. Relationships between North and South American Smilodon. Journal of Vertebrate Paleontology 10(2): 158–169.

Kurtev, M., P. Iankov and I. Angelov. 2005. National Action plan for Conservation of the Egyptian Vulture (*Neophron percnopterus*) in Bulgaria.

Lambertucci, S.A. 2010. Size and spatio-temporal variations of the Andean Condor *Vultur gryphus* population in north-west Patagonia, Argentina: communal roosts and conservation. Oryx 44(3): 441–447.

Lambertucci, S.A. and A. Ruggiero. 2013. Cliffs used as communal roosts by Andean Condors protect the birds from weather and predators. PLoS One 8(6): e67304. doi: 10.1371/journal.pone. 0067304.

Lambertucci, S.A., A. Trejo, S. Di Martino, J.A. Snchez-Zapata, J.A. Donzar and F. Hiraldo. 2009. Spatial and temporal patterns in the diet of the Andean Condor: ecological replacement of native fauna by exotic species. Animal Conservation 12: 338–345.

Land, M.F. and D.-E. Nilsson. 2002. Animal Eyes. Oxford University Press, Oxford.

Landberg, L.C.W. 1965. The Chumash Indians of Southern California. Southwest Museum Papers 19: 1–158.

Lasch, U., S. Zerbe and M. Lenk. 2010. Electrocution of raptors at power lines in Central Kazakhstan. Waldökologie, Landschaftsforschung und Naturschutz 9: 95–100.

Latta, F.F. 1949. Handbook of Yokuts Indians. 1st edn. Bear State Books, Oildale, California.

Latta, F.F. 1977. Handbook of Yokuts Indians. 2nd edn. Bear State Books, Santa Cruz, California.

Laybourne, R. 1974. Collision between a vulture and an aircraft at an altitude of 37000 feet. Wilson Bulletin 86: 461–2.

Laybourne, R.C. 1974. Collision between a Vulture and an Aircraft at an Altitude of 37,000 Feet. The Wilson Bulletin (Wilson Ornithological Society) 86(4): 461–462.

Le Xuan Canh, Pham Trong Anh, J.W. Duckworth, Vu Ngoc Thanh and Lic Vuthy. 1997. A survey of large mammals in Dak Lak Province, Vietnam. WWF/IUCN, Hanoi.

Ledger, J.A. and H.J. Annegarn. 1981. Electrocution Hazards to the Cape Vulture (*Gyps coprotheres*) in South Africa. Biol. Conservation 20: 15–24.

Ledger, J.A. and J.C.A. Hobbs. 1999. Raptor use and abuse of powerlines in southern Africa. Journal of Raptor Research 33(1): 49–52.

Ledger, J.A. and J.C.A. Hobbs. 1999. Raptor use and abuse of powerlines in southern Africa. Journal of Raptor Research 33: 49–52.

Lee, G. and S. Horne. 1978. The painted Rock site (SBa-502 and SBa-526): Sapaksi, the house of the sun. Journal of California Anthropology 5: 216–224.

Lee, K.S., M.W.-N. Lau, J.R. Fellowes and B.P.L. Chan. 2006. Forest bird fauna of South China: notes on current distribution and status. Forktail 22: 23–38.

LeFrank, M., Jr. and W. Clark. 1983. Working Bibliography of the Golden Eagle and the Genus Aquila. National Wildlife Federation—Scientific and Technical Series 7, Washington, D.C.

Lehman, R.N. 2001. Raptor electrocution on power lines: current issues and outlook. Wildlife Society Bulletin 29: 804–813.

Lehman, R.N., P.L. Kennedy and J.A. Savidge. 2007. The state of the art in raptor electrocution research: a global review. Biological Conservation 136: 159–174.

Leighton, A.H. 1928. The Turkey Vulture's eyes. Auk 45: 352–355.

Lekagul, B. and P.D. Round. 1991. A Guide to the Birds of Thailand. Saha Karn Bhaet, Bangkok.

Lekuona, J.M. 2001. Uso del espacio por la avifauna y control de la mortalidad de aves en los parques eolicos de Navarra. Gobierno de Navarra, Pamplona.

Lekuona, J.M. and C. Ursula. 2007. Avian mortality in wind power plants of Navarra (North Spain). pp. 177–192. In: M. De Lucas, G.F.E. Jonas and M. Ferrer (eds.). Birds and Wind Farms. Risk Assessment and Mitigation. Quercus, Madrid.

Lemke, T., J.A. Mack and D.B. Houston. 1998. Winter range expansion by the Northern Yellowstone elk herd. Intermountain Journal of Sciences 4: 1–9.

Lemon, W.C. 1991. Foraging behavior of a guild of Neotropical vultures. The Wilson Bulletin 103(4): 698–702.

Lemus, J.A., G. Blanco, J. Grande, B. Arroyo, M. García-Montijano and F. Martínez. 2008. Antibiotics threaten wildlife: circulating quinolone residues and disease in avian scavengers. PLoS One 1–6.

Leonard, L., G.J. Dyke and M. Van Tuinen. 2005. A new specimen of the fossil palaeognath Lithornis from the Lower Eocene of Denmark. American Museum Novitates 491(3491): 1–11.

Lerner, H.R. and D.P. Mindell. 2005. Phylogeny of eagles, old world vultures, and other Accipitridae based on nuclear and mitochondrial DNA. Molecular Phylogenetics and Evolution 37: 327–346.

Lerner, H.R.L. and D.P. Mindell. 2005. Molecular Phylogenetics and Evolution 37: 327–346.

Leshem, Y. 1984. The rapid population decline of Israel's Lappet-faced Vulture *Torgos tracheliotus negevensis*. International Zoological Yearbook 23: 41–46.

Leshem, Y. and Y. Yom-Tov. 1998. Routes of migrating soaring birds. Ibis 140(1): 41–52.

Levy, N. 1996. Present status, distribution and conservation trends of the Egyptian vulture (Neophron percnopterus) in the Mediterranean countries and adjacent arid regions. In: J. Muntaner and J. Mayol (eds.). Biology and Conservation of Mediterranean Raptors (Monografia 4), 13–33. Sociedad Espanola de Ornitologia, Palma de Mallorca.

Levy, N. 1996. Present status, distribution and conservation trends of the Egyptian vulture (Neophron percnopterus) in the Mediterranean countries and adjacent arid regions. J. Muntaner and J. Mayol (eds.). Biology and Conservation of Mediterranean Raptors (Monographic 4). Sociedad Espanola de Ornitologia, Palma de Mallorca.

Lewis, A. and D. Pomeroy. 1989. A Bird Atlas of Kenya. AA Balkema, Rotterdam.

Lewis, J.B. 1928. Sight and scent in the Turkey Vulture. Auk 45: 467–470.

Lewis, M.E. and L. Werdelin. 2000. Carnivora from the South Turkwel hominid site, northern Kenya. Journal of Paleontology 74: 1173–1180.

Li, Y.D. and C. Kasorndorkbua. 2008. The status of the Himalayan Griffon Gyps himalayensis in South-East Asia. Forktail 24: 57–62.

Liechti, F. 2006. Birds: blowin' by the wind? Journal of Ornithology 147: 202–211.

Ligon, J.D. 1967. Relationships of cathartid vultures. Occasional Papers of the Museum of Zoology, University of Michigan 651: 1–26.

Ligue pour la Protection des Oiseaux (LPO). 2008. Les vautours: Alliés indispensables et fragiles du pastoralisme. Ligue pour la Protection des Oiseaux.

Likoff, L. 2007. King Vulture. The Encyclopedia of Birds. Infobase Publishing, pp. 557–6.

Lindsay, N. 2008. South Asia vulture recovery programme: WAZA Project 08001. WAZA News 15.

Lindsey, P.A., R. Alexander, L.G. Frank, A. Mathieson and S.S. Romañach. 2006. Potential of trophy hunting to create incentives for wildlife conservation in Africa where alternative wildlifebased land uses may not be viable. Animal Conservation 9: 283–291.

Lindsey, P.A., P.A. Roulet and S.S. Romañach. 2007. Economic and conservation significance of the trophy hunting industry in sub-Saharan Africa. Biological Conservation 134: 455–469.

Linnaeus, C. 1758. Systema naturae per regna tria naturae, secundum classes, ordines, genera, species, cum characteribus, differentiis, synonymis, locis. Tomus I. Editio decima, reformata. Holmiae (Laurentii Salvii). p. 86. "V. naribus carunculatis, vertice colloque denudate."

Litvaitis, J.A. and R. Villafuerte. 1996. Intraguild predation, mesopredator release, and prey stability. Conservation Biology 10: 676–677.

Livezey, B.C. and R.L. Zusi. 2007. Higher-order phylogeny of modern birds (Theropoda, Aves: Neornithes) based on comparative anatomy. II. Analysis and discussion. Zoological Journal of the Linnean Society 149: 1–95.

Loeb, E.M. 1932. The Western Kuksu cult. University of California Publications in American Archaeology and Ethnology 33: 1–137.

Loeb, E.M. 1933. The Eastern Kuksu cult. University of California Publications in American Archaeology and Ethnology 33: 139–232.

Long, B., S. Swan and Kry Masphal. 2000. A Wildlife Survey of north-eastern Mondulkiri Province. Fauna and Flora International, Hanoi.

López-López, P., M. Ferrer, A. Madero, E. Casado and M. McGrady. 2011. Solving man-induced large-scale conservation problems: the Spanish Imperial Eagle and power lines. PLoS One 6: e17196.

Louette, M. 1981. The Birds of Cameroon. An Annotated Check-list. Verhandelingen Koninklijke Academie Brussel (Klasse Wetenschappen) 43(163): 1–29.

Lovaszi, P. 1998. Status of the White Stork (Ciconia ciconia) in Hungary: results of national censuses between 1941 and 1994. Ornis Hungarica 8: 1–8.

Lovell, C.D. and R.A. Dolbeer. 1999. Validation of the U.S. Air Force bird avoidance model. Wildland Society Bulletin 27: 161–71.

Lowery, G.H., Jr. and W.W. Dalquest. 1951. Birds from the State of Veracruz, Mexico. University of Kansas Publications, Museum of Natural History 3: 531–649.

Lowery, G.H. and W.W. Dalquest. 1951. Birds from the State of Veracruz, Mexico. University of Kansas Publications, Museum of Natural History 3: 531–649.

Lowney, M.S. 1999. Damage by black and turkey vultures in Virginia, 1990–1996. Wildland Society Bulletin 27: 715–719.

Lozano, J., E. Virgós, S. Cabezas-Díaz and J.G. Mangas. 2007. Increase of large game species in Mediterranean areas: is the European wildcat (Felis silvestris) facing a new threat? Biological Conservation 138: 321–329.

Lu, X., D. Ke, X. Zeng, G. Gong and R. Ci. 2009. Status, ecology, and conservation of the Himalayan griffon Gyps himalayensis (Aves, Accipitridae) in the Tibetan plateau. Ambio 38(3): 166–73.

Lucio, A. and FJ. Purroy. 1992. Caza y conservación de aves en España. Ardeola 39: 85–98.

Lumeij, J.T. 1994. Nephrology. pp. 538–555. *In*: B.W. Ritchie, G.J. Harrison and I.R. Harrison (eds.). Avian Medicine: Principles and Application. Wingers Publishing, Lake Worth.

Lutz, R.L. 2002. Patagonia: At the Bottom of the World. DIMI Press, Salem.

Ma, Z., Y. Cai, B. Li and J. Chen. 2010. Managing wetland habitats for waterbirds: An international perspective. Wetlands 30: 15–27.

Maanen, E. van, I. Goradze, A. Gavashelishvili and R. Goradze. 2001. Opinion: Trapping and hunting of migratory raptors in western Georgia. Bird Conservation International 11(2): 77–92.

Macdonald, K.C. 1906. A list of birds found in the Myingyan district of Burma. Journal of the Bombay Natural History Society 17: 184–194, 492–502.

Mackinney, L.C. 1942. The vulture in ancient medical lore; vulture medicine in the Medieval world; vulture medicine in the modern world. Ciba Symposium (Ciba Pharmaceutical Products) 4(3): 1258–1271.

Mackinney, L.C. 1943. An Unpublished Treatise on Medicine and Magic from the Age of Charlemagne. Speculum 18: 4.

MacKinnon, B. 1999. Database for Bird and Wildlife Collissions with Aircraft in Canada. Internet Report http://www.tc.gc.ca/aviation/aerodrme/birdstke/info/about.htm. Aerodrome Safety Branch, Ottawa.

Maclean, G.L. 1996. Roberts' Birds of South Africa, 6th Edition. John Voelcker Bird Book Fund, Cape Town.

Macqueen, J. 1978. Secondary burial at Catal Huyuk. Numen 25(3): 226.

Madrid, J., H. Madrid, S. Funes, J. López, R. Botzoc and A. Ramos. 1991. Reproductive biology and behavior of the Ornate Hawk-Eagle in Tikal National Park (*Spizaetus ornatus*). *In*: D.F. Whitacre, W.A. Burnham and J.P. Jenny (eds.). Maya Project: use of birds of prey and other fauna as environmental indicators for design, management, and monitoring of protected areas and for building local capacity for conservation in Latin America, Progress Report IV, 93–113. The Peregrine Fund, Inc., Boise, ID.

MAGRAMA (Ministerio de Agricultura, Alimentacion y Medio Ambiente). 2012. Http://www.magrama.gob.es/es/ [accessed February 2014].

Mahood, S. 2012. Preventing Poisoning of Cambodia's Vultures. Final report to CEPF from Wildlife Conservation Society, Phnom Penh, Cambodia.

Maina, S. 2007. Furadan, a cheap and deadly weapon in the human-wildlife conflict. Swara 30: 16–18.

Majors, H.M. 1975. Exploring Washington. Van Winkle Publishing Co. p. 114.

Malan, G. 2009. Raptor Survey and Monitoring: A Field Guide for African Birds of Prey. Briza Publications, Pretoria.

Mandel, J.T. and K.L. Bildstein. 2007. Turkey vultures use anthropogenic thermals to extend their daily activity period. Wilson Journal of Ornithology 119: 102–105.

Mandel, J.T., G. Bohrer, D.W. Winkler, D.R. Barber, C.S. Houston and K.L. Bildstein. 2011. Migration path annotation: cross-continental study of migrationflight response to environmental conditions. Ecological Applications 21: 2258–2268.

Mander, M. 2007. Survey of the trade in vultures for the traditional health industry in South Africa. Future Works Report for Ezemvelo KZN Wildlife, Kwazulu Natal.

Mander, M., N. Diedericks, L. Ntuli, K. Mavundla, V. Williams and S. McKean. 2007. Survey of the trade in vultures for the traditional health industry in South Africa. Future Works Report for Ezemvelo KZN Wildlife.

Mander, M., N. Diederichs, L. Ntuli, K. Mavundla, V. Williams and S. McKean. 2007. Survey of the trade in vultures for the traditional health industry in South Africa. KZN Wildlife, Kwazulu Natal.

Manville, A.M., II. 2005. Bird strikes and electrocutions at power lines, communication towers, and wind turbines: state of the art and state of the science—next steps toward mitigation. Pacific Southwest Research Station, Forest Service, U.S. Department of Agriculture. Albany, California.

Maphisa, D. 2001. British Schools Exploration Society's (BSES) survey of vultures at selected sites in Lesotho—July to August 1998. Vulture News 45: 11–19.

Maransky, B., L. Goodrich and K. Bildstein. 1997. Seasonal shifts in the effects of weather on the visible migration of Red-Tailed Hawks at Hawk Mountain, Pennsylvania, 1992–1994' The Wilson Bulletin 109(2): 246–252.

Marchant. 1960. The breeding of some S.W. Ecuadorian birds. Ibis 102(4): 584–599.

Margalida, A. and J. Bertran. 2001. Function and temporal variation in the use of ossuaries by the Bearded Vulture (*Gypaetus barbatus*) during the nestling period. Auk 118: 785–789.

Margalida, A. 2008a. Presence of bone remains in the ossuaries of Bearded Vultures *Gypaetus barbatus*: storage or nutritive rejection? Auk 125: 560–564.

Margalida, A. 2008b. Bearded Vultures (*Gypaetus barbatus*) prefer fatty bones. Behavioral Ecology and Sociobiology 63: 187–193.

Margalida, A. 2012. Baits, budget cuts: a deadly mix. Science 338: 192.

Margalida, A. and M.A. Colomer. 2012. Modelling the effects of sanitary policies on European vulture conservation. Scientific Reports 2, 753.

Margalida, A., D. García, J. Bertran and R. Heredia. 2003. Breeding biology and success of the bearded vulture (*Gypaetus barbatus*) in the eastern Pyrenees. Ibis 145: 244–252.

Margalida, A., D. García and A. Cortés–Avizanda. 2007a. Factors influencing the breeding density of Bearded Vultures, Egyptian Vultures and Eurasian Griffon Vultures in Catalonia (NE Spain): management implications. Animal Biodiversity and Conservation 30(2): 189–200.

Margalida, A., S. Mañosa, J. Bertran and D. García. 2007b. Biases in studying the diet of the Bearded Vulture *Gypaetus barbatus*. Journal of Wildlife Management 71: 1621–1625.

Margalida, A., J.A. Donazar, J. Bustamante, F.J. Hernandez and M. Romero-Pujante. 2008a. Application of a predictive model to detect long-term changes in nest-site selection in the Bearded Vulture *Gypaetus barbatus*: conservation in relation to territory shrinkage. Ibis 150: 242–249.

Margalida, A., R. Heredia, M. Razin and M. Hernández. 2008b. Sources of variation in mortality of the Bearded Vulture *Gypaetus barbatus* in Europe. Bird Conservation International 18: 1–10.

Margalida, A., J.A. Donazar, M. Carrete and J.A. Sanchezzapata. 2010. Sanitary versus environmental policies: fitting together two pieces of the puzzle of European vulture conservation. Journal of Applied Ecology 47: 931–935.

Margalida, A., D. Campion and J.A. Donazar. 2011a. European vultures' altered behaviour. Nature 480: 457.

Margalida, A., M.A. Colomer and D. Sanuy 2011b. Can wild ungulate carcasses provide enough biomass to maintain avian scavenger populations? An empirical assessment using a bioinspired computational model. PLoS One 6: e20248.

Margalida, A., D. Campion and J. Donazar. 2014. Vultures vs livestock: conservation relationships in an emerging conflict between humans and wildlife. Fauna & Flora International Oryx 1–5. doi:10.1017/S0030605312000889.

Markandya, A., T. Taylor, A. Longo, M. Murty, S. Murty and K. Dhavala. 2008. Counting the cost of vulture decline—an appraisal of the human health and other benefits of vultures in India. Ecological Economics 67: 194–204.

Markus, M.B. 1972. Mortality of vultures caused by electrocution. Nature 238: 228.

Marquez, C., F. Gast, V.H. Vanega and M. Bechard. 2005. Aves rapaces diurnas de Colombia. Bogotá, Colombia: Instituto de Investigación de Recursos Biológicos Alexander von Humboldt, Ministerio del Medio Ambiente.

Marr, N.V., W.D. Edge, R.G. Anthony and R. Valburg. 1995. Sheep carcass availability and use by bald eagles. Wilson Bulletin 107: 251–257.

Marti, C. 1998. Effects of power lines on birds: Documentation (in German with English summary). Schriftenreihe Umwelt Nr. 292. Bundesamt für Umwelt, Wald und Landschaft (BUWAL), Bern.

Marti, R. and J.C. Del Moral. 2000. Atlas de las aves reproductoras de Espana. Direccion General de Conseracion de las Naturaleza—SEO/BirdLife, Madrid.

Martin, A. 1987. Atlas de las aves nidificantes en la isla de Tenerife. Tenerife, Canaries: Instituto de estudios canarios.

Martin, D.P. 1996. On the cultural ecology of sky burial on the Himalayan Plateau. East and West 46(3–4): 353–370.

Martin, G.R. 2011a. Through birds' eyes: insights into avian sensory ecology. Journal of Ornithology doi: 10.1007/s10336-011- 0771-5.

Martin, G.R. 2011b. Understanding bird collisions with manmade objects: a sensory ecology approach. Ibis 153: 239–254.

Martin, G.R. and S.J. Portugal. 2011. Differences in foraging ecology determine variation in visual field in ibises and spoonbills (Threskiornithidae). Ibis 153: 662–671.

Martin, G.R. and J.M. Shaw. 2010. Bird collisions with power lines: failing to see the way ahead? Biological Conservation 143: 2695–2702.

Martin, L.D. 1989. Fossil history of the terrestrial Carnivora. *In*: J.L. Gittleman (ed.). Carnivore Behaviour, Ecology, and Evolution. Cornell University Press, Ithaca 1: 536–568.

Martin, P., D. Campbell, K. Hughes and T. McDaniel. 2008. Lead in the tissues of terrestrial raptors in southern Ontario, Canada, 1995–2001. Science of the Total Environment 391: 96–103.

Martínez-Sánchez, J.C. and T. Will. 2010. Thomas R. Howell's Check-list of the birds of Nicaragua as of 1993. Ornithological Monographs 68: 1–107.

Martínez-Sánchez, J.C. and T. Will (eds.). 2010. Thomas R. Howell's check-list of the birds of Nicaragua as of 1993. Ornithological Monographs 68. American Ornithologists' Union, Washington, D.C.

Martínez, J.E. 2003. Impact of power lines on raptor populations in the Sierra Espuña Regional Park (Murcia). Proceedings of the III International Conference on Prevention Strategies for Fires in Southern Europe, Barcelona.

Marzluff, J.M., B. Heinrich and C.S. Marzluff. 1996. Raven roosts are mobile information centres. Animal Behaviour 51: 89–103.

Maslow, J.E. 1939. Feeding habits of Black Vulture. Auk 56: 472–474.

Maslow, J.E. 1986. Bird of Life, Bird of Death. Simon and Schuster, New York.

Maslowski, K.H. 1934. An aerial nest of the Turkey Vulture. Auk 51: 229–230.

Masphal, K. and B. Vorsak. 2007. Vulture restaurants across Cambodia. The Babbler: BirdLife in Indochina 24.

Massa, B. 1999. New and less known birds from Libya. Bulletin of the British Ornithologists' Club 119: 129–133.

Mateo-Tomás, P. and P.P. Olea. 2010. When hunting benefits raptors: a case study of game species and vultures. European Journal of Wildland Resources 56: 519–528.

Mateo, R. 1998. La Intoxicacion por Ingestion de Perdigones de Plomo en Aves Silvestres: Aspectos Epidemiologicos y Propuestas para su Prevencion En Espana. Tesis doctoral, Universidad Autonoma de Barcelona, Barcelona.

Mateo, R., R. Molina, J. Grifols and R. Guitart. 1997. Lead poisoning in a free ranging Griffon Vulture (Gyps fulvus). Veterinary Record 140: 47–48.

Mateo, R., M. Taggart and A.A. Meharg. 2003. Lead and arsenic in bones of birds of prey from Spain. Environmental Pollution 126: 107–14.

Matsyna, A.I. and E.L. Matsyna. 2011. Protection of birds on the power lines in Russia. Poster presented on International Conference on Power Lines and Bird Mortality in Europe, Budapest.

Matthee, C.A. and S.K. Davis. 2001. Molecular insights into the evolution of the family Bovidae: a nuclear DNA perspective. Molecular Biology and Evolution 18: 1220–1230.

Mauldin, R.E., B.A. Kimball, J.J. Johnston, J.C. Hurley and M.L. Avery. 2003. Development of a synthetic materials mimic for vulture olfaction research. pp. 430–435. *In*: K.A. Fagerstone and G.W. Witmer (eds.). Proceedings of the 10th Wildlife Damage Management Conference. The Wildlife Damage Management Working Group of the Wildlife Society, Fort Collins.

Mayell, H. 2001. Shrinking African Lake Offers Lesson on Finite Resources. National Geographic News. Retrieved 25 July 2013.

Maynard, M.C.J. 1881. The Birds of Eastern North America. C.J. Maynard & Co., Newtonville.

Mayr, E. and G.W. Cottrell (eds.). 1979. Check-list of birds of the world. Volume 1 (2 edn.): Museum of Comparative Zoology, Cambridge.

Mayr, E. and G.W. Cottrell (eds.). 1979. Check-list of the Birds of the World Volume 1. 2nd edition. Harvard University Press, Cambridge, MA.

Mayr, G. 2008. First substantial Middle Eocene record of the Lithornithidae (Aves): A postcranial skeleton from Messel (Germany). Annales de Paléontologie 94: 29–37.

Mayr, G. and J. Clarke. 2003. The deep divergences of neornithine birds: a phylogenetic analysis of morphological characters. Cladistics 19: 527–553.

McCrary, J.K., W.J. Arendt, L. Chavarría, L.J. López, P.A. Somarriba, P.-O. Boudrault, A.L. Cruz, F.J. Muñoz and D.G. Mackler. 2009. A contribution to Nicaraguan ornithology, with a focus on the pine-oak region. Cotinga 31: 72–7.

McCulloch, G. 2006a. Lappet-faced vulture—a social hunter? Africa —Birds & Birding 11: 32–34.

McCulloch, G. 2006b. Lappet-faced vultures—social hunters? Vulture News 10–13.

McGahan, J. 1973. Gliding flight of the Andean condor in nature. J. Exp. Biol. 58: 225–237.

McGrady, M.J. and A. Gavashelishvili. 2006. Tracking vultures from the Caucasus into Iran. Podoces 1(1/2): 21–26.

McHargue, L. 1977. Nesting of turkey and black vultures in Panama. Wilson Bulletin 89(2): 328–329.

McIlenny, E.A. 1945. An unusual feeding habit of the Black Vulture. Auk 62: 136–137.

McIlhenny, E.A. 1939. Feeding habits of the black vulture. Auk 56: 472–474.

Mckean, S. 2007. Vultures and traditional medicine. Fact Sheet: Birds of Prey Working Group and Endangered Wildlife Trust, Modderfontein.

Mckean, S. and A. Botha 2007. Traditional medicine demand threatens vultures in Southern Africa. Media release for Ezemvelo KZN Wildlife, Endangered Wildlife Trust and Future Works.

Mckern, W.C. 1922. Functional families of the Patwin. University of California Publications in American Archaeology and Ethnology 13: 235–258.

McMillan, I. 1968. Man and the California Condor: The embattled history and uncertian future of North America's largest free-living bird. E.P. Dutton and Company, New York.

McShea, W.J., E.G. Reese, T.W. Small and P.J. Weldon. 2000. An experiment on the ability of free-ranging Turkey Vultures (*Cathartes aura*) to locate carrion by chemical cues. Chemoecology 10: 49–50.

McVey, K.J., P.D.B. Skrade and T.A. Sordah. 2008. Use of a communal roost by Turkey Vultures in northeastern Iowa. Journal of Field Ornithology 79(2): 170–175.

Meade, G.E. 1961. The saber-toothed cat Dinobastis serus. Bulletin of the Texas Memorial Museum 2(II): 23–60.

Mech, L.D., D.W. Smith, K.M. Murphy and D.R. MacNulty. 2001. Winter severity and wolf predation on a formerly wolf-free elk herd. Journal of Wildlife Management 65: 998–1003.

Medina, F.M. 1999. Alimentacion del alimoche, *Neophron percnopterus* (L.), en Fuerteventura, islas Canarias (Aves, Accipitridae).Vieraea 27: 77–86.

Mee, A., B.A. Rideout, J.A. Hamber, J.N. Todd, G. Austin, M. Clark and M.P. Wallace. 2007. Junk ingestion and nestling mortality in a reintroduced population of California Condors Gymnogyps californianus. Bird Conservation International 17(2): 1–13.

Meinertzhagen, R. 1957. Kenya Diary. Collins, London.

Meiri, S. and T. Dayan. 2003. On the validity of Bergmann's rule. Journal of Biogeography 30(3): 331–351.

Melcher, C. and L. Suazo. 1999. Raptor electrocutions: the unnecessary losses continue. Journal of the Colorado Field Ornithologists 33: 221–224.

Melero de Blas, M. 2007b. Respiro para las carroñeras. Nuevo Real Decreto para la alimentación de aves rapaces en muladares. Panda 99: 20–22.

Mellaart, J. 1967. Catal Huyuk: A Neolithic town in Anatolia. McGraw Hill, New York.

Meloro, C. and G.J. Slater. 2012. Covariation in the skull modules of cats: the challenge of growing saber-like canines. Journal of Vertebrate Paleontology 32(3): 677–685.

Mendelsohn, J., C. Brown, M. Mendelsohn and M. Diekmann. 2005. Observations on the movements of adult Cape Vultures in central Namibia. Lanioturdus 38: 16–20.

Mendelssohn, H. and Y. Leshem. 1983. Observations on reproduction and growth of old world vultures. pp. 214–244. In: S.R. Wilber and J.A. Jackson (eds.). Vulture Biology and Management. University of California Press, Berkeley.

Mendelssohn, H. and U. Marder. 1989. Reproduction of the Lappet-faced vulture Torgos tracheliotus negevensis at Tel Aviv University Research Zoo. In: P.J.S. Olney and P. Ellis (eds.). International Zoo Yearbook. Zoological Society of London, London 28: 229–234.

Meretsky, V.J., N.F.R. Snyder, S.R. Beissinger, D.A. Clendenen and J.W. Wiley. 2000. Demography of the California Condor: implications for reestablishment. Conservation Biology 14: 957–967.

Mesta, R. 1996. California Condor Recovery Program, Ventura.

Meyburg, B.-U., M. Gallardo, C. Meyburg and E. Dimitrova. 2004. Migrations and sojourn in Africa of Egyptian vultures (Neophron percnopterus) tracked by satellite. Journal of Ornithology 145(4): 273–280.

Meyburg, B., O. Manowsky and C. Meyburg. 1996. The Osprey in Germany: its adaptation to environments altered by man. pp. 125–135. In: D.M. Bird, D.E. Varlan and J.J. Negro (eds.). Raptors in Human Landscapes: Adaptations to Built and Cultivated Environments. Academic Press, New York.

Meyer, A. 1994. Shortcomings of the cytochrome b gene as a molecular marker. Trends in Ecology & Evolution 9: 278–280.

Mijele, D. 2009. Incidences of poisoning of vultures and lions in Masai Mara National Reserve. Kenya Wildlife Service Masai Mara Veterinary Report. Nairobi: Kenya Wildlife Service.

Mijele, D. 2009. Incidences of Poisoning of Vultures and Lions in the Masai Mara National Reserve. Kenya Wildlife Service Masai Mara Veterinary Report, Nairobi, Kenya.

Mikami, O.K. and M. Kawata. 2004. Does interspecific territoriality reflect the intensity of ecological interactions? A theoretical model for interspecific territoriality. Evolutionary Ecological Research 6: 765–775.

Miller, A.H. 1942. A California Condor bone from the coast of southern Oregon. Murrelet 23: 77.

Miller, A.H. 1957. Bird remains from an Oregon Indian midden. Condor 59: 59–63.

Miller, A.H. and L.V. Compton. 1939. Two fossil birds from the lower Miocene of South Dakota. Condor 41: 153–156.

Miller, A.H., I. McMillan and E. McMillan. 1965. The current status and welfare of the California Condor. National Audubon Research Report 6: 1–61.

Miller, L. 1910. The condor-like vultures of Rancho La Brea. University of California Bulletin of the Department of Geology 6: 1–19.

Miller, L. 1911. Avifauna of the Pleistocene cave deposits of California. University of California Bulletin of the Department of Geology 6: 385–400.

Miller, L. 1931. The California Condor in Nevada. Condor 33(1): 32.

Miller, L. 1957. Bird remains from a prehistoric cave deposit in Grant County, New Mexico. Condor 59(1): 59–63.

Miller, L. 1957. Bird Remains from an Oregon Indian Midden. Condor 59: 59–63.

Miller, L. 1960. Condor Remains from Rampart Cave, Arizona. Condor 62(1): 70.

Miller, L. 1963. Birds and Indians in the West. Bulletin of the Southern California Academy of Sciences 62: 178–191.

Miller, L.H. and I. DeMay. 1942. The fossil birds of California. University of California Publications in Zoology 47(4): 47–142.

Miller, L.H. 1909. Teratornis, a new avian genus from Rancho La Brea. University of California Publications, Bulletin of the Department of Geology 5: 305–317.

Miller, M.J.R., M.E. Wayland, E.H. Dzus and G.R. Bortolotti. 2000. Availability and ingestion of lead shotshell pellets by migrant bald eagles in Saskatchewan. Journal of Raptor Research 34: 167–174.

Mills, M.G.L. 1993. Social systems and behaviour of the African wild dog Lycaon pictus and the spotted hyaena *Crocuta crocuta* with special reference to rabies. Onderstepoort Journal of Veterinary Research 60: 405–409.

Mills, M.G.L. and H. Hofer. 1998. Hyaenas. Status Survey and Conservation Action Plan. IUCN/SSC Hyaena Specialist Group, IUCN, Gland, Switzerland and Cambridge.

Millsap, B.A., M.E. Cooper and G. Holroyd. 2007. Legal Considerations 25. pp. 437–449. *In*: D.M. Bird and K.L. Bildstein (eds.). Raptor Research and Management Techniques. Hancock House, Surrey, British Columbia.

Milton, J. Patradise Lost [1667] (Menston, Yorkshire 1968), 273–278.

Mineau, P., M.R. Fletcher, L.C. Glaser, N.J. Thomas, C.A.N.D.A.C.E. Brassard, L.K. Wilson and S.L. Porter. 1999. Poisoning of raptors with organophosphorus and carbamate pesticides with emphasis on Canada, the United States and the United Kingdom. Journal Raptor Research 33: 1–37.

Mingozzi, T. and R. Esteve. 1997. Analysis of a Historical Extirpation of the Bearded Vulture *Gypaetus barbatus* (L.) in the Western Alps (France-Italy): former distribution and causes of extirpation. Biological Conservation 79: 155–171.

Moccia, J. and A. Arapogianni. 2011. Pure power: Wind energy targets for 2020 and 2030. Brussels: European Wind Energy Association.

Modi, J.J. 1928. The Funeral Ceremonies of the Parsees: their Origins and Explanation. Bombay.

Modi, J.J. 1979. The Funeral Ceremonies of the Parsees: their Origins and Explanation. Garland Publications, Bombay.

Moir, J. 2009. A rare chance to view nesting condors. Winging It 21(4): 19.

Mol, D., W. van Logchem, K. van Hooijdonk and R. Bakker. 2008. The Saber-Toothed Cat of the North Sea. Norg, Uitgeverij DrukWare.

Monadjem, A. and D.K. Garcelon. 2005. Nesting distribution of vultures in relation to land use in Swaziland. Biodiversity and Conservation 14: 2079–2093.

Monadjem, A. 2001. Observations on the African White-backed Vulture Gyps africanus nesting at Mlawula Nature Reserve, Swaziland. Vulture News 45: 3–10.

Monadjem, A. 2003a. Nest site selection by African White-backed Vulture in Swaziland. Vulture News 48: 24–26.

Monadjem, A. 2003b. Nesting distribution and status of vultures in Swaziland. Vulture News 48: 12–19.

Monadjem, A., M.D. Anderson, S.E. Piper and A. Boshoff (eds.). 2004. Vultures of southern Africa—Quo Vadis? Proceedings of a Workshop on Vulture Research and Conservation in Southern Africa. Birds of Prey Working Group, Johannesburg.

Monadjem, A., K. Wolter, W. Neser and A. Kane. 2013. Effect of rehabilitation on survival rates of endangered Cape Vultures. Animal Conservation. DOI: 10.1111/acv.12054.

Monroe, B.L. 1968. A Distributional Survey of the Birds of Honduras. Ornithological Monographs 7: 1–458.

Montejo Díaz, J. and E. Ruelas Inzunza. 1997. Notes on spring migration in the Pacific and Atlantic slopes of Guatemala, with observations on Wood Storks and Laughing Gulls. Hawk Migration Association of North America, Hawk Migration Studies 22(2): 6–8.

Moore, J.E. and R.K. Swihart. 2005. Modeling site occupancy by forest rodents: Incorporating detectability and spatial autocorrelation with hierarchically structured data. Journal of Wildlife Management 69: 933–949.

Moreno-Opo, R., A. Margalida, A. Arredondo, F. Guil, M. Martın, R. Higuero, C. Soria and J. Guzman. 2010. Factors influencing the presence of the cinereous vulture *Aegypius monachus* at carcasses: food preferences and implications for the management of supplementary feeding sites. Wildlife Biology 16: 25–34.

Moreno-Opo, R., A. Arredondo and F. Guil. 2010. Foraging range and diet of Cinereous Vulture *Aegypius monachus* using livestock resources in Central Spain. Ardeola 57: 111–119.

Moritzi, M., R. Spaar and O. Biber. 2001. Causes of death of White Storks (*Ciconia ciconia*) ringed in Switzerland (1947–1997). Vogelwarte 41: 44–52.

Morris, R.C. 1934. Death of an elephant Elephas maximus Linn. while calving. Journal of the Bombay Natural History Society 37(3): 722.

Morrison, M.L., K.H. Pollack, A.L. Oberg and K.C. Sinclair. 1998. Predicting the Response of Bird populations to Wind Energy-Related Deaths. National Renewable Energy Laboratory, Golden Colorado.

Mortimore, M. 1989. Adapting to Drought: Farmers, Famines, and Desertification in West Africa. Cambridge University Press, Cambridge.

Mossman, M.J. 1989. Black and Turkey Vultures. Proceedings Midwest Raptor Management Symposium and Workshop, Chicago.

Mrosovsky, N. 1971. Black vultures attack live turtle hatchlings. Auk 88: 672–673.

Mudur, G. 2001. Human anthrax in India may be linked to vulture decline. British Medical Journal 322: 320.

Mueller, H.C. and D.D. Berger. 1967. Turkey Vultures attack living prey. Auk 94: 430–431.

Mukherjee, A.K. 1995. Birds of arid and semi-arid tracts. Records of the Zoological Survey of India Occasional Paper. 142.

Muller-Schwarze, D. 2006. Chemical Ecology of Vertebrates. Cambridge University Press, Cambridge.

Mundy, P., D. Butchart, J. Ledger and S. Piper. 1992. The Vultures of Africa. London: Academic Press.

Mundy, P. 1982. The Comparative Biology of Southern African Vultures. Vulture Study Group, Johannesburg, South Africa.

Mundy, P. 1997. Whiteheaded Vulture. *In*: J.A. Harrison, D.G. Allan and L.G. Underhill.

Mundy, P.J. 1997. Hooded Vulture Necrosyrtes monarchus. pp. 156–157. *In*: J.A. Harrison, D.G. Allan, L.G. Underhill, M. Herremans, A.J. Tree, V. Parker and C.J. Brown (eds.). The Atlas of Southern African Birds, Vol. 1: Non-passerines. BirdLife South Africa, Johannesburg.

Mundy, P.J. 2000. More on the Hooded Vultures in Ghana. Vulture News 43: 64.

Mundy, P.J., P.C. Benson and D.G. Allan. 1997. Cape vulture Kransaalvoël Gyps coprotheres. pp. 158–159. *In*: J.A. Harrison, D.G. Allan, L.G. Underhill, M. Herremans, A.J. Tree, V. Parker and C.J. Brown (eds.). The Atlas of Southern African Birds. Vol. 1: Non-passerines. BirdLife South Africa, Johannesburg.

Mundy, P., D. Butchart, J. Ledger and S. Piper. 1992. The Vultures of Africa. Acorn Books C.C. Randburg & Russel Friedman Books C.C. Halfway House, South Africa.

Mundy, P.J., P.C. Benson and D.G. Allan. 2007. Cape vulture. *In*: The South African Bird Atlas Project. Birdlife South Africa, Blairgowrie, Animal Demography Unit, Cape Town, South African National Biodiversity Institute, Pretoria. 2: 158–159.

Mundy, P.J. 1997. Hooded Vulture Necrosyrtes monarchus. *In*: J.A. Harrison, D.G. Allan, L.G. Underhill, M. Herremans, A.J. Tree, V. Parker and C.J. Brown (eds.). The Atlas of Southern African Birds, Vol. 1: Non-passerines, 156–157. BirdLife South Africa, Johannesburg.

Mundy, P.J. 1997. Lappetfaced Vulture. pp. 162–163. *In*: J.A. Harrison et al. (eds.). The Atlas of South African birds. Volume 1: Non-passerines. BirdLife South Africa and Avian Demography Unit, Johannesburg, South Africa.

Mundy, P.J. 1997. White-headed Vulture. pp. 164–165. *In*: J.A. Harrison, D.G. Allan, L.G. Underhill, M. Herremans, A.J. Tree, V. Parker and C.L. Brown (eds.). The Atlas of South African Birds. Volume 1: Non-passerines. BirdLife South Africa and Avian Demography Unit, Johannesburg.

Mundy, P.J. 2000. The status of vultures in Africa during the 1990s. pp. 151–164. *In*: R.D. Chancellor and B.-U. Meyburg (eds.). Raptors at Risk. World Working Group on Birds of Prey, Berlin, and Hancock House, Blaine, WA.

Mundy, P.J. 2001. On the vultures of Africa. *In*: Wings Over Africa—Proceedings of the International Seminar on Bird Migration: Research, Conservation, Education and Flight Safety, 110–115, Tel Aviv University, Tel Aviv.

Mundy, P.J. and J.A. Ledger. 1977. The plight of the Cape Vulture. Endangered Wildlife 1(4): 2–3.

Mundy, P.J., A.S. Robertson, J. Komen and T.G. O'Connor. 1986. Attacks by Black Eagles on vultures. Journal of Raptor Research 20:2: 61–64.

Munn. 1899. Cited in BirdLife International 2001. Threatened Birds of Asia: the BirdLife International Red Data Book. Cambridge, UK: BirdLife International.

Muntifering, J.R., A.J. Dickman, L.M. Perlow, T. Hruska, P.G. Ryan, L.L. Marker and R.M. Jeo. 2006. Managing the matrix for large carnivores: a novel approach and perspective from cheetah (*Acinonyx jubatus*) habitat suitability modelling. Animal Conservation 9: 103–112.

Murie, A. 1940. Ecology of the Coyote in the Yellowstone. US National Park Service Fauna Series No. 4. Washington, D.C.

Murillo. 2003. Environmental impact and preventive and corrective measures for power lines and substations. Proceedings of the 4th Technical Session on Power Lines and the Environment. Red Eléctrica de España, Madrid.

Murn, C., M.D. Anderson and A. Anthony. 2002. Aerial survey of African white-backed vulture colonies around Kimberley, Northern Cape and Free State provinces, South Africa. South African Journal of Wildlife Research 32: 145–152.

Murn, C. and M.D. Anderson. 2008. Activity patterns of African whitebacked vultures Gyps africanus in relation to different land-use practices and food availability. Ostrich 79(2): 191–198.

Murn, C., U. Khan and F. Farid. 2008. Vulture populations in Pakistan and the Gyps vulture restoration project. Vulture News 35–43.

Murphy, R.K., S.M. McPherron, G.D. Wright and K.L. Serbousek. 2009. Effectiveness of avian collision averters in preventing migratory bird mortality from powerline strikes in the Central Platte River, Nebraska. University of Nebraska-Kearney, Kearney.

Murray, M., W.I. Morrison and D.D. Whitelaw. 1982. Host susceptibility to African Trypanosomiasis: Trypanotolerance. Annals of Tropical Medicine and Parasitology 21: 1–68.

Musters, C.J.M., M.A.W. Noordervliet and W.J. Terkeus. 1996. Bird casualties caused by a wind energy project in an estuary. Bird Study 43: 124–126.

Muzzolini, A. 1983. L'art rupestre du Sahara central: classification et chronologie. Le bœuf dans la préhistoire africaine (2 vols). Université de Provence. Aix-en-Provence and Marseille.

Myers, P., R. Espinosa, C.S. Parr, T. Jones, G.S. Hammond and T.A. Dewey. 2008. Family Cathartidae University of Michigan Animal Diversity Web.

Nadeem, M.S., M. Asif, T. Mahmood and G. Mujtaba. 2007. Reappearance of red-headed Vulture Sarcogyps calvus in Tharparker, Southeast Pakistan. Podoces 2(2): 146–147.

Naidoo, V. 2007. Diclofenac in Gyps vultures: A molecular mechanism of toxicity. Unpublished PhD Thesis. University of Pretoria.

Naidoo, V. and G.E. Swan. 2009. Diclofenac toxicity in Gyps vulture is associated with decreased uric acid excretion and not renal portal vaso constriction. Comparative Biochemistry and Physiology 149: 269–274.

Naidoo, V., K. Wolter, D. Cromarty, M. Diekmann, N. Duncan, A.A. Meharg, M.A. Taggart, L. Venter and R. Cuthbert. 2010. Toxicity of non-steroidal anti-inflammatory drugs to Gyps vultures: a new threat from ketoprofen. Biology Letters 6(3): 339–341.

Naidoo, V., K. Wolter, R. Cuthbert and N. Duncan. 2009. Veterinary diclofenac threatens Africa's endangered vulture species. Regulatory toxicology and pharmacology 53: 205–208.

Naidoo, V., K. Wolter, I. Espie and A. Kotze. 2011. Vulture rescue and rehabilitation in South Africa: an urban perspective. Journal of the South African Veterinary Association 82(1): 24–31.

Naidoo, V., K. Wolter, D. Cromarty, M. Diekmann, N. Duncan, A.A. Meharg, M.A. Taggart, L. Venter and R. Cuthbert. 2010. Toxicity of non-steroidal anti-inflammatory drugs to Gyps vultures: a new threat from ketoprofen. Biology Letters 6(3): 339–341.

Namgail, T and Y. Yom-Tov. 2009. Elevational range and timing of breeding in the birds of Ladakh: the effects of body mass, status and diet. Journal of Ornithologist 150(2): 505–510.

Naoroji, R. 2006. Birds of prey of the Indian subcontinent. Christopher Helm, A&C Black Publishers Ltd., London.

Naoroji, R. 2006. Birds of Prey of the Indian Subcontinent. Christopher Helm/A & C Black, London.

Navarro, B.M. and P. Palmquist. 1996. Presence of the African Saber-toothed Felid Megantereon whitei (Broom 1937) (Mammalia, Carnivora, Machairodontinae) in Apollonia-1 (Mygdonia Basin, Macedonia, Greece). Journal of Archaeological Science 23: 869–872.

Navarro, B.M. and P. Palmqvist. 1995. Presence of the African machairodont Megantereon whitei (Broom 1937) (Felidae, Carnivora, Mammalia) in the Lower Pleistocene site of Venta Micena (Orce, Granada, Spain), with some considerations on the origin, evolution and dispersal of the genus. Journal of Archaeological Science 22: 569–582.

Negro, J.J. 1999. Past and future research on wildlife interaction with power lines. pp. 21–28. *In*: M. Ferrer and G.F. Janss (eds.). Birds and Power Lines: Collision, Electrocution, and Breeding. Quercus, Madrid, Spain.

Nelson, M.W. 1979. Power lines progress report on eagle protection research. Idaho Power Company, Boise, Idaho, USA. Available from the U.S. Geological Survey, Richard R. Olendorff Memorial Library, 970 Lusk St., Boise, ID 83706.

Nelson, M.W. 1980. Update on eagle protection practices. Idaho Power Company. Boise, Idaho, USA. Available from the U.S. Geological Survey, Richard R. Olendorff Memorial Library, 970 Lusk St., Boise, ID 83706.

Nelson, M.W. and P. Nelson. 1976. Power lines and birds of prey. Idaho Wildlife Review 28: 3–7.

New Scientist, 2007-JUN-01: Starving vultures switch to live Retrieved 2007-JUN-20.

Newby, J., J. Grettenberger and J. Wakins. 1986. The Birds of Northern Air, Niger. Malimbus 9: 4–9.

Newton, I. 1990. Birds of Prey. Merehurst, London.

Newton, I. 1979. Population ecology of raptors. In T. and A.D. Posyer, Berkhempsted, United Kingdom.

Newton, I. 2008. The migration ecology of birds. Academic Press, Oxford.

Newton, I., J.A. Bogan and P. Rothery. 1986. Trends and effects of organochlorine compounds in Sparrowhawk eggs. Journal of Applied Ecology 23: 461–478.

Newton, I. ed. 1990. Birds of Prey. Facts on File, New York.

Newton, I., J. Bogan, E. Meek and B. Little. 1982. Organochlorine compounds and shell-thinning in British merlins (Falco columbarius). Ibis 124: 238–335.

Newton, S.F. and M. Shobrak. 1993. The Lappet-faced vulture Torgos tracheliotos in Saudi Arabia. pp. 111–117. *In*: R.T. Wilson (ed.). Proceedings of the eighth Pan-African Ornithological Congress: birds and the African environment. Musée Royal de l'Afrique Centrale, Tervuren, Belgium.

Newton, S.F. and A.V. Newton. 2008. Breeding biology and seasonal abundance of Lappet-faced Vultures Torgos tracheliotus in western Saudi Arabia. Ibis 138(4): 675–683.

Nielsen, J. 2006. Condor: To the Brink and Back—The Life and Times of One Giant Bird. Harper Perennial, New York.

Niklaus, G. 1984. Large numbers of birds killed by electric power line. Scopus 8: 42.

Nikolaus, G. 2001. Bird exploitation for traditional medicine in Nigeria. Malimbus 23: 45–55.

Nikolaus, G. 1984. Distinct status changes of certain Palearctic migrants in the Sudan. Scopus 8: 36–38.

Nikolaus, G. 1984. Large numbers of birds killed by electric power line. Scopus 8(42).

Nikolaus, G. 1987. Distribution atlas of Sudan's birds with notes on habitat and status. Bonner Zoologische Monographien 25: 1–32a.

Nikolaus, G. 2006. Commentary: where have the African vultures gone? Vulture News 65–67.

Nikolov, S., C. Nikolov and I. Angelov. 2013. First Record on Ground Nesting of Egyptian Vulture Neophron percnopterus (Aves: Accipitriformes) in Continental Europe. Acta Zoologica Bulgarica 65(2): 417–419.

Njilima, F., N. Kyonjola and J. Wolstencroft. 2010. Tanzania vultures baseline survey report. Unpublished report to the Royal Society for the Protection of Birds.

Norgaard, R.B. 1994. Development Betrayed: The End of Progress and a Coevolutionary Revisioning of the Future. Routledge, London.

Norgaard, R.B. 1984. Economic Development and Cultural Change. Routledge, London.

Novaes, W.G. and R. Cintra. 2013. Factors influencing the selection of communal roost sites by the Black Vulture *Coragyps atratus* (Aves: Cathartidae) in an urban area in Central Amazon. ZOOLOGIA http://dx.doi.org/10.1590/S1984-46702013005000014.

Nunez-Iturri, G., O. Olsson and H.F. Howe. 2008. Hunting reduces recruitment of primate-dispersed trees in Amazonian Peru. Biological Conservation 141: 1536–1546.

O'Neil, T. 1988. An analysis of bird electrocutions in Montana. Journal of Raptor Research 22: 27–28.

Oaks, J.L., M. Gilbert, M.Z. Virani, R.T. Watson, C.U. Meteyer, B.A. Rideout, H.L. Shivaprasad, S. Ahmed, M.J.I. Chaudhry, M. Arshad, S. Mahmood, A. Ali and A.A. Khan. 2004. Diclofenac residues as the cause of vulture population decline in Pakistan. Nature 427(6975): 630–633.

Oaks, J.L., C.U. Meteyer, B.A. Rideout, H.L. Shivaprasad, M. Gilbert, M.Z. Virani, R.T. Watson and A.A. Khan. 2004. Diagnostic investigation of vulture mortality: the anti-inflammatory drug diclofenac is associated with visceral gout. Falco 13–14.

Oaks, L., B.A. Rideout, M. Gilbert, R. Watson, M. Virani and A. Ahmed Khan. 2001. Summary of diagnostic investigation into vulture mortality: Punjab Province, Pakistan, 2000–2001. Reports from the workshop on Indian Gyps vultures. pp. 12–13. *In*: T. Katzner and J. Parry-Jones (eds.). Proceedings of the 4th Eurasian Congress on Raptors. Estación Biológica Donaña, Raptor Research Foundation, Seville.

Oatley, T.B., H.D. Oschadleus, R.A. Navarro and L.G. Underhill. 1998. Review of ring recoveries of birds of prey in southern Africa: 1948–1998. Endangered Wildlife Trust, Johannesburg.

Oatley, T.B., H.D. Oschadleus, R.A. Navarro and L.G. Underhill. 1998. Review.

Oehler, J.D. and J.A. Litvaitis. 1996. The role of spatial scale in understanding responses of medium-sized carnivores to forest fragmentation. Canadian Journal of Zoology 74: 2070–2079.

Ogada, D.L., M.E. Torchin, M.F. Kinnaird and V. Ezenwa. 2012. Effects of vulture declines on facultative scavengers and potential implications for mammalian disease transmission. Conservation Biology 26(3): 453–460.

Ogada, D.L. and R. Buij. 2011. Large declines of the Hooded Vulture Necrosyrtes monachus across its African range. Ostrich: Journal of African Ornithology 82(2): 101–113.

Ogada, D.L. and F. Keesing. 2010. Decline of Raptors over a three-year period in Laikipia, Central Kenya. Journal of Raptor Research 44(2): 129–135.

Ogada, D.L., F. Keesing and M.Z. Virani. 2012. Dropping dead: causes and consequences of vulture population declines worldwide. Annals of New York Academy of Sciences 1–15. doi: 10.1111/j.1749-6632.2011.06293.

Ogada, D.L., S. Thomsett, M.Z. Virani, C. Kendall and M. Odino. 2010. Raptor road counts in Kenya: with emphasis on vultures. Unpublished report to the Royal Society for the Protection of Birds.

Ogada, D.L., M.E. Torchin, M.F. Kinnaird and V.O. Ezenwa. 2012. Effects of vulture declines on facultative scavengers and potential implications for mammalian disease transmission. Conservation Biology 26: 453–460.

Ohmart, R.D. and R.G. Clark. 1985. Spread-winged posture of Turkey Vulture: single or multiple function? Condor 87: 350–355.

Olden, J.D. and N.L. Poff. 2003. Toward a mechanistic understanding and prediction of biotic homogenization. American Naturalist 162: 442–460.

Olea, P.P. and P. Mateo-Tomas. 2013. Assessing species habitat using google street view: A case study of cliff-nesting vultures. PLoS One 8(1): e54582. doi:10.1371/journal.pone.0054582.

Olea, P.P. and P. Mateo-Tomás. 2009. The role of traditional farming practices in ecosystem conservation: the case of transhumance and vultures. Biological Conservation 142(8): 1844–1853.

Olea, P.P., J. García and J. Falagán. 1999. Expansión del buitre leonado Gyps fulvus: tamaño de la población y parámetros reproductores en un área de reciente colonización. Ardeola 46: 81–88.

Olendorff, R.R., A.D. Miller and R.N. Lehman. 1981. Suggested practices for raptor protection on power lines—the state-of-the-art in 1981. Raptor Research Report No. 4. Raptor Research Foundation, Inc., St. Paul, Minnesota.

Olrog, C.C. 1985. Status of Wet Forest Raptors in Northern Argentina. ICBP Technical Publication 191–197.

Olsen, P. and L. Joseph. 2011. Stray Feathers: Reflections on the Structure, Behavior and Evolution of Birds. CSIRO Publishing, Collingwood.

Olsen, S.L. 1985. The fossil record of birds. pp. 79–252. *In*: D.S. Farmer, J.R. King and K.C. Parkes (eds.). Avian Biology 8. Academic Press, Orlando.

Olson, D.M., E. Dinerstein, E.M. Wikramanayake, N.D. Burgess, G.V.N. Powell, E.C. Underwood, J.A. D'Amico, I. Itoua, H.E. Strand, J.C. Morrison, C.J. Loucks, T.F. Allnutt, T.H. Ricketts, Y. Kura, J.F. Lamoreux, W.W. Wettengel, P. Hedao and K.R. Kassem. 2001. Terrestrial ecoregions of the world: A new map of life on Earth. BioScience 51(11): 933–938.

Olson, K.A., T.K. Fuller, G.B. Schaller, B. Lhagvasuren and D. Odonkhuu. 2005. Reproduction, neonatal weights, and first-year survival of Mongolian gazelles (*Procapra gutturosa*). Journal of Zoology 265(3): 227–233.

Oniki, Y. and E.O. Willis. 1983. A study of breeding birds of the Belém area, Brazil 1. Tinamidae to Columbidae. Ciencia e Cultura (Sao Paulo) 35: 947–956.

Orians, G.H. and D.R. Paulson. 1969. Notes on Costa Rican birds. Condor 71: 426–431.

Orloff, S. and A. Flannery. 1996. A continued examination of avian mortality in the Altamont Pass Wind Resource Area. California Energy Commission, USA.

Orloff, S.G. and A.W. Flannery. 1993. Wind turbine effects on avian activity, habitat use, and mortality in the Altamont Pass and Solano County Wind Resource Areas. pp. 1–14. *In*: J.W. Huckabee (ed.). Avian Interactions with Utility Structures. Avian Power Line Interactions Committee (APLIC). Electric Power Research Institute, Palo Alto.

Oro, D., A. Margalida, M. Carrete, R. Heredia and J.A. Donázar. 2008. Testing the goodness of supplementary feeding to enhance population viability in an endangered vulture. PLoS One 3: e4084.

Orozco, J. 2008. Sabertooth Cousin Found in Venezuela Tar Pit—A First. National Geographic News. National Geographic Society.

Orr, P.C. 1968. Prehistory of Santa Rosa Island. Santa Barbara Museum of Natural History, Santa Barbara.

Orta, J. 1994. Eurasian Griffon. pp. 127 *In*: J. del Hoyo, A. Elliott and J. Sargatal (eds.). Handbook of Birds of the World. Vol. 2. New World vultures to guineafowl. Lynx Edicions, Barcelona, Spain.

Osborn, R.G., K.F. Higgins, R.E. Usgaard, C.D. Dieter and R.E. Neiger. 2000. Bird mortality associated with wind turbines at the Buffalo Ridge Wind Resource Area, Minnesota. American Midland Naturalist 143: 41–45.

Osborn, R.G., K.F. Higgins, E.R. Usgaard, C.D. Dieter and R.D. Neiger. 2000. Bird mortality associated with wind turbines at the buffalo ridge wind resource area, Minnesota. Am. Midland Nat. 143: 41–52.

Ostende, L.W., H. van den, M. Morlo and D. Nagel. 2006. Fossils explained 52 Majestic killers: the sabre-toothed cats. Geology Today 22(4): 150.

Otieno, P.O., J.O. Lalah, M. Virani, I.O. Jondiko and K. Schramm. 2010. Carbofuran and its toxic metabolites provide forensic evidence for Furadan exposure in vultures (*Gyps africanus*) in Kenya. Bulletin of Environmental Contamination and Toxicology 84: 536–544.

Owre, O.T. and P.O. Northington. 1961. Indication of the sense of smell in the Turkey Vulture, *Cathartes aura* (Linnaeus), from feeding tests. American Midland Naturalist 66: 200–105.

Pain, D.J., A.A. Cunningham, P.F. Donald, J.W. Duckworth, D. Houston, T. Katzner, J. Parry Jones, C. Poole, V. Prakash, P. Round and R. Timmins. 2003. Gyps vulture declines in Asia: temperospatial trends, causes and impacts. Conservation Biology 17: 661–671.

Pain, D.J., C.G.R. Bowden, A.A. Cunningham, R. Cuthbert, D. Das, M. Gilbert, R.D. Jakati, Y. Jhala, A.A. Khan, V. Naidoo, J.L. Oaks, J. Parry-Jones, V. Prakash, A. Rahmani, S.P. Ranade, H.S. Baral, K.R. Senacha, S. Saravanan, N. Shah, G. Swan, D. Swarup, M.A. Taggart, R.T. Watson, M.Z. Virani, K. Wolter and R.E. Green. 2008. The race to prevent the extinction of South Asian vultures. Bird Conservation International 18: S30–S48.

Pain, D.J., V. Prakash, A.A. Cunningham and G.R. Ghalsasi. 2002. Vulture declines in India: patterns, causes and spread. pp. 4–7. *In*: T. Katzner and J. Parry-Jones (eds.). Conservation of Gyps Vulture in Asia. Third North American Ornithological Congress, September 24–28, 2002. New Orleans, Louisinana.

Pain, D.J. and C. Amiard-Triquet. 1993. Lead poisoning in raptors in France and elsewhere. Ecotoxicology and Environmental Safety 25: 183–192.

Pain, D.J. and M.W. Pienkowski (eds.). 1997. Farming and Birds in Europe: The Common Agricultural Policy and its Implications for Bird Conservation. Academic Press, London.

Pain, D.J., A.A. Cunningham, P.F. Donald, J.W. Duckworth, D.C. Houston, T. Katzner, J. Parry-Jones, C. Poole, V. Prakash, P. Round and R. Timmins. 2003. Causes and effects of temporospatial declines of Gyps vultures in Asia. Conservation Biology 17: 661–671.

Palacios, C.J. 2000. Decline of the Egyptian Vulture (*Neophron percnopterus*) in the Canary Islands. J. Raptor Res. 34: 61.

Palacios, C.J. 2000. Decline of the Egyptian vulture (*Neophron percnopterus*) in the Canary Islands. Journal of Raptor Research 34: 61.

Palacios, M.J. 2003. Power lines in Extremadura: conservation action and bird conservation. Proceedings of the National Conference on Power lines and Bird Conservation in Protected Areas, Dirección General de Medio Ambiente, Murcia.

Palacios, M.J. and M.J. García-Baquero. 2003. Power lines in Extremadura: Conservation and Protection of BirdLife (in Spanish). Proceedings of the 4th technical session on Power Lines and the Environment, Red Eléctrica de España, Madrid.

Palgrave, K.C. 1977. Trees of Southern Africa. C. Struik, Cape Town.

Palmqvist, P. and S.F. Vizcaíno. 2003. Ecological and reproductive constraints of body size in the gigantic Argentavis magnificens (Aves, Theratornithidae) from the Miocene of Argentina. Ameginiana 40(3): 379–385.

Parish, C.N., W.R. Heinrich and W.G. Hunt. 2007. Lead exposure, diagnosis, and treatment in California Condors released in Arizona. pp. 97–108. *In*: A. Mee, L.S. Hall and J. Grantham (eds.). California Condors in the 21st Century. American Ornithologists Union, McLean, VA, U.S.A.

Parish, C.N., W.R. Heinrich and W.G. Hunt. 2007. Lead exposure, diagnosis, and treatment in California Condors released in Arizona. pp. 97–108. *In*: A. Mee, L.S. Hall and J. Grantham (eds.). California Condors in the 21st Century. American Ornithologists Union, McLean.

Parker, V. 2004. The status of vultures in Mozambique. pp. 137–138. *In*: A. Monadjem, M.D. Anderson, S.E. Piper and A.F. Boshoff (eds.). The Vultures of Southern Africa—Quo Vadis? Birds of Prey Working Group, Johannesburg.

Parker, P. 1987. Recruitment to food in black vultures: evidence for following from communal roosts. Animal Behavior 35: 1775–1785.

Parker, P.G., T.A. Waite and M.D. Decker. 1995. Kinship and association in communally roosting black vultures. Animal Behavior 49: 395–401.

Parker, T.A., D.F. Stotz and J.W. Fitzpatrick. 1996. Ecological and distributional databases. pp. 113–436. *In*: D.F. Stotz, J.W. Fitzpatrick, T.A. Parker and D.K. Moskovits (eds.). Neotropical Bird Ecology and Conservation. University of Chicago Press, Chicago.

Parker, V. 1994. Swaziland Bird Atlas 1985–1991. Webster's, Mbabane.

Parker, V. 1999. The Atlas of the Birds of Sul do Save, Southern Mozambique. Avian Demography Unit and Endangered Wildlife Trust., Cape Town and Johannesburg.

Parker, V. 2005a. Endangered Wildlife Trust and Avian Demography Unit, Johannesburg, South Africa.

Parker, V. 2005b. The Atlas of the Birds of Central Mozambique. Avian Demography Unit and Endangered Wildlife Trust, Cape Town and Johannesburg.

Parmalee, P.W. 1969. California Condor and other birds from Stanton Cave, Arizonia. Journal of the Arizona Academy of Sciences 5: 204–206.

Parmenter, R.R. and J.A. MacMahon. 2009. Carrion decomposition and nutrient cycling in a semiarid shrub-steppe ecosystem. Ecological Monographs 79: 637–661.

Parra, J. and J.L. Telleria. 2004. The increase in the Spanish population of Griffon Vulture Gyps fulvus during 1989–1999: effects of food and nest site availability. Bird Conservation International 14: 33–41. DOI: 10.1017/S0959270903000000.

Patte, O.H. and S.K. Hennes. 1983. Bald eagles and waterfowl: the lead shot connection. Transactions of the North American Wildlife and Natural Resources Conference 48: 230–237.

Pattee, O., P. Bloom, J. Scott and M. Smith. 1990. Lead hazards within the range of the California Condor. Condor 92: 931–937.

Paudel, S. 2008. Vanishing vultures and diclofenac prevalence in Lumbini IBA. Danphe 17(2): 1–3.

Pavokovic, G. and G. Susic. 2006. Population viability analysis of (Eurasian) Griffon Vulture Gyps fulvus in Croatia. pp. 75–86. In: D.C. Houston and S.E. Piper (eds.). Proceedings of the International Conference on Conservation and Management of Populations. Thessaloniki, Greece. Natural History Museum of Crete & WWF Greece.

Payen, L.A. and R.E. Taylor. 1976. Man and Pleistocene fauna at Potter Creek Cave, California. Journal of California Anthropology 3(1): 51–8.

Paynter, W.P. 1924. Lesser White Scavenger Vulture N. ginginianus nesting on the ground.— Journal of the Bombay Natural History Society 30(1): 224–225.

Pearman, M. 2003. In BirdLife International 2014. Species factsheet: Vultur gryphus. Downloaded from http://www.birdlife.org on 21/02/2014.

Pearson, D.L., D. Tallman and E. Tallman. 1977. The Birds of Limoncocha. Instituto Linguistico, Quito.

Pearson, T.C. 1919. Turkey Vulture. Bird-Lore 21: 319–322.

Peet, R. and M. Watts. 1996. Liberation Ecologies: Environment, Development, Social Movements. Routledge, London.

Pendleton, E. 1978. To save raptors from electrocution. Defenders 53: 18–21.

Pennycuick, C.J. 1971. Gliding flight of the white-backed vulture Gyps africanus. Journal of Experimental Biology 55: 39–46.

Pennycuick, C.J. 1972. Soaring behaviour and performance of some east African birds, observed from a motor-glider. Ibis 114: 178–218.

Pennycuick, C.J. 1972b. Animal Flight. Edward Arnold, London.

Pennycuick, C.J. 1976. Breeding of the Lappet-faced and White-headed vultures (Torgos tracheliotus Forster and Trigonoceps occipitalis Burchell) on the Serengeti Plains, Tanzania. East African Wiidlife Journal 14: 67–84.

Pennycuick, C.J. 1983. Effective nest density of Rüppell's Griffon vulture in the Serengeti-Rift Valley Area of Northern Tanzania. pp. 172–184. In: S.R. Wilbur and J.A. Jackson (eds.). Vulture Biology and Management. University of California Press, Berkeley.

Pennycuick, C.J. 2008. Modelling the Flying Bird. Elsevier, Burlington.

Pennycuick, C.J. and K.D. Scholey. 1984. Flight behavior of Andean condors Vultur gryphs and turkey vultures Cathartes aura around the Paracas Peninsula, Peru. Ibis 126: 253–256. doi: 10.1111/j.1474-919x.1984.tb08005.x.

Pennycuick, C.J., T. Alerstam and B. Larsson. 1979. Soaring migration of the common crane Grus grus observed by radar and from an aircraft. Ornis Scandinavica 10: 241–51.

Pennycuick. 1972. Soaring behavior and performance of some East African birds, observed from a motor-glider. Ibis 144: 178–218.

Penry, H. 1994. Bird Atlas of Botswana. University of Natal Press, Pietermaritzburg.

Peregrine Fund. 1998. Peregrine Fund Annual Report.

Pérez de Ana, J.M. 2007. Los buitres de Sierra Salvada sufren la recogida de ganado muerto. Quercus 261: 30–32.

Perfecto, I., J.H. Vandermeer, G.L. Bautista, G.I. Nunez, R. Greenberg and P. Bichier. 2004. Greater predation in shaded coffee farms: The role of resident neotropical birds. Ecology 85: 2677–2681.

Perry, P. 2007. Lappetfaced Vulture attacking Golden Jackal at Gnu carcass Locality Serengeti, Tanzania. http://ibc.lynxeds.com/photo/lappet-faced-vulture-torgos-tracheliotus/lappetfaced-vulture-attacking-golden-jackal-gnu-carca. Accessed 30/03.2014.

Peters, J.L., E. Mayr and W. Cottrell. 1979. Check-list of Birds of the World. Museum of Comparative Zoology, Cambridge, Massachusetts.

Peters, J. and K. Schmidt. 2004. Animals in the symbolic world of Pre-Pottery Neolithic Göbekli Tepe, south-eastern Turkey: a preliminary assessment. Anthropozoologica 39(1): 179–218.

Peters, J.L. 1931. Check-list of the Birds of the World Volume 1. Harvard University Press, Cambridge.

Peterson, A.P. 2007. Richmond Index—Genera Aaptus—Zygodactylus. The Richmond Index. Division of Birds at the National Museum of Natural History.

Peterson, R.O. and P. Ciucci. 2003. The wolf as a carnivore. *In*: L.D. Mech and L. Boitani (eds.). Wolves Behavior, Ecology and Conservation. The University of Chicago Press, Chicago.

Peterson, R.T. 2001. A Field Guide to Western Birds. Houghton Mifflin Field Guides.

Pettersson, J. and T. Stalin. 2003. Influence of Offshore Windmills on Migration Birds in Southeast Coast of Sweden. Report to GE Wind Energy, Fairfield, Connecticut.

Pettersson, J. and T. Stalin. 2003. Influence of offshore windmills on migration birds in southeast coast of Sweden. Report to GE Wind Energy. Fairfield.

Phillips, R.L. 1986. Current issues concerning the management of Golden Eagles in Western U.S.A. pp. 149–156. *In*: R.D. Chancellor and B.-U. Meyburg (eds.). Birds of Prey Bulletin No. 3. Proceedings of the Western Hemisphere Meeting of the World Working Group on Birds of Prey, Berlin, Germany.

Phillips, S.J. and P.W. Comus. 2000. A Natural History of the Sonoran Desert. University of California Press, Berkeley.

PIB (Press Information Bureau). 2005. Saving the Vultures from Extinction (Press release). Press Information Bureau, Government of India. 2005-05-16. Retrieved 2006-05-12.

Pickford, B. and P. Pickford. 1989. Cape Vulture. Southern African Birds of Prey. Struik Publishers, Cape Town.

Pienaar, U.V. 1969. Predator-prey relationships amongst the larger mammals of the Kruger National Park. Koedoe 12: 108–176.

Pinto, O. 1965. Dos frutos da palmeira Elaeis guineenszs na dieta de Cathartes aura ruficollis. Hornero 10: 276–277.

Pinto, O.M.O. 1938. Catalogo das aves do Brasil e lista dos exemplares que as representam no Museu Paulista. Revista do Museu Paulista 22: 1–566.

Piper, S.E. 1994. Mathematical Demography of the Cape Vulture. MSc. Thesis. Witwatersrand University.

Piper, S.E. 1994. Mathematical demography of the Cape Vulture *Gyps coprotheres*. PhD thesis. University of Cape Town, Cape Town.

Piper, S.E. 2004a. Vulture restaurants. pp. 220–227. *In*: A. Monadjem, M.D. Anderson, S.E. Piper and A.F. Boshoff (eds.). Vultures in the Vultures of Southern Africa—Quo Vadis? Proceedings of a workshop on vulture research and conservation in southern Africa. Birds of Prey Working Group, Johannesburg.

Piper, S.E. 2004b. Vulture restaurants—conflict in the midst of plenty. pp. 341–349. *In*: R.D. Chancellor and B.-U. Meyburg (eds.). Raptors Worldwide. World Working Group on Birds of Prey and Owls, Budapest.

Piper, S.E. 2006. Supplementary feeding programmes: how necessary are they for the maintenance of numerous and healthy populations. pp. 41–50. *In*: D.C. Houston and S.E. Piper (eds.). Proceedings of the International Conference on Conservation and Management of Vulture Populations. Natural History Museum of CreteThessaloniki.

Piper, S.E., A.F. Boshoff and H.A. Scott. 1999. Modelling survival rates in the Cape Griffon Gyps coprotheres, with emphasis on the effects of supplementary feeding. Bird Study 46(Supplement): S230–238.

Piper, S.E., P.J. Mundy and J.A. Ledger. 1981. Estimates of survival in the Cape Vulture Gyps corprotheres. Journal of Animal Ecology 50: 815–825.

Pitman, R.S. 1960. An unusual case of predation by Aquila verreauxi. Bulletin of the British Ornithological Club 80: 67.

Plutarch. 1952. The Lives of the Noble Grecians and Romans, translated by John Dryden (1631–1700) Great Books of Western World Vol. 14. William Benton, Chicago.

Poharkar, A., P.A. Reddy, V.A. Gadge, S. Kolte, N. Nurkure and S. Shivaji. 2009. Is malaria the cause for decline in the Indian White-backed Vulture (*Gyps bengalensis*)? Current Science 96(4): 553–558.

Pollo, C.J., L. Robles, A. García-Miranda, R. Otero and J.R. Obeso. 2003. Variaciones en la densidad y asociaciones espaciales entre ungulados silvestres y urogallo cantábrico. Ecología 17: 199–206.

Pomeroy, D.E. 1975. Birds as scavengers of refuse in Uganda. Ibis 117: 69–81.

Poole, A.F. 1994. Family Pandionidae (Osprey). *In*: J. del Hoyo, A. Elliott and J. Sargatal (eds.). Handbook of Birds of the World. Vol 2. New World Vultures to Guinea Fowl, 42–51. Lynx Edicions, Barcelona.

Porter, R. and A.S. Suleyman. 2012. The Egyptian vulture *Neophron percnopterus* on Socotra, Yemen: population, ecology, conservation and ethno-ornithology. Sandgrouse 34.

Poulakakis, N., A. Antoniou, G. Mantziou, A. Parmakelis, T. Skartsi, D. Vasilakis, J. Elorriaga, J. De La Puente, A. Gavashelishvili, M. Ghasabyan, T. Katzner, M. McGrady, N. Batbayar, M. Fuller and T. Natsagdor. 2008. Population structure, diversity, and phylogeography in the near-threatened Eurasian black vultures *Aegypius monachus* (Falconiformes; Accipitridae) in Europe: insights from microsatellite and mitochondrial DNA variation. Biological Journal of the Linnean Society, pp. 1–14.

Pounds, J.A., M.P.L. Fogden and J.H. Campbell. 1999. Biological response to climate change on a tropical mountain. Nature 398: 611–615.

Powers, S. 1877. Tribes of California. Contributions to North American Ethnology 3. U.S. Geographical and Geological Survey of the Rocky Mountain Region, Washington.

Prakash, V., R.E. Green, D.J. Pain, S.P. Ranade, S. Saravanan et al. 2007. Recent changes in populations of resident Gyps vultures in India. Journal of the Bombay Natural History Society 104: 129–135.

Prakash, V. 1989. The general ecology of raptors (Families: Accipitridae, Strigidae, Class: Aves) in Keoladeo National Park, Bharatpur. Ph.D. thesis. Bombay Natural History Society, Bombay University, Mumbai.

Prakash, V. 1999. Status of vultures in Keoladeo National Park, Bharatpur, Rajasthan with special reference to population crash in Gyps species. Journal of the Bombay Natural History Society 96: 365–378.

Prakash, V. and A.R. Rahmani. 1999. Notes about the decline of Indian Vultures, with particular reference to Keoladeo National Park. Vulture's News 41: 6–13.

Prakash, V., M.C. Bishwakarma, A. Chaudhary, R. Cuthbert and R. Dave. 2012. The population decline of Gyps Vultures in India and Nepal has slowed since veterinary use of Diclofenac was banned. PLoS One 7(11): e49118. doi:10.1371/journal.pone.0049118.

Prakash, V., D.J. Pain, A.A. Cunningham et al. 2003. Catastrophic collapse of Indian white-backed *Gyps bengalensis* and long-billed *Gyps indicus* vulture populations. Biological Conservation 109: 381–390.

Prakash, V., R.E. Green, D.J. Pain, S.P. Ranade, S. Saravanan and N. Prakash. 2007. Recent changes in populations of resident Gyps vultures in India. Journal of the Bombay Natural History Society 104(2): 127–133.

Prakash, V., R.E. Green, D.E. Pain, S.P. Ranade, S. Saravanan, N. Prakash, R. Venkitachalam, R. Cuthbert, A.R. Rahmani and A.A. Cunningham. 2007. Recent changes in populations of resident Gyps vultures in India. Journal of the Bombay Natural History Society 104: 129–135.

Prakash, V., D.J. Pain, A.A. Cunningham, P.F. Donald, N. Prakash, A. Verma, R. Gargi, S. Sivakumar and A.R. Rahmani. 2003. Catastrophic collapse of Indian white-backed *Gyps bengalensis* and longbilled *Gyps indicus* vulture populations. Biological Conservation 109: 381–390.

Prange, S. and S.D. Gehrt. 2004. Changes in mesopredator community structure in response to urbanization. Canadian Journal of Zoology 82: 1804–1817.

Principado de Asturias. 2007. Plan de caza de las Reservas Regionales de Caza del Principado de Asturias 2007–2008. Informe inédito.

Prinsen, H.A.M., G.C. Boere, N. Píres and J.J. Smallie. 2011. Review of the conflict between migratory birds and electricity power grids in the African-Eurasian region. CMS Technical Series No. XX, AEWA Technical Series No. XX Bonn.

Priolo, A. 1967. Distrutti i Grifoni delle Caronie? Rivista Italiana di Ornitologia Bologna 37: 7–11.

Prostov, A. 1955. New data for the avifauna of the Bulgarian Black sea coast. Bulletin of the Zoological Institute Sofia 4-5: 451–460.

Prugh, L.R., C.J. Stoner, C.W. Epps, W.T. Bean, W.J. Ripple and A.S. Laliberte. 2009. The rise of the mesopredator. Bioscience 59: 779–791.

Putman, R.J. 1983. Carrion and Dung: The Decomposition of Animal Wastes. Edward Arnold, London.

Pyle, P. and S.N.G. Howell. 1993. An aggregation of Peregrine Falcons on the Mexican winter grounds. Euphonia 2: 92–94.

Rabenold, P.P. and M.D. Decker. 1990. Black vultures in North Carolina: statewide population surveys and analysis of Chatham County population trends. Final report, North Carolina Wildlife Resources Commission, Nongame and Endangered Wildlife Program, Contract 1989SG07, Raleigh.

Rabenold, P.P. 1983. The communal roost in black and turkey vultures—an information center? pp. 303–321. *In*: S.R. Wilbur and J.R. Jackson (eds.). Vulture Biology and Management. University of California Press, Berkley.

Rabenold, P.P. and M.D. Decker. 1989. Black and turkey vultures expand their ranges northward. Eyas 2: 11–5.

Ragg, J.R., C.G. Mackintosh and H. Moller. 2000. The scavenging behaviour of ferrets (*Mustelo furo*), feral cats (*Felis domesticus*), possums (*Trichosurus vulpecula*), hedgehogs (*Erinaceus europaeus*) and harrier hawks (*Circus approximans*) on pastoral farmland in New Zealand: implications for bovine tuberculosis transmission. New Zealand Veterinary Journal 48: 166–175.

Rainey, H. 2008. In BirdLife International 2014. Species factsheet: *Sarcogyps calvus*. Downloaded from http://www.birdlife.org on 15/03/2014.

Ralls, K. and J.D. Ballou. 2004. Genetic status and management of California Condors. Condor 106: 215–228.

Rahmani, A. 1998. A possible decline of vultures in India. Oriental Bird Club Bulletin 28: 40–41.

Rand, A.L. and R.L. Fleming. 1957. Birds of Nepal. Fieldiana Zoology 41: 1–218.

Rapp, W.F. 1943. Turkey vulture feeding habits. Auk 60: 95.

Rapp, W.F. 1943. Turkey Vulture feeding habits. Auk 60: 95.

Rasmussen, P.C. and S.J. Parry. 2000. On the specific distinctness of the Himalayan Long-billed Vulture Gyps [indicus] tenuirostris. Newsletter of the World Working Group on Birds of Prey and Owls 29/32: 70–71.

Rasmussen, P.C. and S.J. Parry. 2000. On the specific distinctness of the Himalayan Longbilled Vulture *Gyps [indicus] tenuirostris*. 118th Meeting of the American Ornithologists' Union (AOU), Memorial University of Newfoundland, St. John's, Newfoundland.

Rasmussen, P.C. and J.C. Anderton. 2005. Birds of South Asia: The Ripley Guide. Volume 2. Smithsonian Institution and Lynx Edicions, Washington DC and Barcelona.

Rasmussen, P.C. and S.J. Parry. 2000. On the specific distinctness of the Himalayan Longbilled Vulture *Gyps [indicus] tenuirostris*, 16. 118th stated meeting of the American Ornithologists' Union (AOU). AOU, Memorial University of Newfoundland, St. John's, Newfoundland.

Rasmussen, P.C. and S.J. Parry. 2001. The taxonomic status of the 'Longbilled' Vulture *Gyps indicus*. Vulture News 44: 18–21.

Rasmussen, P.C., W.S. Clark, S.J. Parry and J. Schmitt. 2001. Field identification of 'Long-billed' Vultures (Indian and Slender-billed Vultures). Oriental Bird Club Bulletin 34: 24–29.

Ratcliffe, D.A. 1967a. Decrease in eggshell weight in certain birds of prey. Nature 215: 208–210.

Ratcliffe, D.A. 1967b. The peregrine situation in Great Britain: 1965–1966. Bird Study 14: 238–246.

Rattner, B.A., M.A. Whitehead, G. Gasper, C.U. Meteyer, W.A. Link, M.A. Taggart, A.A. Meharg, O.H. Pattee and D.J. Pain. 2008. Apparent tolerance of turkey vultures (*Cathartes aura*) to the non-steroidal anti-inflammatory drug diclofenac. Environmental Toxicology and Chemistry 27: 2341–2345.

Rawn-Schatzinger, V. 1992. The scimitar cat Homotherium serum Cope. Osteology, functional morphology, and predator behavior. Illinois State Museum Reports of Investigation 47: 1–80.

Razin, M., I. Rebours and C. Arthur. 2008. Le Vautour fauve Gyps fulvus dans les Pyrénées françaises: status récent et tendance. Ornithos 5-6: 385–393.

Rea, A.M. 1980. New World vultures: diminishing and misunderstood. Environment Southwest 12–13.

Rea, A.M. 1983. Cathartid affinities: a brief overview. pp. 26–54. In: S.R. Wilbur and J.A. Jackson (eds.). Vulture Biology and Management, University of California, Berkeley.

Reading, R.P., S. Amgalanbaatar, D. Kenny and B. Dashdemberel. 2005. Cinereous Vulture Nesting Ecology in Ikh Nartyn Chuluu Nature Reserve, Mongolia. Mongolian Journal of Biological Sciences 3(1): 13–19.

Reading, R.P. and B. Miller. 2000. Endangered Animals: A Reference Guide to Conflicting Issues. Greenwood Press, Westport.

Regidor, S., C. Santos, M. Ferrer and J.J. Negro. 1988. An experiment with modified electric pylons in Doñana National Park. Ecología 2: 251–256.

Reid, S.B. 1989. Flying behaviour and habitat preferences of the king vulture Sarcorarnphus papa in the western Orinoco Basin of Venezuela. Ibis 131: 301–303.

Reilly, J.R. and R.J. Reilly. 2009. Bet-hedging and the orientation of juvenile passerines in fall migration. Journal of Animal Ecology 78: 990–1001.

Reiser, O. 1905. Materialien zu einer Omis Balcanica III. Griechenland.

Reiter, A.S. 2000. Casualties of Great Bustards (*Otis tarda* L.) on overhead power lines in the western Weinviertel (Lower Austria). Egretta 43: 37–54.

Remsen, J.V., Jr., C.D. Cadena, A. Jaramillo, M. Nores, J.F. Pacheco, M.B. Robbins, T.S. Schulenberg, F.G. Stiles, D.F. Stotz and K.J. Zimmer. 2007. A classification of the bird species of South America. American Ornithologists' Union, Bato, C.A.

Remsen, J.V., Jr. (Chair). 2008. A Classification of the Bird Species of South America (Part 2). South American Classification Committee, American Ornithologists' Union, Baton Rouge, LA.

Restall, R., C. Rodner and M. Lentino. 2006. Birds of Northern South America: An Identification Guide. Vol. 1. Christopher Helm/A&C Black Publishers Ltd., London.

Reumer, J.W.F., L. Rook, K. Van Der Borg, K. Post, D. Mol and J. De Vos. 2003. Late Pleistocene survival of the saber-toothed cat Homotherium in northwestern Europe. Journal of Vertebrate Paleontology 23: 260–262.

Rexer-Huber, K. and K.L. Bildstein. 2012. Winter diet of striated caracara Phalcoboenus australis (Aves, Polyborinae) at a farm settlement on the Falkland Islands. Polar Biology DOI 10.1007/s00300-012-1266-4.

Rhys, D. 1980. Argentavis magnificens: World's Largest Flying Bird. Origins 7(2): 87–88.

Rich, P.V. 1980. New World vultures with Old World affinities? Contributions to Vertebrate Ecology 5: 1–115.

Rich, T.D., C.J. Beardmore, H. Berlanga, P.J. Blancher, M.S.W. Bradstreet, G.S. Butcher, D.W. Demarest, E.H. Dunn, W.C. Hunter, E.E. Inigo-Elias, J.A. Kennedy, A.M. Martell, A.O. Panjabi, D.N. Pashley, K.V. Rosenburg, C.M. Rustay, J.S. Wendt and T.C. Will. 2004. Partners in Flight North American Landbird Conservation Plan. Cornell Laboratory of Ornithology, Ithaca.

Richards, D.K. 1982. The birds of Conakry and Kakulima, Democratic Republic of Guinee. Malimbus 4: 93–104.

Richards, S.A. 2008. Dealing with overdispersed count data in applied ecology. Journal of Applied Ecology 45: 218–227.

Richardson, W.J. 1978. Timing and amount of bird migration in relation to weather: a review. Oikos 30: 224–72.

Richardson, W.J. 1990. Timing of bird migration in relation to weather: updated review. pp. 78–101. *In*: E. Gwinner (ed.). Bird Migration. Springer-Verlag, Berlin.

Richardson, W.J. 1994. Serious birdstrikes-related accidents to Military aircraft of ten countries: Preliminary analysis of circumstances. Proceedings of the BSCE 22, 129–152, Vienna. 29 August to 2 September. WP 21.

Richardson, W.J. 1996. Serious Birdstrike-related Accidents to Military Aircraft of Europe and Israel. Lists and Analysis of Circumstances. Proceedings of the IBSC 23, 33–56. London. 13–17 May WP-2.

Richford, A.S. 1976. Black Vultures in Mallorca. Oryx 13(4): 383–386.

Richman, A.D. and T. Price. 1992. Evolution of ecological differences in the Old World leaf warblers. Nature 355: 817–821.

Rideout, B.A., I. Stalis, R. Papendick, A. Pessier, B. Puschner, M.E. Finkelstein, D.R. Smith, M. Johnson, M. Mace, R. Stroud, J. Brandt, J. Burnett, C. Parish, J. Petterson, C. Witte, C. Stringfield, K. Orr, J. Zuba, M. Wallace and J. Grantham. 2012. Patterns of mortality in free-ranging California Condors (Gymnogyps californianus). Journal of Wildlife Diseases 48(1): 95–112.

Ridgely, R.S. and P. Greenfield. 2001. The Birds of Ecuador. Cornell University Press, Ithaca.

Ridgely, R.S. and J.A. Gwynne. 1989. A Guide to the Birds of Panama: with Costa Rica, Nicaragua, and Honduras. Princeton University Press, Princeton.

Rincón, A., F. Prevosti and G. Parra. 2011. New saber-toothed cat records (Felidae: Machairodontinae) for the Pleistocene of Venezuela, and the Great American Biotic Interchange. Journal of Vertebrate Paleontology 31(2): 468–478.

Rincon, P. 2008. Big cat fossil found in North Sea. BBC News 18 November 2008 accessed 18 November 2008.

Ríos-Uzeda, B. and R.B. Wallace, 2007. Estimating the size of the Andean Condor population in the Apolobamba Mountains of Bolivia. Journal of Field Ornithology 78(2): 170–175.

Ristow, D. 2003. Can the sense of smell in the Turkey Vulture be put to any conservation use? International Hawkwatcher 7: 18–19.

Ritchie, E.G. and C.N. Johnson. 2009. Predator interactions, mesopredator release and biodiversity conservation. Ecology Letters 12: 982–998.

Ritter, L.V. 1983. Growth, development, and behavior of nestling Turkey Vultures in Central California. pp. 287–302. *In*: S.R. Wilbur and J.A. Jackson (eds.). Vulture Biology and Management. Univ. of California Press, Berkeley.

Rivers, I.L. 1941. The Mormon cricket as food for birds. Condor 43: 65–69.

Rivers, J.W., J.M. Johnson, S.M. Haig, C.J. Schwarz, J.W. Glendening, L.J. Burnett, D. George and J. Grantham. 2014. Resource selection by the California Condor (Gymnogyps californianus) relative to terrestrial-based habitats and meteorological conditions. PLoS One 9(2): e88430. doi: 10.1371/journal.pone.0088430.

Roach, J. 2004. Peru's Andean Condors are rising tourist attraction. National Geographic News 07-22-2004. National Geographic Society, Washington D.C.

Roberts, T.J. 1991–1992. The Birds of Pakistan. Oxford University Press, Karachi.

Robinson, S.K. 1994. Habitat selection and foraging ecology of raptors in Amazonian Peru. Biotropica 26: 443–458.

Robson, C. 2000. A Guide to the Birds of Southeast Asia. Princetown University Press, Princeton, New Jersey.

Robson, C.R., H. Buck, D.S. Farrow, T. Fisher and B.F. King. 1998. A birdwatching visit to the Chin Hills, West Burma (Myanmar), with notes from nearby areas. Forktail 13: 109–120.

Roche, C. 2006. Breeding records and nest site preference of Hooded Vultures in the greater Kruger National Park. Ostrich 77: 99–101.

Rock, P. 2005. Urban gulls: problems and solutions. British Birds 98: 338–355.

Rodriguez-Estella, R. and L. Rivera-Rodriguez. 1992. Kleptoparasitism and other interactions of Crested Caracara in the Cape Region, Baja California, Mexico. Journal of Field Ornithology 63(2): 177–180.

Rodriguez-Ramos, J., V. Gutierrez, U. Höfle, R. Mateo, L. Monsalve, E. Crespo and J.M. Blanco. 2009. Lead in Griffon and Cinereous Vultures in Central Spain: Correlations between clinical signs and blood lead levels. Extended abstract in Watson, R.T., Fuller, M., Pokras, M. and Hunt, W.G. (eds.) Ingestion of Lead from Spent Ammunition: Implications for Wildlife and Humans, 236. The Peregrine Fund, Boise. DOI 10.4080/ilsa.2009.0213.

Roelke-Parker, M.E. et al. 1996. A canine distemper outbreak in Serengeti lions (Panthera leo). Nature 379: 441–445.

Roelke-Parker, M.E., L. Munson, C. Packer, R. Kock, S. Cleaveland, M. Carpenter, S.J. O'Brien, A. Pospischil, R. Hofmann-Lehmann, H. Lutz, G.L. Mwamengele, M.N. Mgasa, G.A. Machange, Summers, B.A. and Appel, M.J. 1996. A canine distemper virus epidemic in Serengeti lions (Panthera leo). Nature 379: 441–445.

Rondeau, G. and J.M. Thiollay. 2004. West African vulture decline. Vulture News 51: 13–33.

Rondeau, G. 2008. Tagged vultures in Fouta Djallon, Guinea. Vulture News 58: 56.

Rondeau, G., M.M. Condeé, B. Ahon, O. Diallo and D. Pouakouyou. 2008. Survey of the occurrence and relative abundance of raptors in Guinea subject to international trade. JNCC Report no. 412. Joint Nature Conservation Conmmittee, Peterborough.

Rondeau, G., P. Pilard, B. Ahon and M. Conde. 2006. Tree-nesting Rüppell's Griffon Vultures. Vulture News 55: 14–22.

Rose, M.D. and G.A. Polis. 1998. The distribution and abundance of coyotes: The effects of allochthonous food subsidies from the sea. Ecology 79: 998–1007.

Round, P. 2006. BirdLife International 2014. Species factsheet: Sarcogyps calvus. Downloaded from http://www.birdlife.org on 04/02/2014.

Round, P.D. 1988. Resident forest birds in Thailand. Monograph 2. International Council for Bird Preservation, Cambridge.

Round, P.D. 2000. Field check-list of Thai birds. Bird Conservation Society of Thailand, Bangkok.

Round, P.D. and V. Chantrasmi. 1985. A status report on birds of prey in Thailand. Proceedings of the Third East Asian Bird Protection Conference 291–297.

Rowley, J.S. 1984. Breeding records of land birds in Oaxaca, Mexico. Proceedings of the Western Foundation for Vertebrate Zoology 2: 73–224.

Rubolini, D., M. Gustin, G. Bogliani and R. Garavaglia. 2005. Birds and powerlines in Italy: an assessment. Bird Conservation International 15(2): 131–145.

Rudnick, J.A., T.E. Katzner, E.A. Bragin and J.A. Dewoody. 2008. A noninvasive genetic evaluation of population size, natal philopatry, and roosting behavior of non-breeding eastern Imperial Eagles (Aquila heliaca) in central Asia. Conservation Genetics 9: 667–676.

Russell, S.M. 1964. A distributional study of the birds of British Honduras. Ornithological Monographs 1: 1–195.

Rutledge, L.Y., B.N. White, J.R. Row and B.R. Patterson. 2012. Intense harvesting of eastern wolves facilitated hybridization with coyotes. Ecology and Evolution 2(1): 19–33.

Ruxton, G.D. and D.C. Houston. 2004. Obligate vertebrate scavengers must be large soaring fliers. Journal of Theoretical Biology 228: 431–436.

Ryser, F.A. 1985. Birds of the Great Basin: A Natural History. University of Nevada Press, Reno.

Saikia, P. and P.C. Bhattacharjee. 1990. Indian Longbilled Vulture nesting in Assam. Newsletter for Birdwatchers 30(1-2): 9.

Saitou, N. and M. Nei. 1987. The neighbor-joining method, a new method for reconstructing phylogenetic trees. Molecular Biology and Evolution 4: 406–425.

Saleem, A., M. Husheem, P. Härkönen and K. Pihlaja. 2002. Inhibition of cancer cell growth by crude extract and the phenolics of Terminalia chebula retz. Fruit. Journal of Ethnopharmacology 81(3): 327–336.

Salesa, M. et al. 2012. A rich community of Felidae (Mammalia, Carnivora) from the late Miocene (Turolian, MN 13) site of Las Casiones (Villalba Baja, Teruel, Spain). Journal of Vertebrate Paleontology 32(3): 658–676.

Salisbury, C. 2013. 600 vultures killed by elephant poachers in Namibia. http://news.mongabay.com/2013/0910-salisbury-vultures-killed-by-elephant-poachers.html. Accessed 05/03/2014.

Salvadori, T. and E. Festa. 1900. Not Cathartes burrovianus Cassin 1845 (Rio Peripa, Ecuador). Bollettino del Musei di Zoologia ed Anatomia Comparata, Torino 15(368): 27.

Sanchez, F. 2008. Saber-toothed cat fossils discovered in Venezuela. Associated Press 22 August 2008. Accessed 22 August 2008.

Sapir, N. 2009. The effect of weather on Bee-eater (Merops apiaster) migration. Hebrew University, Jerusalem.

Sarà, M. and M. Di Vittorio. 2003. Factors influencing the distribution, abundance and nest-site selection of an endangered Egyptian vulture (*Neophron percnopterus*) population in Sicily. Animal Conservation 6(4): 317–328.

Sara, M., S. Grenci and M. Di Vittorio. 2009. Status of Egyptian Vulture (Neophron percnopterus) in Sicily. Journal of Raptor Research 43(1): 66–69.

Sarker, S.U. and N.J. Sarker. 1985. Migratory raptorial birds of Bangladesh. pp. 291–293. *In*: I. Newton and R.D. Chancellor (eds.). Conservation Studies on Raptors. International Council for Bird Preservation, Cambridge.

Sarrazin, F., C. Bagnolini, J.-L. Pinna, E. Danchin and J. Clobert. 1994. High survival estimates of griffon vultures (Gyps fulvus fulvus) in a reintroduced population. The Auk 111: 853–862.

Satheesan, S.M. 1989a. Birds at vulture feeding sites in Agra. Vulture News 21: 25.

Satheesan, S.M. 1989b. On the differences in feeding between Scavenger and Indian Whitebacked vultures. Vulture News 22: 49–50.

Satheesan, S.M. 1989c. Behaviour of the Indian Whitebacked Vulture in the presence of man. Vulture News 22: 52–53.

Satheesan, S.M. 1990. Scavengers on the wing. Sanctuary Asia X(4): 26–37.

Satheesan, S.M. 1992a. Solutions to the Kite Hazard at Indian Airports. ICAO Journal of International Civil Aviation Organization.

Satheesan, S.M. 1992b. An updated list of birds and bat species involved in collision with aircraft in India. Journal of the Bombay Natural History Society 89(1): 129–132.

Satheesan, S.M. 1992c. Bird and bat collisions with aircraft. Journal of the Bombay Natural History Society 89(3): 379–380.

Satheesan, S.M. 1992d. Ecology and behaviour of Pariah Kite Milvus migrans govinda Sykes as a problem bird at some Indian aerodromes. Ph.D. Thesis. University of Bombay. Bombay.

Satheesan, S.M. 1994. The more serious vulture hits to Military Aircraft in India between 1980 and 1994. Proc. 22 BSCE Meeting Vienna, Austria, 163–168. 29 August–2 September, WP 23.

Satheesan, S.M. 1996. Raptors Associated with Airports and Aircraft. pp. 315–323. *In*: D. Bird, D. Varland and J. Negro (eds.). Raptors in Human Landscapes. Academic Press, U.K. and Raptor Research Foundaton.

Satheesan, S.M. 1998a. Sighting of the Bearded Vulture *Gypaetus barbatus* at Katra, near Jammu, India. Vulture News 38: 23.

Satheesan, S.M. 1998b. The role of vultures in the disposal of the human corpses in India and Tibet. Vulture News 39: 32–33.

Satheesan, S.M. 1998c. Need for imparting training at national level to bird controllers at Civil and Military Airports. Proceedings of the IBSC 24, 3–8. Stara Lesna, Slovakia, 14–18 September. WP1.

Satheesan, S.M. 1998d. Sound and light can control bird activity at Indian Airports. Proceedings of the IBSC 24, 223–226. Stara Lesna, Slovakia, 14–18 September. WP19. Ibid, WP 19.

Satheesan, S.M. 1998e. Vultures in Asia Proceedings of the fifth Meeting of the World Working Group on Birds of Prey and Owls (WWGBP), Midrand, South Africa, August 4–11, 1998.

Satheesan, S.M. 1998f. Vulture-eating communities in India. Vulture News 41: 15–17.

Satheesan, S.M. 1999a. The vanishing skylords. WWF-India Network Newsletter 9(4): 13–18. WWF-India, New Delhi.

Satheesan, S.M. 1999b. The decline of vultures in India. Vulture News 40: 35–36.

Satheesan, S.M. 1999c. Zero Bird Strike Rate—An Achievable Target, Not a Pipedream! Proceedings of the Bird Strike 1999, the First Combined Meeting of the BSCUSA and BSCC, Richmond, Vancouver, Canada. May 9–13, 1999.

Satheesan, S.M. 2000a. The role of poisons in the Indian vulture population crash. Vulture News 42: 3–4.

Satheesan, S.M. 2000b. Vultures in Asia. pp. 165–174. In: R. Chancellor and B.-U. Meyburg (eds.). Raptors at Risk. Hancock House, Berlin/Blaine.

Satheesan, S.M. and M. Satheesan. 2000. Serious vulture-hits to aircraft over the world. International Bird Strike Committee IBSC25/WP-SA3. IBSC, Amsterdam.

Sauer, E.G.F. 1973. Notes on the behaviour of Lappetfaced Vultures and Cape Vultures in the Namib Desert of South West Africa. Madoqua II 2: 43–62

Sauer, J.R., J.E. Hines and J. Fallon. 2005. The North American Breeding Bird Survey, results and analysis, 1966–2005. Version 6.2.2006. U.S. Geological Survey, Patuxent Wildlife Research Center, Laurel.

Sauer, J.R., S. Schwartz and B. Hoover. 1996. The Christmas Bird Count home page. Version 95.1. Patuxent Wildlife Research Center, Laurel, Maryland, USA. http://www.mbr-pwr.usgs.gov/bbs/cbc.html/. Accessed 12 Dec 2013.

SAVE. 2012. Report from the 1st meeting of Saving Asia's Vultures from Extinction (SAVE).

Sayad, A. 2007. Causes of extinction and the decline of birds in Morocco]. Ostrich 78: 170.

Sayles, I. 1887. The sense of smell in Cathartes aura. Auk 4: 51–56.

Schaefer, E. 1938. Ornithologisch Eergebnisse Zweier Forsc-hungrseisen nach Tibet. Journal of Ornitology 86: 1–79.

Schaller, G.B. 1972. The Serengeti Lion. University of Chicago Press, Chicago.

Schalock, R.L., M.A.V. Alonso and D.L. Braddock. 2002. Handbook on Quality of Life for Human Service Practitioners. American Association on Mental Retardation, Washington, D.C.

Schalock, R.L. 1997. Can the concept of quality of life make a difference? pp. 245–267. In: R.L. Schalock (ed.). Quality of Life Volume II: Application to Persons with Disabilities. American Association on Mental Retardation, Washington D.C.

Schalock, R.L. 2000. Three decades of quality of life. Focus on Autism and Other Developmental Disabilities 15(2): 116–128.

Schaltegger, S. and U. Beständig. 2012. Corporate Biodiversity Management Handbook: A Guide for Practical Implementation. Federal Ministry of the Environnment, Berlin.

Schaub, M., A. Aebischer, O. Gimenez, S. Berger and R. Arlettaz. 2010. Massive immigration balances high anthropogenic mortality in a stable eagle owl population: Lessons for conservation. Biological Conservation 143: 1911–1918.

Schaub, M., F. Liechti and L. Jenni. 2004. Departure of migrating European robins, Erithacus rubecula, from a stopover site in relation to wind and rain. Animal Behaviour 67: 229–237.

Schenk, W.E. and E.J. Dawson. 1929. Archaeology of the northern San Joaquin Valley. University of California Publications in American Archaeology and Ethnology 25: 289–413.

Schlee, M. 1995. Nest records for the king vulture (*Sarcoramphus papa*) in Venezuela. Journal of Raptor Research 29(4): 269–272.

Schlee, M. 2005. King vultures (*Sarcoramphus papa*) forage in moriche and cucurit palm stands. Journal of Raptor Research 39(4): 458–61.

Schlee, M.A. 1988. Breeding Rüppell's griffon vulture Gyps rueppellii at the Paris Menagerie. International Zoo Yearbook 27(1): 252–257.

Schlee, M.A. 1989. Breeding the Himalayan griffon *Gyps himalayensis* at the Paris menagerie. International Zoo Yearbook 28: 234–240.

Schlee, M.A. 1995. Nest records for the King Vulture (*Sarcorarnphus papa*) in Venezuela. Journal of Raptor Research 29(4): 269–272.

Schlee, M.A. 2000. The status of vultures in Latin America. pp. 191–206. *In*: R.D. Chancellor and B.U. Meyburg (eds.). Raptors at Risk. WWGBP/Hancock House, London.

Schmaljohann, H., F. Liechti and B. Bruderer. 2009. Trans-Sahara migrants select flight altitudes to minimize energy costs rather than water loss. Behavioral Ecology and Sociobiology 63: 609–619.

Schochat, E., P.S. Warren and S.H. Faeth. 2006. Future directions in urban ecology. Trends in Ecology and Evolution 21: 661–662.

Schoener, T.W. 1971. Theory of feeding strategies. Annual Review of Ecological Systems 2: 369–404.

Scholte, P. 1998. Status of vultures in the Lake Chad Basin, with special reference to Northern Cameroon and Western Chad. Vulture News 39: 3–19.

Schulenberg, T.S. (ed.) 2010. Greater Yellow-headed Vulture (*Cathartes melambrotus*), Neotropical Birds Online). Cornell Lab of Ornithology, Ithaca. Retrieved from Neotropical Birds Online: http://neotropical.birds.cornell.edu/portal/species/overview?p_p_spp=118556.

Schultz, P. 2007. Does bush encroachment impact foraging success of the critically endangered Namibian population of the Cape Vulture *Gyps coprotheres*? MSc thesis, University of Cape Town.

Schulz, P.D. and D.D. Simons. 1973. Fish species diversity in prehistoric central California Indian midden. California Fish and Game 59: 107–113.

Schürenberg, B., R. Schneider and H. Jerrentrup. 2010. Implementation of recommendation No. 110/2004 on minimising adverse effects of above ground electricity transmission facilities (power lines) on birds. Report by the NGOs to the 30th meeting of the Standing Committee of the Berne Convention, Strasbourg. T-PVS/Files (2010) 21. Council of Europe, Strasbourg.

Schüz, E. and C. König. 1983. Old world vultures and man. pp. 461–469. *In*: S.R. Wilbur and J.A. Jackson (eds.). Vulture Biology and Management. University of California Press, Berkely.

Sclater, P.L. and O. Salvin. 1859. On the ornithology of Central America. Part. III. Ibis I: 213–234.

Scott, M.P. 1998. The ecology and behavior of burying beetles. Annual Review of Entomology 43: 595–618.

Scully, J. 1879. A contribution to the ornithology of Nepal. Stray Feathers 8: 204–368.

Seamans, T.W. 2004. Response of roosting turkey vultures to a vulture effigy. Ohio Journal of Science 104: 136–138.

Seibold, I. 1994. Untersuchungen zur molekularen Phy-logenie der Greifvogel anhand von DNA-Sequenzen des mitochondriellen Cytochrom b-Gens. PhD Dissertation, University of Heidelberg, Heidelberg.

Seibold, I., A.J. Helbig and M. Wink. 1993. Molecular systematics of falcons (family Falconidae). Naturwis senschaften 80: 87–90.

Seibold, I. and A. Helbig. 1996. Phylogenetic relationships of the sea eagles (genus Haliaeetus). Reconstructions based on morphology, allozymes and mitochondrial DNA sequences. Journal of Zoological Systematics and Evolutionary Research 34: 103–112.

Seibold, I. and A.J. Helbig. 1995. Evolutionary history of new and old world vultures inferred from nucleotide sequences of the mitochondrial cytochrome b gene. Philosophical Transactions of the Royal Society B: Biological Sciences 350(1332): 163–178.

Seibold, I. and A.J. Helbig. 1995. Zur systematischen Stellung des Fischadlers Pandion haliaetus nach mitochondriellen DNA-Sequenzen. Vogelwelt 116: 209–217.

Seibold, I., A.J. Helbig, B.U. Meyburg, J.J. Negro and M. Wink. 1996. Genetic differentiation and molecular phylogeny of European Aquila eagles according to cytochrome b nucleotide sequences. *In*: B.U. Meyburg and R.D. Chancellor (eds.). Eagle Studies. World Working Group on Birds of Prey (WWGBP), London.

Sekercioglu, C.H. 2010. Ecosystem functions and services. pp. 45–72. *In*: N.S. Sodhi and P.R. Ehrlich (eds.). Conservation Biology for All. Oxford University Press, Oxford.

Sekercioglu, C.H., G.C. Daily and P.R. Ehrlich. 2004. Ecosystem consequences of bird declines. Proceedings of the National Academy of Sciences of the United States of America 101: 18042–18047.

Sekercioglu, C.H. 2006. Ecological significance of bird populations. pp. 15–51. *In*: J. del Hoyo, A. Elliott and D.A. Christie (eds.). Handbook of the Birds of the World, Vol. 11. Lynx Edicions, Barcelona.

Sekercioglu, C.H. 2006. Increasing awareness of avian ecological function. Trends in Ecology and Evolution 21: 464–471.

Selva, N. and M.A. Fortuna. 2007. The nested structure of a scavenger community. Proceedings of the Royal Society B: Biological Sciences 274: 1101–1108.

Selva, N., B. Jedrzejewska, W. Jedrzejewski and A. Wajrak. 2005. Factors affecting carcass use by a guild of scavengers in European temperate woodland. Canadian Journal of Zoology 83: 15901601.

Serra, G., M. Abdallah, A. Al Assaed, G. Al Qaim and A.K. Abdallah. 2005. A long-term bird survey in the central Syrian desert (2000–2003) Part 1. Sandgrouse 27: 9–23.

Sesé, J.A., R.J. Antor, M. Alcántara, J.C. Ascaso and J.A. Gil. 2005. La alimentación.suplementaria en el quebrantahuesos: estudio de un comedero del Pirineo occidental aragonés. pp. 279–304. *In*: A. Margalida and R. Heredia (eds.). Biología de la Conservación del Quebrantahuesos Gypaetus barbatus en España. Organismo Autónomo Parques Nacionales, Madrid.

Seshadri, B. 1969. The Twilight of India's Wildlife. John Barker, London.

Shamoun-Baranes, J., W. Bouten and E.V. Loon. 2010. Integrating Meteorology into Research on Migration Integrative and Comparative Biology 50(3): 280–292.

Shamoun-Baranes, J., Y. Leshem, Y. Yom-Tov and O. Liechti. 2003. Differential use of thermal convection by soaring birds over central Israel. Condor 105: 208–218.

Shamoun-Baranes, J., E. van Loon, D. Alon, P. Alpert, Y. Yom-Tov and Y. Leshem. 2006. Is there a connection between weather at departure sites, onset of migration and timing of soaring-bird autumn migration in Israel? Global Ecology and Biogeography 15: 541–52.

Sharp, A., M. Norton and A. Marks. 2001. Breeding activity, nest site selection and nest spacing of wedge-tailed eagles, Aquila audax, in western New South Wales. Emu 101: 323–328.

Sharpe, B. 1874. Catalogue of Birds of British Museum.1. British Museum, London.

Sharpe, C.J., F. Rojas-Suárez and D. Ascanio. 2008. Cóndor Vultur gryphus. *In*: J.P. Rodríguez and F. Rojas-Suárez (eds.). Libro Rojo de la fauna Venezolana. Tercera Edición, 128. Provita & Shell Venezuela, S.A., Caracas.

Sharpe, R.B. 1873. On a new species of Turkey Vulture from the Falkland Islands and a new genus of Old-World vultures. Annals and Magazines of Natural History 11: 133.

Sharpe, R.B. 1874. Catalogue of the Accipitres, or diurnal birds of prey, in the collection of the British Museum London: The Trustees.

Shaw, A.P.M. and C.H. Hoste. 1987. Trypanotolerant cattle and livestock development in West and Central Africa. FAO Animal Production and Health Paper No. 67(2).

Shaw, C. and S. Cox. 2006. The large carnivorans: wolves, bears, and big cats. pp. 187–188. *In*: G. Jefferson and L. Lindsay (eds.). Fossil Treasures of the Anza-Borrego Desert: The Last Seven Million Years. Sunbelt Publications, San Diego.

Shaw, J.M., A.R. Jenkins, P.G. Ryan and J.J. Smallie. 2010. A preliminary survey of avian mortality onpower lines in the Overberg, South Africa. Ostrich 81: 109–113.

Sherrod, S.K. 1978. Diets of North American Falconiformes. Journal of Raptor Research 12: 49–121.

Shields, G.F. and A.C. Wilson. 1987. Calibration of mitochondrial DNA evolution in geese. Journal of Molecular Evolution 24: 212–217.

Shimelis. 2007. In BirdLife International 2014. Species factsheet: Torgos tracheliotos. Downloaded from http://www.birdlife.org on 08/02/2014. R

Shimelis, A., E. Sande, S. Evans and P. Mundy. 2005. International Species Action Plan for the Lappet-faced Vulture, *Torgos tracheliotus*. BirdLife International, Nairobi, Kenya and Royal Society for the Protection of Birds, Sandy, Bedfordshire, UK. Available at: http://www.birdlife.org/action/science/species/species_action_plans/africa/lappet-faced_vulture_sap.pdf.

Shimelis, A., E. Sande, S. Evans and P. Mundy (eds.). 2005. International Action Plan for Lappet-faced Vulture, Torgos tracheliotus. BirdLife International, Nairobi, Kenya and Royal Society for the Protection of Birds, Sandy, Bedfordshire.

Shirihai, H. 1987. Field characters of the Negev Lappet-faced Vulture. In International Bird Identification: Procedings of the 4th International Meet Eilat 1st–8th November 1986, 8–11. Internation Birdwatching Center, Eliat.

Shirihai, H. 1996. The Birds of Israel. Academic Press, San Diego, London.

Shivik, J.A. 2006. Tools for the edge: what's new for conserving carnivores. BioScience 56: 253–259.

Shobrak, M. 1996. Ecology of Lappet-faced Vulture Torgos tracheliotus in Saudi Arabia. Unpublished PhD Thesis, University of Glasgow, Glasgow.

Shobrak, M. 1999. Status and conservation of vultures in Saudi Arabia. Proceedings of the International Conference on Conservation and Management of Griffon, Black and Egyptian Vultures, 24–27 June, 1999.

Shrestha, B.P. and B.P. Devkota. 2011. Status of critically endangered vultures in Dang Deukhuri Foothill Forests and West Rapti Wetlands. The Initiation 4: 28–34.

Schüz, E. and C. König. 1983. Old world vultures and Man. pp. 461–469. In: S.R. Wilbur and J.A. Jackson (eds.). Vulture Biology and Management. University of California Press, Berkely.

Shultz, S., H.S. Baral, S. Charman, A.A. Cunningham, D. Das, G.R. Ghalasi, M.S. Goudar, R.E. Green, A. Jones, P. Nighot, D.J. Pain and V. Prakash. 2004. Diclofenac poisoning is widespread in declining vulture populations across the Indian subcontinent. Proceedings of the Royal Society of London Series B 271: S458–S460.

Sibley, C.G. 1994. On the phylogeny and classification of living birds. Journal Avian Biology 25: 87–92.

Sibley, C.G. and B.L. Monroe. 1990a. Distribution and Taxonomy of Birds of the World. Yale University Press, New Haven, London.

Sibley, S.G. and J.E. Ahlquist. 1990b. Phylogeny and Classification of Birds. A Study in Molecular Evolution. Yale University Press, Yale, New Haven.

Sick, H. 1993. Birds in Brazil. Princeton University Press, Princeton.

Siebold, I. and A.J. Helberg. 1995. Evolutionary history of new and old world vultures inferred from nucleotide sequences of the mitochondrial cytochrome b gene. Philosophical Transactions of the Royal Society of London 350B: 163–178.

Sigismondi, A. and E. Politano. 1996. Unusually high concentrations of the Egyptian vulture Neophron percnopterus in a border area of the Dancalia region of Ethiopia. It may be one of the most important wintering areas known, 88–89. Abstracts 2nd International Conference on Raptors, Urbino, Italy 2–5 October 1996. Raptor Research Foundation—University of Urbino, Urbino.

Sikes, D.S. and C.J. Raithel. 2002. A review of hypotheses of decline of the endangered American burying beetle (Silphidae: *Nicrophorus americanus Olivier*). Journal of Insect Conservation 6: 103–113.

Simmons, R. 1986. Delayed breeding and the non-adaptive significance of delayed maturity in vultures: a fruit fly perspective. Vulture News 15: 13–18.

Simmons, R. 1995. Mass poisoning of lappet faced vultures in Namibia. Journal of African Raptor Biology 10: 3.

Simmons, R.E. 2002. A helicopter survey of Cape Vultures Gyps coprotheres, Black Eagles Aquila verreauxii and other cliff-nesting birds of the Waterberg Plateau, Namibia, 2001. Vulture News 46: 3–7.

Simmons, R.E. and A.R. Jenkins. 2007. Is climate change influencing the decline of Cape and Bearded Vultures in southern Africa? Vulture News: 41–51.

Simmons, R.E., C. Boix-Hinzen, K.N. Barnes, A.M. Jarvis and A. Robertson. 1998. Important Bird Areas of Namibia. pp. 295–332. *In*: K.N. Barnes (ed.). The Important Bird Areas of Southern Africa. BirdLife South Africa, Johannesburg.

Simmons, R.E. and P. Bridgeford. 1997. The status and conservation of vultures in Namibia. pp. 67–75. *In*: A.F. Boshoff, M.D. Anderson and W.D. Borello (eds.). Vultures in the 21st Century: Proceedings of a workshop on vulture research and conservation in southern Africa, Vulture Study Group, Johannesburg.

Simmons, R.E. and C.J. Brown. 2006a. Birds to watch in Namibia: red, rare and endemic species. National Biodiversity Programme, Windhoek, Namibia.

Simmons, R.E. and C.J. Brown. 2006b. In BirdLife International 2014. Species factsheet: Trigonoceps occipitalis. Downloaded from http://www.birdlife.org on 03/02/2014.

Simmons, R.E. and C.J. Brown. 2007. Cape Vulture—Critically endangered. Birds to watch in Namibia: red, rare and endemic species. National Biodiversity Programme. Namibia Nature Foundation, Windhoek.

Simmons, R.E. and A.R. Jenkins. 2007. Is climate change influencing the decline of Cape and Bearded Vultures in southern Africa? Vulture News 56: 41–51.

Simons, D.D. 1983. Interactions between California condors and humans in prehistoric far western North America. pp. 470–494. *In*: S.R. Wilbur and J.A. Jackson (eds.). Vulture Biology and Management. University of California Press, Berkeley, Los Angeles, London.

Sinclair, A.R.E. 1977. The African Buffalo: A Study of Resource Limitation of Populations. University of Chicago Press, Chicago.

Singh, D., B. Singh and R.K. Goel. 2011. Traditional uses, phytochemistry and pharmacology of Ficus religiosa: A review. Journal of Ethnopharmacology 134(3): 565–583.

Singh, R.B. 1999. Ecological strategy to prevent vulture menace to aircraft in India. Defence Science Journal 49(2): 117–121.

Skinner, N.J. 1997. The breeding seasons of birds in Botswana. III: Nonpasserine families (Ostrich to Skimmer). Babbler 32: 10–23.

Skutch, A. 1945. The migration of Swainson's and Broad-winged hawks through Costa Rica. Northwest Science 19: 80–89.

Slack, K.E., F. Delsuc, P.A. McClenachan, U. Arnason and D. Penny. 2007. Resolving the root of the avian mitogenic tree by breaking up long branches. Molecular Phylogenetics and Evolution 42: 1–13.

Slater, G.J. and B. Van Valkenburgh. 2008. Long in the tooth: evolution of sabertooth cat cranial shape. Paleobiology 34: 403–419.

Slotta-Bachmayr, L., R. Bögel and A. Camina-Cardenal (eds.). 2005. The Eurasian Griffon Vulture (Gyps fulvus) in Europe and the Mediterranean—Status Report and Action Plan. The European Griffon Vulture Working Group EGVWG.

Slud, P. 1964. The birds of Costa Rica. Distribution and ecology. Bulletin of the Amer. Museum of Natural History 128: 1–430.

Smallie, J. and M. Virani. 2010. A preliminary assessment of the potential risks from electrical infrastructure to large birds in Kenya. Scopus 30: 32–39.

Smallie, J. and L. Strugnell. 2011. Use of camera traps to investigate Cape Vulture roosting behaviour on power lines in South Africa. Unpublished report to Eskom, Sunninghill.

Smallie, J., M. Diamond and A. Jenkins. 2009. Lighting up the African continent—what does it mean for our birds? pp. 38–43. *In*: D.M. Harebottle, A.J.F.K. Craig, M.D. Anderson, H. Rakotomanana and M. Muchai (eds.). Proceedings of the 12th Pan-African Ornithological Congress, 2008, Cape Town. Animal Demography Unit, Cape Town.

Smit, G.N. 2004. Changing land-use patterns in southern Africa with emphasis on bush encroachment and tree removal in savanna. pp. 166–176. *In*: A. Monadjem, M.D. Anderson, S.E. Piper and A.F. Boshoff (eds.). The Vultures of Southern Africa—Quo Vadis? Proceedings of a workshop on vulture research and conservation in southern Africa, Birds of Prey Working Group, Johannesburg.

Smith, A.B. 1980. Domesticated cattle in the Sahara and their introduction into West Africa. *In*: M.A.J. Williams and H. Faure (eds.). The Sahara and the Nile: Quaternary Environments and Prehistoric Occupation in Northern Africa. Ch. 20. Balkema, Rotterdam.

Smith, D. 1978. Condor journal: The history, mythology, and reality of the California California. Capra Press/Santa Barbara Museum of Natural History, Santa Barbara.

Smith, D. and R. Easton. 1964. California Condor: Vanishing American. McNally and Loftin, Santa Barbara.

Smith, D.G. and J.R. Murphy. 1972. Unusual causes of raptor mortality. Raptor Research 6: 4–5.

Smith, H.R., R.M. DeGraaf and R.S. Miller. 2002. Exhumation of food by Turkey Vultures. Journal of Raptor Research 36: 144–145.

Smith, K.D. 1957. An annotated check list of the birds of Eritrea. Ibis 99: 1–26, 307–337.

Smith, M.D., E. Brantley and J.B. Armstrong. 2013. Proper Disposal of Deer Carcasses. Alabama Cooperative Extension System (Alabama A&M University and Auburn University).

Smith, N.G. 1980. Hawk and vulture migrations in the neotropics. pp. 51–65. *In*: A. Keast and E.S. Morton (eds.). Migrant Birds in the Neotropics. Smithsonian Institution Press, Washington D.C.

Smith, N.G. 1985a. Dynamics of the trans-isthmian migration of raptors between Central and South America. pp. 271–290. *In*: I. Newton and R.D. Chancellor (eds.). Conservation Studies on Raptors. ICBP Tech. Publication 5, International Council for Bird Preservation, Cambridge, UK.

Smith, N.G. 1985b. Thermals, cloud streets, trade winds, and tropical storms: how migrating raptors make the most of atmospheric energy in Central America. pp. 51–65. *In*: M. Harwood (ed.). Proceedings of Hawk Migration Conference IV. Hawk Migration Association of North America, Lynchburg.

Smith, N.G. 1970. Nesting of King Vulture and Black Hawk-eagle in Panama. Condor 72(2): 247–248.

Smith, N.G. 1980. Hawk and vulture migrations in the neotropics. pp. 51–65. *In*: A. Keast and E.S. Morton (eds.). Migrant Birds in the Neotropics. Smithsonian Institution Press, Washington, DC.

Smith, N.G. 1985a. Dynamics of the transist migration of raptors between Central and South America. 271–290. *In*: I. Newton and R.D. Chancellor (eds.). Conservation Studies on Raptors. ICBP Technical Publication No. 5, International Council for Bird Preservation, Cambridge.

Smith, N.G. 1985b. Thermals, cloud streets, trade winds, and tropical storms: how migrating raptors make the most of atmospheric energy in Central America. *In*: M. Harwood (ed.). Proceedings of Hawk Migration Conference 4, 51–65. Hawk Migration Association of North America, Lynchburg.

Smith, S.A. and R.A. Paselk. 1986. Olfactory sensitivity of the Turkey Vulture (*Cathartes aura*) to three carrion associated odorants. Auk 103: 586–592.

Smyser, T.J., J.C. Beasley, Z.H. Olson and O.E. Rhodes, Jr. 2010. Use of rhodamine B to reveal patterns of interspecific competition and bait acceptance in raccoons. Journal of Wildlife Management 74: 1405–1416.

Smythies, B.E. 1953. The Birds of Burma 2nd edition. Oliver and Boyd, Edinburgh.

Smythies, C.E. 1986. The Birds of Burma. Nimrod Press, Liss.

Snow, D.W. 1978. An Atlas of Speciation in African Non-passerine Birds. British Museum of Natural History, London.

Snow, D.W. and C.M. Perrins. 1998. The Birds of the Western Palearctic. Oxford University Press, Oxford.

Snyder, N.F.R. and H. Snyder. 2000. The California Condor. Academic Press, San Diego.

Snyder, N. and H. Snyder. 2000. The California Condor: a saga of natural history and conservation. Academic Press, San Diego.

Snyder, N.F.R. and N.J. Schmitt. 2002. California Condor (Gymnogyps californianus). *In*: A. Poole (ed.). The Birds of North America Online, No. 610. Cornell Lab of Ornithology, Ithaca.

Snyder, N.F.R. and H. Snyder. 2006. Raptors of North America: Natural History and Conservation. Voyageur Press.

Sodeinde, S.O. and D.A. Soewu. 1999. Pilot study of the traditional medicine trade in Nigeria. Traffic Bulletin 18: 35–40.

Sorenson, K., J. Burnett, A. Mee and L. Hall. 2007. Lead concentrations in the blood of Big Sur California condors. California condors in the 21st century. Series in Ornithology No. 2., 185–195. Nuttall Ornithological Club and American Ornithologists' Union, Cambridge, MA.

Sorkin, B. 2008. A biomechanical constraint on body mass in terrestrial mammalian predators. Lethaia 41(4): 333–347.

Spaar, R. 1997. Flight strategies of migrating raptors; a comparative study of interspecific variation in flight characteristics. Ibis 139(3): 523–535.

Spaar, R. and B. Bruderer. 1997. Migration by flapping or soaring: Flight strategies of Marsh, Montagu's and Pallid Harriers in southern Israel. Condor 99: 458–69.

Spaar, R., H. Stark and F. Liechti. 1998. Migratory flight strategies of Levant sparrowhawks: time or energy minimization? Animal Behaviour: 56: 1185–1197.

Sparkman, P.S. 1908. The culture of the Luseno Indians. University of California Publications in American Archaeology and Ethnology 8: 187–234.

Speakman, J.R. and E. Król. 2010. Maximal heat dissipation capacity and hyperthermia risk: Neglected key factors in the ecology of endotherms. Journal of Animal Ecology 79: 726–746.

Spiegel, O., W.M. Getz and R. Nathan. 2013. Factors influencing foraging search efficiency: why do scarce Lappet-Faced Vultures outperform ubiquitous White-Backed Vultures? The American Naturalist 181(5): E102–E115.

Spier, I. 1923. Southern Diegueno customs. University of California Publications in American Archaeology and Ethnology 20: 295–358.

Srikosamatara, S. and V. Suteethorn. 1995. Populations of Gaur and Banteng and their management in Thailand. Natural History Bulletin of the Siam Society 43: 55–83.

Ssemmanda, R. and D. Pomeroy. 2010. Scavenging birds of Kampala: 1973–2009. Scopus 30: 26–31.

Ssemmanda, R. 2005. An apparent increase in Hooded Vulture *Necrosyrtes monachus* numbers in Kampala, Uganda. Vulture News 53: 10–14.

Ssemmanda, R. and A.J. Plumptre. 2011. Declining populations: what is the status of vultures in Queen Elizabeth Park-Uganda? Wildlife Conservation Society.

Ssemmanda, R. and D. Pomeroy. 2010. Scavenging birds in Kampala since 1973. Scopus 30: 26–31.

Stager, K.E. 1964. The role of olfaction in food location by the Turkey Vulture and other cathartids. Los Angeles County Museum Contributions in Science 81: 1–63.

Stalmaster, M.V. 1987. The Bald Eagle. Universe Books, New York.

Stanford, J.K. and C.B. Ticehurst. 1935. Notes on the birds of the Sittang-Irrawaddy Plain, Lower Burma. Journal of the Bombay Natural History Society 37: 859–899.

Stanford, J.K. and C.B. Ticehurst. 1938–1939. On the birds of northern Burma. Ibis (14)2: 65–102, 197–229, 391–428, 599–638; (14)3: 1–45, 211–258.

Statistix. 2008. Statistix 9 user's manual. Tallahassee, Florida: Analytical Software.

Steadman, D.W., J. Arroyo-Cabrales, E. Johnson and A.F. Guzman. 1994. New Information on the Late Pleistocene Birds from San Josecito Cave, Nuevo Leon, Mexico. Condor 96(3): 577–589.

Steele, R. 1905. Mediaeval Lore from Bartholomaeus Anglicus, London.

Steele, R. 2004. Mediaeval Lore from Bartholomew Anglicus. Kessinger Publishing, Whitefish, Montana.

Steenhof, K., M.N. Kochert and J.A. Roppe. 1993. Nesting raptors and common ravens on electrical transmission line towers. Journal of Wildlife Management 57: 271–281.

Stevenson, T. and J. Fanshawe. 2001. Field Guide to the Birds of East Africa: Kenya, Tanzania, Uganda, Rwanda, Burundi. Elsevier Science, Amsterdam.

Stewart, P.A. 1978. Behavioral interactions and niche separation in Black and Turkey vultures. Living Bird 17: 79–84.

Stewart, P.A. 1984. Population decline of black vultures in North Carolina. Chat 48: 65–68.

Steyn, P. 1982. Birds of prey of southern Africa: their identification and life histories. David Phillip, Cape Town, South Africa.

Steyn, P. 1982. Birds of Prey of Southern Africa. David Philip, Cape Town.

Stiles, F.G. and A. Skutch. 1989. A Guide to the Birds of Costa Rica. Cornell Universiy. Press, Ithaca.

Stiles, F.G. 1985. Conservation of forest birds in Costa Rica: problems and perspectives. *In*: A.W. Diamond and T.E. Lovejoy (eds.). Conservation of Tropical Forest Birds. ICBP Technical Publications, Cambridge. 4: 141–168.

Stiles, F.G. and A.F. Skutch. 1989. A Guide to the Birds of Costa Rica. Cornell University Press, Ithaca.

Still, C.J., P.N. Foster and S.H. Schneider. 1999. Simulating the effects of climate change on tropical montane cloud forests. Nature 398: 608–610.

Stock, C. 1958. Rancho La Brea: a record of Pleistocene life in California 6th edn. Los Angeles County Museum. Science Series 20: 1–83.

Stolen, E.D. 2003. Black vultures on vegetables and tortoise dung. Florida Field Naturalist 31(2): 21.

Stolen, E.D. 1996. Black and turkey vulture interactions with bald eagles in Florida. Florida Field Naturalist. 24(2): 43–45.

Stone, W. 1890. On birds collected in Yucatan and southern Mexico. Proceedings of the Academy of Natural Sciences of Philadelphia 42: 201–218.

Storch, V., U. Welsch and M. Wink. 2001. Evolutionsbiologie. Springer, Heidelberg.

Storz, J.F. and H. Moriyama. 2008. Mechanisms of hemoglobin adaptation to high altitude hypoxia. High altitude medicine and biology 9(2): 148–157.

Stotz, D.F., J.W. Fitzpatrick, T.A. Parker and D.K. Moskovits. 1996. Neotropical Birds: Ecology and Conservation. University of Chicago Press, Chicago.

Stoychev, S. and T. Karafeisov. 2003. Power Line Design and Raptor Protection in Bulgaria. Sixth World Conference on Birds of Prey and Owls, Budapest.

Strange, I.J. 1996. The Striated Caracara Phalcoboenus australis in the Falkland Islands. Author's Edition, Falkland Islands.

Stresemann, E. and D. Amadon. 1979. Order Falconiformes. *In*: E. Mayr and G.W. Cotrelli (eds.). Checklist of the Birds of the World. Cambridge Museum of Comparative Zoology, Cambridge, MA. 1: 271–425.

Strix. 2012. Developing and testing the methodology for assessing and mapping the sensitivity of migratory birds to wind energy development. BirdLife International, Cambridge.

Strong, W.D. 1929. Aboriginal society in southern California. University of California Publications in American Archaeology and Ethnology 26: 1–358.

Stroud, D.A. 2003. The Status and Legislative Protection of Birds of Prey and their Habitats in Europe. pp. 51–84. *In*: D.B.A. Thompson, S.M. Redpath, A.H. Fielding, M. Marquiss and C.A. Galbraith (eds.). Birds of Prey in a Changing Environment. Scottish Natural Heritage, Edinburgh.

Stucchi, M. and S.D. Emslie. 2005. A new condor (Ciconiiformes, Vulturidae) from the Late Miocene/Early Pliocene Pisco Formation, Peru. Condor 107: (1) 107–113.

Suárez Arangüena, L. 2008. Buitres hambrientos. Panda 102: 23.

Suarez, W. 2004. The identity of the fossil raptor of the genus Amplibuteo (Aves: Accipitridae) from the Quaternary of Cuba. Caribbean Journal of Science 40: (1) 120–125.

Suárez, W. and S.D. Emslie. 2003. New fossil material with a redescription of the extinct condor Gymnogyps varonai (Arredondo 1971) from the Quaternary of Cuba (Aves: Vulturidae). Proceedings of the Biological Society of Washington 116(1): 29–37.

Suazo, L.R. 2000. Powerlines and raptors, using regulatory influence to prevent electrocutions. pp. 773–778. *In*: R.D. Chancellor and B.-U. Meyburg (eds.). Raptors at Risk. World Working Group on Birds of Prey and Owls. Hancock House, Blaine.

Subramanian, M. 2008. Towering silence. Science & Spirit May/June 34–38.

Subramanya, S. and O.C. Naveein. 2006. Breeding of Long-billed Vulture *Gyps indicus* at Ramanagaram hills, Karnataka, India. Indian Birds 2(2): 32–34.

Sudmann, S.R., S. Hüppeler-Borcherding and S. Klostermann. 2000. The Behaviour of Overwintering, Arctic Geese in the Proximity of Marked and Unmarked High-tension Power Lines at the Niederrhein. Naturschutzzentrum im kreis Kleve.

Sullivan, K., R. Sieg and C. Parish. 2007. Arizona's efforts to reduce lead exposure in California Condors.In California Condors in the 21st Century. pp. 109–121. *In*: A. Mee and L.S. Hall (eds.). The Nuttall Ornithological Club and the American Ornithologists' Union. Lancaster, Pennsylvania.

Sutton, C. and P. Sutton. 1999. River of raptors: exploring and enjoying Pronatura Veracruz's raptor conservation project. Birding 31: 229–236.

Sutton, C. and P. Sutton. 1999. River of raptors: exploring and enjoying Pronatura Veracruz's raptor conservation project. Birding 31: 229–236.

Swan, G. 2004. Steroidal Anti-inflammatory Drugs—risk to vultures. pp. 202–207. *In*: A. Monadjem, M.D. Anderson, S.E. Piper and A.F. Boshoff (eds.). Vultures in The Vultures of Southern Africa—Quo Vadis? Proceedings of a workshop on vulture research and conservation in southern Africa. Birds of Prey Working Group, Johannesburg.

Swan, G., V. Naidoo, R. Cuthbert, R.E. Green, D.J. Pain, D. Swarup, V. Prakash, M. Taggart, L. Bekker, D. Das, J. Diekmann, M. Diekmann, E. Killian, A. Meharg, R. Chandra Patra, M. Saini and K. Wolter. 2006. Removing the threat of diclofenac to critically endangered Asian vultures. PLoS Biology 4(3): 395–402.

Swan, G.E., R. Cuthbert, M. Quevedo, R.E. Green, D.J. Pain, P. Bartels, A.A. Cunningham, N. Duncan, A.A. Meharg, J.L. Oaks, J. Parry-Jones, M.A. Taggart, G. Verdoorn and K. Wolter. 2006. Toxicity of diclofenac to Gyps vultures. Biology Letters 279–282.

Swaringen, K., R.J. Wiese, K. Willis and M. Hutchins (eds.). 1995. AZA Annual Report on Conservation and Science. American Association of Zoological Parks and Aquariums, Bethesda.

Swarup, D., R.C. Patra, V. Prakash, R. Cuthbert, D. Das, P. Avari, D.J. Pain, R.E. Green, A.K. Sharma, M. Saini, D. Das and M. Taggart. 2007. Safety of meloxicam to critically endangered Gyps vultures and other scavenging birds in India. Animal Conservation 10(2): 192–198.

Sweeney, R. 2013. Captive management of ground hornbills for a sustainable population. Rochester Institute of Technology, New York.

Swezey, S.L. and R.F. Heizer. 1977. Ritual management of salmonid fish resources in California. Journal of California Anthropology 4: 6–29.

Swezey, S.L. 1975. Ethnographic interpretations 12: The energies of subsistence-assurance ritual in Native California. Contributions of the University of California Archaeological Research Facility 23: 1–46.

Swihart, R.K., T.M. Gehring, M.B. Kolozsvary and T.E. Nupp. 2003. Responses of 'resistant' vertebrates to habitat loss and fragmentation: The importance of niche breadth and range boundaries. Diversity and Distributions 9: 1–18.

Switzer, F. 1977. Saskatchewan Power's experience. Blue Jay 35: 259–260.

Swofford, D.L. 1993. PAUP, Phylogenetic analysis using parsimony. Version 3.1.1, Illinois.

Taberlet P., A. Meyer and J. Bouvet. 1992. Unusual mitochondrial polymorphism in two local populations of blue tit Parus caeruleus. Molecular Ecology 1: 27–36.

Tait, M. 2006. Going, Going, Gone: Animals and Plants on the Brink of Extinction. Sterling Publishing, New York.

Tallman, D.A. and E. Tallman. 1977. [Additions and revisions to the avifaunal list of Limoncocha, Napo Province, Ecuador]. Revista de la Universidad Catolica (Pontificia Universidad Catolica del Ecuador) 5: 217–224.

Tambussi, C.P. and J.I. Noriega. 1999. The fossil record of condors (Ciconiiformes: Vulturidae) in Argentina. Smithsonian Contributions to Paleobiology 89: 177–184.

Tarboton, W.R. and D.G. Allan. 1984. The Status and Conservation of Birds of Prey in the Transvaal. Transvaal Museum Monograph 3, Pretoria, South Africa.

Tarboton, W. 1990. African Birds of Prey. Cornell University Press, Ithaca.

Taylor, W.P. 1923. Note on the sense of smell in the Golden Eagle and certain other birds. Condor 25: 28.

Taylor, W.P. and C.T. Vorhies. 1933. The Black Vulture in Arizona. Condor 35: 205–206.

Te Velde, H. 2008. The Goddess Mut and the vulture. pp. 242–245. *In*: S.H. D'Auria (ed.). Servant of Mut: Studies in Honour of Richard A. Fazzini. Koninklijkc Brill NV, Leiden and Boston.

Tella Escobedo, J.L. 2006. Action is needed now, or BSE crisis could wipe out endangered birds of prey. Vulture News 55: 35–36.

Tella, J.L. 2001. Action is needed now, or BSE crisis could wipe out endangered birds of prey. Nature 410: 408. doi:10.1038/35068717.

Telleria, J.L. 2009a. Overlap between wind power plants and griffon vultures Gyps fulvus in Spain. Bird Study 56: 368–271.

Telleria, J.L. 2009b. Wind power plants and the conservation of birds and bats in Spain: a geographical assessment. Biodiversity Conservation 18: 1781–1791.

Temple-Lang, J. 1982. The European Community Directive on bird conservation. Biological Conservation 22: 11–25.

Terminski, B. 2011. Towards Recognition and Protection of Forced Environmental Migrants in the Public International Law: Refugee or IDPs Umbrella. Policy Studies Organization (PSO), Washington.

Terrasse, J.F., F. Sarrazin, J.P. Choisy, C. Clemente, S. Henriquet, P. Lecuyer, J.L. Pinna and C. Tessier. 2004. A success story: the reintroduction of Griffon Gyps fulvus and Black *Aegypius monachus* Vultures in France. pp. 127–145 *In*: R.D. Chancellor and B.-U. Meyburg (eds.). Raptors Worldwide. World Working Group on Birds of Prey and Owls and MME/BirdLife Hungary, Berlin and Budapest.

Terrasse, M. 2006. Long-term reintroduction projects of Griffon *Gyps fulvus* and Black vultures *Aegypius monachus* in France. pp. 98–107. *In*: D.C. Houston and S.E. Piper (eds.). Proceedings of the International Conference on Conservation and Management of Vulture Populations, 14–16 November 2005, Thessaloniki, Greece. Natural History Museum of Crete & WWF Greece.

Terrasse, M., C. Bagnolini, J. Bonnet, J.L. Pinna and F. Sarrazin. 1994. Reintroduction of the Griffon Vulture Gyps fulvus in the Massif Central, France. pp. 479–491. *In*: B.-U. Meyburg and R. Chancellor (eds.). Raptor Conservation Today. WWGBP/Pica Press, London.

Terrasse, M., F. Sarrazin, J.P. Choisy, C. Clemente, S. Henriquet, P. Lecuyer, J.L. Pinna and C. Tessier. 2004. A success story: The reintroduction of Eurasian Griffon Gyps fulvus and Black *Aegypius monachus* Vultures to France. pp. 127–145. *In*: R. Chancellor and B.U. Meyburg (eds.). Raptors Worldwide. WWGBP, Budapest.

Terres, J.K. 1980. The Audubon Society Encyclopedia of North American Birds. Knopf, New York.

Terres, J.K. and National Audubon Society. 1991. The Audubon Society Encyclopedia of North American Birds. Reprint of 1980 edition. National Audubon Society, New York.

Tewes, E. 1998. The Black Vulture Conservation Programme: a Species Conservation Strategy ill Europe. *In*: E. Tewes, J.J. Sanchez, B. Heredia and M. Bijleveld van Lexmond (eds.). International Symposium on the Black Vulture in South Eastern Europe and Adjacent Regions (Dadia. Greece. 15 16 September 1993), 102–112. FZS/BVCF, Palma de Mallorca.

Thacker, P.D., N. Lubick, R. Renner, K. Christen and J. Pelley. 2006. Condors are shot full of lead. Environmental Science and Technology 40(19): 5826.

Thakur, M.L., R.C. Kataria and K. Chauhan. 2012. Population decline of vultures and their conservation: scenario in India and Himachal Pradesh. International Journal of Science and Nature 3(2): 241–250.

Thaler, E., S. Maschler and V. Steinkellner. 1986. Vergleichende Studien zur Postembryonalentwicklung dreier Altweltgeier. Annalen Naturhistorisches Museum Wien 881898, 361–376.

The Peregrine Fund. 2003. Annual Report. 28–29.

The Peregrine Fund. 2010. Annual Report. 34.

Thelander, C.G., K.S. Smallwood and L. Rugge. 2003. Bird risk behaviors and Fatalities at the Altamont Pass Wind Resource Area. Period of Performance: March 1998–December 2000. National Renewable Energy Laboratory, Colorado.

Thévenot, M., R. Vernon and P. Bergier. 2003. The Birds of Morocco. BOU Checklist Series 20. British Ornithologists' Union & British Ornithologists' Club, Tring.

Thewlis, R.M., R.J. immins, T.D. Evans and J.W. Duckworth. 1998. The conservation status of birds in Laos: a review of key species. Bird Conseration International 8 (supplement): 1–159.

Thiel, R.P. 1976. Activity patterns and food habits of southeastern Wisconsin Turkey Vultures. Passenger Pigeon 38: 137–143.

Thiollay, J.-M. 1980. Spring hawk migration in eastern Mexico. Journal of Raptor Research 14: 13–19.

Thiollay, J.-M. 1994a. Family Accipitridae (Hawks and Eagles). pp. 52–205. *In*: J. del Hoyo, A. Elliott and J. Sargatal (eds.). Handbook of the Birds of the World. Lynx Edicions, Barcelona, Spain.

Thiollay, J.-M. 1994b. Himalayan Griffon. *In*: J. del Hoyo, A. Elliott and J. Sargatal (eds.). Handbook of the Birds of the World, Vol. 2., 127. Lynx Edicions, Barcelona.

Thiollay, J.M. 2000. Vultures in India. Vulture News 42: 36–38.

Thiollay, J.M. 2001. Long-term changes of raptor populations in northern Cameroon. Journal of Raptor Research 35: 173–186.

Thiollay, J.-M. 2006a. The decline of raptors in West Africa: long-term assessment and the role of protected areas. Ibis 148: 240–254.

Thiollay, J.-M. 2006b. Severe declines of large birds in the northern Sahel of West Africa: a long-term assessment. Bird Conservation International 16(4): 353–365.

Thiollay, J.M. 2006c. Large bird declines with increasing human pressure in savanna woodlands (Burkina Faso). Biodiversity and Conservation 15: 2085–2108.

Thiollay, J.M. 2007a. Raptor declines in West Africa: comparisons between protected, buffer and cultivated areas. Oryx 41(3): 322–329.

Thiollay, J.M. 2007b. Raptor population decline in West Africa. Ostrich 78(2): 405–413.

Thiollay, J. 2012. In BirdLife International 2014. Species factsheet: Trigonoceps occipitalis. Downloaded from http://www.birdlife.org on 03/02/2014.

Thiollay, J.-M. and J.C. Bednarz. 2007. Raptor communities in French Guiana: distribution, habitat selection and conservation. Journal of Raptor Research 41(2): 90–105.

Thirgood, S. and S. Redpath. 2008. Hen harriers and red grouse: science, politics and human-wildlife conflict. Journal Applied Ecology 45: 1550–1554.

Thirgood, S., S. Redpath, I. Newton and P. Hudson. 2000. Raptors and red grouse: conservation conflicts and management solutions. Conservation Biology 14: 95–104.

Thorpe, J. 1992. Serious Bird Strikes to Civil Aircraft 1989–1991. Proceedings of BSCE 21, 291–297. Jerusalem 23–27 March. WP31.

Thorpe, J. 1994. Serious Bird Strikes to Civil Aircraft 1992–1993. Proceedings of BSCE 22, 179–188. Vienna 29 August-2 September. WP26.

Thorpe, J. 1996. Fatalities and Destroyed Civil aircraft due to Bird Strikes 1912–1995. Proc. of IBSC 23, 17–31. London 13–17 May.Wp-1.

Thorpe, J. 1998. The Implications of Recent Serious Bird Strike Accidents and Multiple Engine Ingestion's. Proc. IBSC 24, 11–22. Stara Lesna, Slovakia 14–18 September. WP 3.

Thorup, K., T. Alerstam, M. Hake and N. Kjellen. 2003. Bird orientation: compensation for wind drift in migrating raptors is age dependent. Biology Letters 270: 8–11.

Thurber, W., J.E. Serrano, A. Sermero and M. Benitez. 1987. Status of uncommon and previously unreported birds in El Salvador. Proceedings of the Western Foundation of Vertinary Zoology 3: 109–293.

Tilly, F.C., S.W. Hoffman and R.C. Tilly. 1990. Spring hawk migration in southern Mexico, 1989. Hawk Migration Association of North America, Hawk Migration Studies 15(2): 21–29.

Tilson, R.L. and J.R. Henschel. 1986. Spatial arrangement of spotted hyaena groups in a desert environment, Namibia. African Journal of Ecology 24: 173–180.

Timmins, R.J. and Men Soriyun. 1998. A wildlife survey of the Tonle San and Tonle Srepok river basins in north-eastern Cambodia. Fauna and Flora International and Wildlife Protection Office, Phnom Penh.

Timmins, R.J. and Ou Ratanak. 2001. The importance of Phnom Prich Wildlife Sanctuary and adjacent areas for the conservation of tigers and other key species. World Wildlife Foundation, Indochina Programme, Hanoi.

Timmins, R.J., B. Pech and S. Prum. 2003. An assessment of the conservation and importance of the Western Siem Pang area, Stung Treng province, Cambodia. Phnom Penh: WWF Cambodia Program.

Timmins, R.J. and Ou Ratanak. 2001. The importance of Phnom Prich Wildlife Sanctuary and adjacent areas for the conservation of tigers and other key species. WWF Indochina Programme, Hanoi and Phnom Penh.

Todd, W.E. and M.A. Carriker. 1922. The birds of the Santa Marta region of Colombia: A study in altitudinal distribution. Annals of the Carnegie Museum 14: 3–582.

Tonni, E.P. and J.I. Noriega. 1988. Los condores (Ciconiiformes, Vulturidae) de la Regió n Pampeana de la Argentina durante el cenozoico tardıo: distribucion, interacciones y extinciones. Ameghiniana 35: 141–150.

Toops. 2009. In BirdLife International 2014. Species factsheet: Gymnogyps californianus. Downloaded from http://www.birdlife.org on 13/03/2014.

Trail, P.W. 2007. African hornbills: keystone species threatened by habitat loss, hunting and international trade. Ostrich 78(3): 609–613.

Traverso, J.M. 2001. Nidificaciones sobre árbol del buitre Leonado en España. Quercus 180: 23–25.

Trost, C. 1999. Black-billed magpie (*Pica pica*). *In*: A. Poole and F. Gill (eds.). The Birds of North America No. 389 (The Birds of North America Inc., Philadelphia, PA, 1–28.

Tryjanowski, P., J.Z. Kosicki, S. Kuzniak and T.H. Sparks. 2009. Long-term changes and breeding success in relation to nesting structures used by the white stork, *Ciconia ciconia*. Annales Zoologici Fennici 46: 34–38.

Tso-hsin, C. 1987. A synopsis of the avifauna of China. Science Press, Beijing.

Tucker, V.A. 1996. A mathematical model of bird collisions with wind turbine rotors. Journal of Solar Energy Engineering 118: 253–262.

Tucker, G., S. Bassi, J. Anderson, J. Chiavari, K. Casper and M. Fergusson. 2008. Provision of evidence of the conservation impacts of energy production. Institute for European Environmental Policy (IEEP), London.

Tucker, G.M. and F.A. Heath (eds.). 1994. Birds in Europe: Their Conservation Status. Birdlife International Conservation Series No. 3, Cambridge.

Tucker, G.M. and M.F. Heath (eds.). 1994. Birds in Europe: their Conservation Status. BirdLife International, Cambridge.

TÜIK (Turkish Statistical Institute). 2010. Livestock Statistics.—www.turkstat.gov.tr.

Turner, A and M. Antón. 1997. The Big Cats and Their Fossil Relatives: An Illustrated Guide to Their Evolution and Natural History. Columbia University Press, New York.

Turner, A. 1987. Megantereon cultridens (Cuvier) (Mammalia, Felidae, Machairodontinae) from Plio-Pleistocene deposits in Africa and Eurasia, with comments on dispersal and the possibility of a New World origin. Journal of Paleontology 61(6): 1256–1268.

Turner, A. 1990. The Evolution of the guild of larger terrestrial carnivores during the Plio-Pleistocene in Africa. Geobios 23(3): 349–368.

Turner, A. and M. Antón. 1997. The Big Cats and Their Fossil Relatives: An Illustrated Guide to Their Evolution and Natural History. Columbia University Press, Columbia.

Tyler, W.M. 1961. Turkey vulture. *In:* A.C. Bent (ed.). Life Histories of North American Birds of Prey (part 1). Dover Publications, New York.

Tyler, W.M. 1937. Cathartes aura septentrionalis, Wied. Turkey vulture. pp. 12–28. *In*: Life Histories of North American Birds of Prey, Volume 1. United States National Museum Bulletin 167, Washington, DC.

US Fish and Wildlife Service. 1996. California Condor recovery plan. USFWS, Washington DC.

US Fish and Wildlife Service. 2002. California Condors return to Mexico. USFWS, Washington DC.

UNEP United Nations Environment Programme. 1994. United Nations Convention to Combat Desertification. UNEP, Nairobi.

United States Department of Agriculture. 2003. Wildlife Hazard Assessment for Moody Air Force Base, Georgia. US Government Print Shop, Washington DC.

United States Department of Agriculture USDA. 2003. Fact Sheet: Managing Vulture Damage. United States Department of Agriculture Animal and Plant Health Inspection Service Wildlife Services, Pittstown.

Unitt, P. 1984. The birds of San Diego County. San Diego Society of Natural History Memoir 13.

US Fish & Wildlife Service. 2013. List of Migratory Bird Species Protected by the Migratory Bird Treaty Act as of December 2, 2013. http://www.fws.gov/migratorybirds/RegulationsPolicies/mbta/MBTANDX.HTML. Accessed 13-02-2013.

US Fish and Wildlife Service. 2013. California Condor Recovery Program.

Vaassen, E.W.A.M. 2000. Habitat choice, activity pattern, and hunting method of wintering raptors in the Goksu Delta, southern Turkey. De Takkeling 8: 142–162.

Vagliano, C. 1981. Contribution au statut des rapaces diurnes et nocturnes nicheurs en Crete. In Rapaces mediterraneens, 14–16. annales du CROP. 1. Aix-en-Provence.

Valentine, A.F. 1973. Nesting of a turkey vulture in Alpena County, Michigan. Jack-Pine Warbler 51: 92.

Valkama, J., E. Korpimäki, B. Arroyo, P. Beja, V. Bretagnolle, E. Bro, R. Kenward, S. Mañosa, S. Redpath, S. Thirgood and J. Viñuela. 2005. Birds of prey as limiting factors of gamebird populations in Europe: a review. Biological Review 80: 171–203.

Valmiki. The Ramayan of Valmiki: Translated into English Verse. Trans. Ralph T.H. Griffith (London 1870–74), Canto Li.

Van Belle, J., J. Shamoun-Baranes, E. van Loon and W. Bouten. 2007. An operational model predicting autumn bird migration intensities for flight safety. Journal of Applied Ecology 44: 864–874.

Van Buren, A.N. 2012. Trophic Interactions in South West Atlantic Seabird Colonies. Unpublished PhD Thesis, University of Washington, Seattle.

van den Bergh, L.M.J., A.L. Spaans and N.D. van Swelm. 2002. Lijnopstellingen van windturbines geen barrière voor voedselvluchten van meeuwen en sterns in de broedtijd. Limosa 75: 25–32.

Van den Ostende, L.W., M. Morlo and D. Nagel. 2006. Fossils explained 52 Majestic killers: the sabre-toothed cats. Geology Today 22(4): 150.

Van Dooren, T. 2010. Vultures and their people in India: equity and entanglement in a time of extinctions. Manoa 22: 130–146.

Van Dooren, T. 2011. Vulture. Reaktion Books Ltd., London.

Van Rooyen, C.S. 2000. An overview of vulture electrocutions in South Africa. Vulture News. Vulture Study Group: Johannesburg, South Africa 43: 5–22.

Van Rooyen, C.S. and C.J. Vernon. 1997. Impact of vulture restaurants as management tools. pp. 142–144. *In*: A.F. Boshoff, M.D. Anderson and W.D. Borello (eds.). Vultures in the 21st Century: Proceedings of a Workshop on Vulture Research and Conservation in Southern Africa. Vulture Study Group, Johannesburg.

Van Tuinen, M., C.G. Sibley and S.B. Hedges. 2000. The early history of modern birds inferred from DNA sequences of nuclear and mitochondrial ribosomal genes. Molecular Biology & Evolution 17: 451–457.

Van Valkenburgh, B. 2007. Deja vu: the evolution of feeding morphologies in the Carnivora. Integrative and Comparative Biology 47:147–163.

Van Valkenburgh, B. and T. Sacco. 2002. Sexual dimorphism, social behavior, and intrasexual competition in large Pleistocene carnivorans. Journal of Vertebrate Paleontology 22(1): 164–169.

Van Wyk, E., H. Bouwman, H. van der Bank, G.H. Verdoorn and D. Hofmann. 2001. Persistent organochlorine pesticides detected in blood and tissue samples of vultures from different localities in South Africa. Comparative Biochemistry and Physiology. Part C, Comparative Pharmacology and Toxicology 129(3): 243–264.

Vanderburgh, D.C. 1993. Manitoba Hydro accommodates osprey activity. Blue Jay 51: 173–177.

Vanvalkenburgh, B. and F. Hertel. 1993. Tough times at La Brea: tooth breakage in large carnivores of the late Pleistocene. Science 261(5120): 456–59.

Vardanis, Y., R.H.G. Klaassen, R. Strandberg and T. Alerstam. 2011. Individuality in bird migration: routes and timing. Biology Letters, Animal Behaviour 7: 502–505.

Verdoorn, G.H. 1997a. The impact of poison on vultures. pp. 105–108. In: A.F. Boshoff, M.D. Anderson and W.D. Borello (eds.). Vultures in the 21st Century: Proceedings of a Workshop on Vulture Research and Conservation in Southern Africa. Vulture Study Group, Johannesburg.

Verdoorn, G.H. 1997b. Vulture restaurants as a conservation tool. pp. 119–122. In: A.F. Boshoff, M.D. Anderson and W.D. Borello (eds.). Vultures in the 21st Century: Proceedings of a Workshop on Vulture Research and Conservation in Southern Africa. Vulture Study Group, Johannesburg.

Verdoorn, G.H., N. van Zijl, T.V. Snow, L. Komen and E.W. Marais. 2004. Vulture poisoning in southern Africa. pp. 195–201. In: A. Monadjem, M.D. Anderson, S.E. Piper and A.F. Boshoff (eds.). The Vultures of Southern Africa—Quo Vadis? Birds of Prey Working Group, Johannesburg.

Vermeulen, J. 2011. A Report from birdtours.co.uk Ghana, November 7–19th 2011. http://www.birdtours.co.uk/tripreports/ghana/ghana-5/ghana-nov-2011.htm Accessed 3/2/2014.

Vernon, C. 1999. The Cape Vulture at Colleywobbles: 1977–1997. Ostrich 70: 200–202.

Viñuela, J. and B. Arroyo. 2002. Gamebird hunting and biodiversity conservation: synthesis, recommendations and future research priorities. EC Report. hhttp://digital.csic.es/handle/10261/8275ttp://digital.csic.es/handle/10261/8275.

Viñuela, J. and B. Arroyo. 2002. Gamebird hunting and biodiversity conservation: synthesis, recommendations and future research priorities. Report to EC, Brussels. Downloadable from http://digital.csic.es/handle/10261/8275.

Virani, M., M. Gilbert, R. Watson, L. Oaks, P. Benson, A.A. Kham and H.-S. Baral. 2001. Asian vulture crisis project: field results from Pakistan and Nepal for the 2000–2001 field season. Reports from the workshop on Indian Gyps vultures. pp. 461–469. In: T. Katzner and J. Parry-Jones (eds.). Proceedings of the 4th Eurasian Congress on Raptors. Estación Biológica Donaña, RaptorResearch Foundation, Seville.

Virani, M.Z. and M. Muchai. 2004. Vulture conservation in the Masai Mara National Reserve, Kenya: proceedings and recommendations of a seminar and workshop held at the Masai Mara National Reserve, 23 June 2004. Ornithology Research Report. National Museums of Kenya, Nairobi 57: 2–19.

Virani, M.Z. 2006. In Steep Decline. SWARA (Magazine of the East African Wildlife Society) April–June 2006.

Virani, M.Z., C. Kendall, P. Njoroge and S. Thomsett. 2011. Major declines in the abundance of vultures and other scavenging raptors in and around the Masai Mara ecosystem, Kenya. Biological Conservation 144: 746–752.

Virani, M.Z., A. Monadjem, S. Thomsett and C. Kendall. 2012. Seasonal variation in breeding Rüppell's Vultures *Gyps rueppellii* at Kwenia, southern Kenya and implications for conservation. Bird Conservation International 22: 260–269.

Virani, M., P. Kirui, A. Monadjem, S. Thomsett and M. Githiru. 2010. Nesting status of African White-backed Vultures Gyps africanus in the Masai Mara National Reserve, Kenya. Ostrich 81(3): 205–209.

Virani, M., C. Kendall, P. Njoroge and S. Thomsett. 2011. Major declines in the abundance of vultures and other scavenging raptors in and around the Masai Mara ecosystem, Kenya. Biological Conservation 144: 746–752.

Virani, M.Z., J.B. Giri, R.T. Watson and H.S. Baral. 2008. Surveys of Himalayan Vultures (*Gyps himalayensis*) in the Annapurna conservation area, Mustang, Nepal. Journal of Raptor Research 42(3): 197–203.

Voegelin, E.W. 1938. Tubatula ethnography. University of California Anthropological Records 2: 1–84.

Vogt, W. 1941. Food detection by vultures and condors. Auk 58: 571.

Vrba, E.S. 1985. African Bovidae: evolutionary events since the Miocene. South African Journal of Science 81: 263–266.

Wacher, T., J. Newby, I. Houdou, A. Harouna and T. Rabeil. Vulture observations in the Sahelian zones of Chad and Niger. Bull ABC 20(2): 186–199.

Wadley N.J.P. 1951. Notes on the birds of Central Anatolia. Ibis 93: 3–89.

Wallace, M. 2005. Re-introduction of the California Condor to Baja California, Mexico. Re-introduction News 24: 27–28.

Wallace, W.L. and D.W. Lathrap. 1959. Ceremonial bird burials in San Francisco Bay shellmounds. American Antiquities 25: 262–264.

Wallace, W.L. and D.W. Lathrap. 1975. West Berekeley (CA-Ala 307): a culturally stratified shellmound on the east shore of San Francisco Bay. Contributions of the University of California Archaeological Research Facility 29: 1–64.

Wallace, M.P. and S.A. Temple. 1987. Competitive interactions within and between species in a guild of avian scavengers. Auk 104: 290–295.

Walpole, M.J., M. Nabaala and C. Matankory. 2004. Status of the Mara Woodlands in Kenya. African Journal of Ecology 42: 180–188.

Walsh, J.F. 1987. Records of birds seen in north-eastern Guinea in 1984–1985. Malimbus 9: 105–122.

Walters, J.R., S.R. Derrickson, D.M. Fry, S.M. Haig, J.M. Marzluff and J.M. Wunderle, Jr. 2010. Status of the California condor (Gymnogyps californianus) and efforts to achieve its recovery. Auk 127(4): 969–1001.

Ward, J., D.J. McCafferty, D.C. Houston and G.D. Ruxton. 2008. Why do vultures have bald heads? The role of postural adjustment and bare skin areas in thermoregulation. Journal of Thermal Biology 33(3): 168–173.

Ward, D. 2005. Do we understand the causes of bush encroachment in African savannas? African Journal of Range and Forage Science 22: 101–105.

Ward, J., D. McCafferty, D. Houston and G. Ruxton. 2008. Why do vultures have bald heads? The role of postural adjustment and bare skin areas in thermoregulation. Journal of Thermal Biology 33(3): 168–173.

Ward, P. and A. Zahavi. 1973. The importance of certain assemblages of birds as "information centres" for food–finding. Ibis 115: 515–545.

Wardle, D.A. 1999. Is "sampling effect" a problem for experiments investigating biodiversity–ecosystem function relationships? Oikos 87: 403–407.

Warner, R.E. 1968. The role of introduced diseases in the extinction of the endemic Hawaiian avifauna. Condor 70: 101–120.

Waser, P.M. 1980. Small nocturnal carnivores: ecological studies in the Serengeti. African Journal of Ecology 18: 167–185.

Wassink, A. and G.J. Oreel. 2007. The Birds of Kazakhstan. De Cocksdorp, Texel, Netherlands.

Waterman, T.T. 1910. The religious practices of the Diegueno Indians. University of California Publications in American Archaeology and Ethnology 8: 271–358.

Watson, J., S.R. Rae and R. Stillman. 1992. Nesting density and breeding success of golden eagles in relation to food supply in Scotland. Journal of Animal Ecology 61: 543–550.

Watson, R.T. 2000. Flight, foraging and food of the Bateleur *Terathopius ecaudatus*: an aerodynamically specialized, opportunistic forager. *In*: R.D. Chancellor and B.U. Meyburg (eds.). Raptors at Risk. WWGBP/Hancock House, Berlin

Watson, R.T., M. Gilbert, J.L. Oaks and M. Virani. 2004. The collapse of vulture populations in South Asia. Biodiversity 5(3): 3–7.

Watzke, H. 2007. Reproduction and causes of mortaliy in the breeding area of the Great Bustard in the Saratov region of Russia. pp. 53–64. *In*: H. Litzbarski and H. Watzke (eds.). Great Bustards in Russia and Ukraine. Bustard Studies 6. Förderverein Großtrappenschutz e.V., Germany.

Weber, R.E., I. Hiebl and G. Braunitzer. 1988. High altitude and hemoglobin function in the vultures *Gyps rueppellii* and *Aegypius monachus*. Biological Chemistry Hoppe-Seyler 369(4): 233–240.

Weigeldt, C. and H. Schulz. 1992. Counts of Lappet-faced Vultures *Torgis tracheliotus* at Mahazat As Said (Saudi Arabia), with a discussion of the species' taxonomy. Sandgrouse 14: 16–26.

Wells, D.R. 1984. Bird report: 1978 & 1979. Malayan Nature Journal 38: 113–150.

Wells, D.R. 1999. The Birds of the Thai-Malay Peninsula, Vol. 1: Non-Passerines. Academic Press, London.

Wenny, D.G., T.L. DeVault, M.D. Johnson, D. Kelly and C.H. Sekercioglu. 2011. On the need to quantify ecosystem services provided by birds. Auk 128: 1–14.

Wenzel, B.M. and M.H. Sieck. 1972. Olfactory perception and bulbar electrical activity in several avian species. Physiology & Behavior 9: 287–293.

Wernery, U. 2009. A Lappet-faced Vulture nest in eastern Arabia. Phoenix 25: 15.

Wesley-Hunt, Gina D. and J. Flynn, John. 2005. Phylogeny of the Carnivora: basal relationships among the Carnivoramorphans, and assessment of the position of 'Miacoidea' relative to Carnivora. Journal of Systematic Palaeontology 3: 1–28.

West, J.N. 1988. Raptors of El Imposible Forest, El Salvador, C.A. M.Sc. thesis Central Washington University, Ellensburg, WA.

Western, D., S. Russell and I. Cuthill. 2009. The status of wildlife in protected areas compared to non-protected areas of Kenya. PLoS One 4(7): e6140.

Wetmore, A. 1927. Fossil Birds from the Oligocene of Colorado. Proceedings of the Colorado Museum of Natural History 7(2): 1.

Wetmore, A. 1931. The California Condor in New Mexico. Condor 33(2): 76–77.

Wetmore, A. 1932. Additional records of birds from cavern deposits in New Mexico. Condor 34(3): 141–142.

Wetmore, A. and H. Friedmann. 1933. The California Condor in Texas. Condor 35: 37–38.

Wetmore, A. and H. Friedmann. 1938. The California Condor in Texas. Condor 35(1): 37–38.

Wetmore, A. 1950. The identity of the American Vulture described as *Cathartes burroianus* by Cassin. Journal of the Washington Academy of Science 40, No. 12: 415–417.

Wetmore, A. 1960. A classification of the birds of the world. Smithsonian Miscellaneous Collections 139(11): 137.

Wetmore, A. 1962. Systematic notes concerned with the avifauna of Panamá. Smithsonian Miscellaneous Collections 145: 1–14.

Wetmore, A. 1964. A revision of the American vultures of the genus Cathartes. Smithsonian Miscellaneous Collections 146(6): 1–48.

Wetmore, A. 1965. (Review of) The role of olfaction in food location by the Turkey Vulture (Cathartes aura), by Kenneth E. Stager. Auk 82: 661–662.

Whitacre, D.F., A.J. Baker, J.E. Jones, R. Villegas Patraca, J. Sutter and C.M. Swartz. 1991. Results of census efforts in three units of the Maya Biosphere Reserve/Calakmul Biosphere Reserve Complex. pp. 43–58. *In*: D.F. Whitacre, W.A. Burnham and J.P. Penny (eds.). Maya Progress Report. The Peregrine Fund, Boise.

White, C.M. and L.F. Kiff. 2000. Biodiversity, island raptors and species concepts. pp. 633–652. *In*: R.D. Chancellor and B.U. Meyburg (eds.). Raptors at Risk. WWGBP, Hancock House, Surrey.

White, P.J.C., C. Stoate, J. Szczur and K. Norris. 2008. Investigating the effects of predator removal and habitat management on nest success and breeding population size of a farmland passerine: a case study. Ibis 150(Suppl. 1): 178–190.

Whitford, W.G. 2002. Ecology of desert systems. Academic Press, New York.

Whitson, M.A. and P.D. Whitson. 1968. Breeding Behavior of the Andean Condor (Vultur Gryphus). Condor (Cooper Ornithological Society) 71(1): 73–75.

Wilbur, S. and J.A. Jackson. 1983. Vulture Biology and Management, University of California Press, Berkeley.

Wilbur, S.R. and L.F. Kiff. 1980. The California Condor in Baja California, Mexico. American Birds 34: 856–859.

Wilbur, S.R. 1976. Condor: A doomed species? National Parks Conservation Magazine 50(2): 17–19.

Wilbur, S.R. 1978. The California Condor, 1966–1976: a Look at its Past and Future. Fish and Wildlife Service, Washington DC.

Wilcove, D.S. and M. Wilelski. 2008. Going, going, gone: is animal migration disappearing? Plos Biology 6: 1361–1364.

Wildash, P. 1968. Birds of South Vietnam. Tuttle & Company, Rutland, Vermont.

Wildlife Management Institute. 1982. Eagle electrocutions down. Outdoor News Bulletin 36, 3.

Wille, C. 2006. How many weapons are there in Cambodia? Small arms survey. Geneva, Switzerland.

Williams, A. 1997. Zoroastrianism and the body. *In*: S. Coakley (ed.). Religion and the Body (Cambridge Studies in Religious Traditions (8). Cambridge University Press, Cambridge.

Williams, C.B. 1922. Sense of smell in birds. Nature 110: 149–149.

Williams, T. 2000. Zapped! Audubon 102: 32–44.

Willis, E.O. and E. Eisenmann. 1979. A revised list of birds of Barro Colorado Island, Panama. Smithsonian Contributions to Zoology 291: 1–31.

Wilmers, C.C., R.L. Crabtree, D. Smith and K.M. Murphy. 2003a. Tropic facilitation by introduced top predators: gray wolf subsidies to scavengers in Yellowstone National Park. Journal of Animal Ecology 72: 909–916.

Wilmers, C.C., D.R. Stahler, R.L. Crabtree, S. Smith and W.M. Getz. 2003b. Resource dispersion and consumer dominance: scavenging at wolf and hunter-killed carcasses in Greater Yellowstone, USA. Ecology Letters 6: 996–1003.

Wilmers, C.C., D.R. Stahler, R.L. Crabtree, D.W. Smith and W.M. Getz. 2003. Resource dispersion and consumer dominance: scavenging at wolf- and hunter-killed carcasses in Greater Yellowstone, USA. Ecology Letters 6: 996–1003.

Wink, M., P. Heidrich U. Kahl, H.H. Witt and D. Ristow. 1993a. Inter- and intraspecific variation of the nucleotide sequence for cytochrome b in Cory's shearwater (*Calonectris diomedea*), Manx Shearwater (Puffinus/puffinus) and Fulmar (Fulmarus glacialis). Z. Naturforsch. 48C: 504–508.

Wink, M., P. Heidrich and D. Ristow. 1993b. Genetic evidence for speciation of the Manx shearwater (*Puffinus puffinus*) and the Mediterranean shearwater (*P. yelkouan*). Vogelwelt 114: 226–232.

Wink, M. 1994. PCR in der Evolutionsforschung. pp. 166–184. *In*: M. Wink and H. Wehrle (eds.). PCR im rnedizinischen und biologischen Labor. GIT-Verlag, Darmstadt.

Wink, M., U. Kahl and P. Heidrich. 1994. Lassen sich Silber-, Weißkopf- und Heringsmowe (Larus argentatus, L. cachinnans und L. fuscus) molekulargenetisch unterscheiden? Journal of Ornithology 135: 73–80.

Wink, M. 1995. Phylogeny of old and new world vultures (Aves: Accipitridae and Cathartidae) inferred from nucleotide sequences of the mitochondrial cytochrome b gene. Verlag der Zeitschrift für Naturforschung 50c: 868–882.

Wink, M., P. Heidrich and C. Fentzloff. 1996. A mtDNA phylogeny of sea eagles (genus Haliaeetus) based on nucleotide sequences of the cytochrome b-gene. Biochemical Systematics and Ecology 24(7–8): 783–791.

Wink, M. and I. Seibold. 1996. Molecular phylogeny of Mediterranean raptors (Families Accipitridae and Falconidae). *In*: J. Muntaner and J. Mayol (eds.). Biology and Conservation of Mediterranean Raptors. SEO/BirdLife, Madrid Monografia 4: 335–344.

Wink, M., 1. Seibold, F. Lotfikhah and W. Bednarek. 1998. Molecular systematics of holarctic raptors (Ordes falconiformes). pp. 29–48. *In*: R.D. Chancellor, B.M. Meyburg and J.J. Ferrero (eds.). Holarclic Birds of Prey. ADENEX-WWGBP, Calamonte, Spain.

Wink, M. and H. Sauer-Gurth. 2000. Advances in the molecular systematics of African raptors. pp. 135–147. *In*: R.D. Chancellor and B.-U. Meyburg (eds.). Raptors at Risk. WWGBP/ Hancock House, Mérida and Berlin.

Wink, M. and H. Sauer-Gürth. 2004. Phylogenetic relationships in diurnal raptors based on nucleotide sequences of mitochondrial and nuclear marker genes. pp. 483–498. *In*: R.D. Chancellor and B.-U. Meyburg (eds.). Raptors Worldwide. WWGBP/Hancock House, Mérida and Berlin.

Winkelman, J.E. 1990. Verstoring van vogels door de Sep-proefcentrale te Oosterbierum (Fr.) tijdens bouwfase en half-operationale situaties (1984–1989). RIN-rapport 90/9. DLO-Instituut voor Bos- en Natuuronderzoek, Arnhem.

Winkelman, J.E. 1990. Bird Collision Victims in the Experimental Wind Park near Oosterbierum (Fr), During Building and Partly Operative Situations (1986–1989). Rijksinstituut voor natuurbeheer, Anherm, The Netherlands.

Winkelman, J.E. 1992a. De invloed van de Sep-proefwindcentrale te Oosterbierum (Friesland) op vogels, 1: Aanvaringsslachtoffers.RIN-rapport 92/2, IBN-DLO, Arnhem.

Winkelman, J.E. 1992b. De invloed van de Sep-proefwindcentrale te Oosterbierum (Fr.) op vogels, 2: Nachtelijke aanvaringskansen.

Winkelman, J.E. 1995. Bird/wind turbine investigations in Europe. Proceedings of the National Avian-Wind Power Meetings, 20–21 July 1994, 43–48. Denver, Colorado.

Winter, E. and U. Winter. 1995. The vulture in ancient Egypt—classical tradition and modern behaviourist research. pp. 61–67. *In*: E. Winter and U. Winter (eds.). Bearded Vulture Reintroduction into the Alps Annual Report. Foundation for the Conservation of the Bearded Vulture, Vienna.

Winter, I.J. 1985. After the battle is over: The 'Stele of the Vultures' and the beginning of historical narrative in the art of the ancient near East. pp. 11–32. *In*: H.L. Kessler and M.S. Simpson (eds.). Pictorial Narrative in Antiquity and the Middle Ages. Center for Advanced Study in the Visual Arts, Symposium Series IV 16, National Gallery of Art, Washington D.C.

Wobeser, G. 2002. Disease management strategies for wildlife. Revue Scientifique et Technique 21: 159–178.

Wolfe, L.R. 1938. Eggs of the Falconiformes. Oologists' Record, XVIII.

Wollaston, A.F.R. 1922a. The natural history of south-western Tibet. The Geographical Journal 40(1): 12.

Wollaston, A.F.R. 1922b. The natural history of south-western Tibet. The Geographical Journal 60(1): 5–14.

Wolter, K. 2014. In BirdLife International (2014. Species factsheet: *Gyps coprotheres*. Downloaded from http://www.birdlife.org on 14/03/2014.

Wolter, K., V. Naidoo, C. Whittington-Jones and P. Bartels. unpublished. Does the presence of vulture restaurants influence the movement of Cape Vultures (*Gyps coprotheres*) in the Magaliesberg? Unpublished abstract.

Wood, J.G. 1862. The Illustrated Natural History. Routledge, Warne and Routledge, London.

Wood, G. 1983. The Guinness Book of Animal Facts and Feats.

Woodford, M.H., C.G.R. Bowden and N. Shah. 2008. Diclofenac in Asia and Africa—repeating the same mistake? Harmonisation and improvement of registration and quality control of Veterinary Medicinal Products in Africa. OIE World Organisation for Animal Health, Paris.

Woodgate, G.R. and M. Redclift. 1998. From a 'sociology of nature' to environmental sociology: beyond social construction. Evironmental Values 7: 3–24.

Woods, R.W. 1988. Guide to birds of the Falkland Islands. Anthony Nelson, Oswestry.

Woods, C.P., W.R. Heinrich, S.C. Farry, C.N. Parish, S.A.H. Osborn and T.J. Cade. 2007. Survival and reproduction of California Condors released in Arizona. pp. 57–78. *In*: A Mee, L.S. Hall and J. Grantham (eds.). California Condors in the 21st century. American Ornithologists' Union, McLean.

Woods, R.W. and A. Woods. 2006. Birds and mammals of the Falkland Islands. WILDGuides, Hampshire.

World Health Organization (WHO). 1998. World Survey for Rabies No. 34 for the year 1998. WHO, Geneva.

Wormington, H.A. 1957. Ancient Man in North America 4th edn. Denver Museum of Natural History Denver Museum of Natural History, Denver.

WWF Greece. 1999. Dadia project report.

Wynne-Edwards, V.C. 1955. Low reproductive rates in the birds, especially sea-birds. Acta XI. International Ornithological Congress 1954: 540–547.

Xiao Ti, Y. 1991. Distribution and status of the Cinereous Vulture *Aegypius monachus* in China. Birds of Prey Bulletin 4: 51–56.

Xirouchakis, S. 2007. Seasonal and activity pattern in Griffon Vulture (*Gyps fulvus*) colonies on the island of Crete (Greece). Ornis Fennica 84: 39–46.

Xirouchakis, S. and M. Mylonas. 2004. Griffon vulture (Gyps fulvus distribution and density in Crete. Israeli Journal of Zoology 50: 341–354.

Xirouchakis, S.M. and M. Mylonas. 2005. Status and structure of the griffon vulture (*Gyps fulvus*) population in Crete. European Journal Wildland Research 51: 223–231.

Xirouchakis, S.M. and Tsiakiris. 2009. Status and population trends of vultures in Greece. Munibe (supplement) 29: 154–171.

Yamaç, E. and E. Günyel. 2010. Diet of the Eurasian Black Vulture, *Aegypius monachus* Linnaeus 1766, in Turkey and implications for its conservation (Aves: Falconiformes). Zoology in the Middle East 51: 15–22.

Yang, L. and D. An. 2005. Handbook of Chinese Mythology. Oxford University Press, New York.

Yanosky, A.A. 1987. On the nest of the lesser yellow-headed vulture (Aves, Cathartidae). Nótulas Faunísticas 5(1): 1.

Ye Xiao-Ti. 1991. Distribution and status of the Cinerous Vulture *Aegypius monachus* in China. Birds of Prey Bulletin 4: 51–56.

Yosef, R. and D. Alon. 1997. Do immature Palearctic Egyptian Vultures *Neophron percnopterus* remain in Africa during the northern summer? Vogelwelt 118: 285–289.

Yosef, R. and O. Hatzofe. 1997. Conservation aspects and former nest-site selection of Lappet-faced Vulture Torgos tracheliotos in Israel.Vulture News 37: 2–9.

Young, C.G. 1929. A contribution to the ornithology of the coastland of British Guiana. Ibis 71: 1–38.

Zalles, J. and K.L. Bildstein. 2000. Raptor Watch: a Global Directory of Raptor Migration Sites. BirdLife International, Cambridge.

Zarudny, V. and M. Härms. 1902. Neue Vogelarten. Ornithologische Monatsberichte (4): 49–55.

Zedler, J.H. 1731–54. Grosses vollständiges Universal-Lexicon aller Wissenschafften und Künste. Halle Foundation, Leipzig.

Zhang, Z., Y. Huang, H.F. James and L. Hou. 2012a. Two old world vultures from the Middle Pleistocene of northeastern China and their implications for interspecific competition and biogeography of Aegypiinae. Journal of Vertebrate Paleontology 32(1): 117–124.

Zhang, Z., A. Feduccia and H.F. James. 2012b. A late Miocene Accipitrid (Aves: Accipitriformes) from Nebraska and its implications for the divergence of old world vultures. Plos One 7(11): e48842.

Zilio, F., A. Bolzan, A. de Mendonça-Lima, A.O. da Silva, L. Verrastro and M. Borges-Martins. 2013. Raptor assemblages in grasslands of Southern Brazil: species richness and abundance and the influence of the survey method. Zoological Studies 52: 27.

Zim, H.S., Robbins, S. Chandler and B. Bruun. 2001. Birds of North America: A Guide to Field Identification Golden Publishing.

Zimmerman, D.A., D.A. Turner and D.J. Pearson. 1996. Birds of Kenya and Northern Tanzania. Christopher Helm, London.

Zuberogoitia, I., K. Álvarez, M. Olano, A. Rodríguez and R. Arambarri. 2009. Avian scavenger populations in the Basque Country: status, distribution and breeding parameters. pp. 34–65. *In*: J.A. Donázar, A. Margalida and D. Campión (eds.). Vultures, Feeding Stations and Sanitary Legislation: A Conflict and its Consequences from the Perspective of Conservation Biology. Munibe 29 (Suppl.), Sociedad de Ciencias Aranzadi, Donostia.

Zuberogoitia, I., J.E. Martínez, A. Margalida, I. Gómez, A. Azkona and J.A. Martínez. 2010. Reduced food availability induces behavioural changes in Griffon Vulture Gyps fulvus. Ornis Fennica 87: 52–60.

Zuberogoitia, I., J. Zabala, J.A. Martínez, J.E. Martínez and A. Azkona. 2008. Effect of human activities on Egyptian Vulture breeding. Animal Conservation 11(4): 313–320.

Index